Introduction to the
Petroleum Geology of the North Sea

Introduction to the
Petroleum Geology of the North Sea

EDITED BY K. W. GLENNIE

Shell International Petroleum Mij,
The Hague, The Netherlands

BLACKWELL SCIENTIFIC PUBLICATIONS

OXFORD LONDON EDINBURGH

BOSTON PALO ALTO MELBOURNE

© 1984 by Blackwell Scientific Publications
Editorial offices:
Osney Mead, Oxford, OX2 0EL
8 John Street, London, WC1N 2ES
9 Forrest Road, Edinburgh, EH1 2QH
52 Beacon Street, Boston
    Massachusetts 02108, USA
706 Cowper Street, Palo Alto
    California 94301, USA
99 Barry Street, Carlton
    Victoria 3053, Australia

First published 1984

Set by DMB (Typesetting)
and printed and bound
in Great Britain by
Butler & Tanner Ltd, Frome, Somerset

DISTRIBUTORS

USA and Canada
    Blackwell Scientific Publications Inc
    P O Box 50009, Palo Alto
    California 94303

Australia
    Blackwell Scientific Book Distributors
    31 Advantage Road, Highett
    Victoria 3190

British Library
Cataloguing in Publication Data
Introduction to the petroleum geology of the North Sea.
    1. Gas, Natural, in submerged lands—North Sea
    2. Petroleum in submerged lands—North Sea
    I. Glennie, K.W.
    553.2′8′0916336   TN882.N6
    ISBN 0-632-01267-6
    ISBN 0-632-01268-4 Pbk

Contents

4 Late Permian—Zechstein, 61

J.C.M. TAYLOR *V.C. Illing and Partners, Cuddington Croft, Ewell Road, Cheam, Surrey*

5 Triassic, 85

M.J. FISHER *Area Exploration Manager, Britoil plc, 150 St Vincent Street, Glasgow*

9 Source Rocks and Hydrocarbons of the North Sea, 171

CHRIS CORNFORD *Independent Consultant, Integrated Geochemical Interpretation Ltd, Blair Mill, Dalry, Ayrshire*

10 North Sea Hydrocarbon Plays, 205

A.J. PARSLEY *Manager, International Exploration, Britoil plc, 150 St Vincent Street, Glasgow*

Foreword

R. STONELEY

This book is the outcome of a two-day short course held annually in London, and is based on the manual distributed to the course participants.

The course is arranged by the Joint Association for Petroleum Exploration Courses (UK) (JAPEC), an organisation that came into existence in 1980 in response to demands from UK-based petroleum exploration companies for assistance with their training programmes. JAPEC is sponsored and supported jointly by the Geological Society of London, the Petroleum Exploration Society of Great Britain and the Department of Geology at Imperial College of Science and Technology, University of London. It is run by an honorary committee drawn from the petroleum industry and from university geology departments, and *inter alia* now stages approximately eight short courses each year on a variety of topics.

The 'Introduction to the Petroleum Geology of the North Sea' was the first course to be arranged by JAPEC and was presented initially in the spring of 1981 under the energetic overall direction of K.W. Glennie. The contributors had all had considerable first-hand knowledge of the area. The course was designed as a rapid 'state of the art' overview for exploration geologists and geophysicists with little direct experience of the North Sea, but was also considered useful for those who wished to place detailed local knowledge in a basin-wide context. It thus provided the first systematic review of a basin which is undergoing rapid exploration and development. No attempt has been made to cover areas of the north-west European continental shelf outside the North Sea, and indeed the Norwegian, Danish, German and Dutch sectors are treated less fully than the British areas: this largely reflects the fact that none of the contributors has worked extensively in those sectors, but their descriptions are such that there are believed to be no significant gaps in the coverage.

With each annual repeat of the course, the content has been updated and modified in the light of constructive criticisms from the participants. This book may therefore be regarded as up to date to about the end of 1983, in respect of publicly released information and ideas.

The book falls into three broad sections. The first sets the scene with a summary of the history of exploration in the North Sea, which is followed by a review of the structural framework and pre-Permian development of the region.

The meat of the book is a series of descriptions, in stratigraphical order, of the depositional history and hydrocarbon-related rock units from the Permian to the Tertiary. The final section homes in on petroleum exploration with a review of that all-important factor, the oil and gas source rocks. The last chapter brings the preceding material together with a discussion of the various exploration 'plays': why are the oil and gas where they are, and how have they been found? What of the future?

Although both the JAPEC course and this book have been designed primarily for the benefit of those concerned with North Sea exploration, the book should have a wider appeal. As pointed out by Brennand in the opening chapter, our geological knowledge of the North Sea area was exceedingly sketchy when offshore exploration began in 1964. This then is a record of the geological knowledge gained almost entirely by the industry in a relatively short space of time, and it is a well-informed description of one of the world's major petroleum-bearing basins. It describes the geology of a large part of the north-west European region and therefore will be of significance to geologists in a broader range of disciplines: because the North Sea has been essentially an area of subsidence from the Late Palaeozoic to the present, its geological record has also proved to be important in unravelling some of the history of much of the surrounding, more positive, land areas.

With the passage of time, and with the modifications that have been made to the JAPEC course since its original presentation, there have inevitably been changes in the arrangement and authorship of certain sections. Our thanks, then, are due to those who participated in the past and who, whilst making no direct contribution to this book, have helped to make whatever success it may have: K.W. Barr (Consultant), C.E. Deegan (Hydrocarbons Unit, Institute of Geological Sciences), J.G.C.M. Fuller (Amoco Europe Inc.) and R.C. Selley (Imperial College of Science and Technology).

We sincerely thank the employing organisations of all contributors, past and present, for permitting their staff to give freely of their time and expertise. JAPEC was founded on the basis of industry self-help, and the industry has indeed been generous in its response.

Acknowledgements

This book has been produced in time for distribution at the 1984 JAPEC course of the same title. To achieve this aim, once it was decided to make it available to the public, many factors had to fit into a relatively tight schedule. A publisher had to be found; Bob Stoneley must be thanked for finding one who was prepared to meet all our specifications. Thanks are, of course, also due to my co-authors, all of whom kept within the originally planned timetable for revision of their earlier texts (Chris Cornford had to write his from scratch) and completed their proof-reading. Publication on time would not have been possible without the full cooperation of Blackwell Scientific Publications and the cordial working arrangement that developed between us. Finally, I thank my family for bearing with me during the book's nine-months gestation and my wife, Marjorie, and youngest daughter, Elizabeth, for helping to complete the Index.

Chapter 1

Petroleum Geology in North Sea Exploration 1964–1983

T.P. BRENNAND

1.1 Introduction

Because the oil industry already had the technical ability to undertake offshore exploration by the late 1950s, exploration of the continental shelf beneath the North Sea was inevitable. The spur that finally jolted it into action was the discovery of the giant Groningen gas field near the northern coast of The Netherlands, and the prospect of finding other gas fields of similar size offshore.

The Groningen field (Fig. 1.1) was discovered in 1959 with the well Slochteren-1. Its size was not realised until 1963, however, when the well Ten Boer-1 was deepened to penetrate the gas-rich Rotliegend sandstones near the Field's gas/water contact (Stheeman and Thiadens, 1969).* It is now known to have proven ultimate recoverable reserves in the order of 2425×10^9 m³ (86×10^{12} ft³) of gas.

In petroleum geology, the case history approach is often used with good effect to illustrate, through the practice of the subject, the nature of its principles. The set of papers to which this first contribution forms an introduction, provides an up to date review of a selection of petroleum geological elements from the case history of the North Sea. The result is a concise yet comprehensive picture of petroleum geology in action addressing a great diversity of problems. Indeed, there are few hydrocarbon provinces in the world that can compare with the North Sea for the variety of its petroleum geology.

The theme of this short introduction is not to summarise the papers which follow, but to describe briefly aspects of the growth process by which we arrive at today's view of the petroleum geology of the North Sea. The succeeding papers may then, perhaps, be seen in the context of the overall effort of discovery which began only 20 years before publication of this book. The emphasis in this chapter is on exploration in the U.K. sector of the North Sea. This is very reasonable because not only is half the North Sea under British jurisdiction so far as petroleum regulations are concerned, but well over half the exploration wells have been drilled in British waters. The exploration effort in other national sectors of the North Sea will not be neglected, however, because a find in one sector very

soon stimulates exploration in similar geological settings of other sectors, as will become apparent below.

After a brief summary of some of the general factors which affected exploration over the past 20 years, the development of the main 'plays' is considered by emphasising the successive contributions to knowledge made by the main discoveries of oil and gas. The details of individual plays are given in the last chapter. For convenience, this review is treated in three periods, from 1964-1970, 1971-1976, and the period from 1977 to 1983. Each period tends to be dominated by particular exploration pre-occupations, and each is roughly bounded by important UK events in the realm of acreage licensing (Fig. 1.2).

1.2 Licensing of offshore acreage

The effective start of offshore exploration in different parts of the North Sea depended upon enactment of petroleum legislation by the countries claiming sovereignty over those waters.

'Territorial waters' have, for almost two centuries, been limited by many western countries to a distance of 3 miles (5 km) from their coasts, and by up to 12 miles (20 km) by some others. Since 1945, however, many states have laid claim to the oil and other resources that lay between their coastline and the edge of the continental shelf, which is generally taken at a water depth of 200 m (~ 600 ft). In the North Sea area, this precedent resulted in sovereignty over the mineral rights being extended by bilateral agreement between the countries concerned to a median line half way between them (Fig. 1.1). Boundary settlements between Germany and her neighbours were not agreed until after some wells had already been drilled in waters claimed by The Netherlands and Denmark. A compromise solution was found for these marine boundaries.

Although the acquisition of geophysical data was able to proceed prior to the award of offshore exploration concession areas, the governments involved realised in 1963/4 that drilling could not proceed without the enactment of petroleum legislation. This legislation was to lay down the terms and conditions under which oil and gas could be searched for and exploited when found. These vary from country to country, and within countries have changed with the passage of time. Details will not be reviewed here, but reference can be made to Walmsley (1983) for the main conditions

---

* References appear in subsections 1.7.1-1.7.5, grouped by subject and each in alphabetical order.

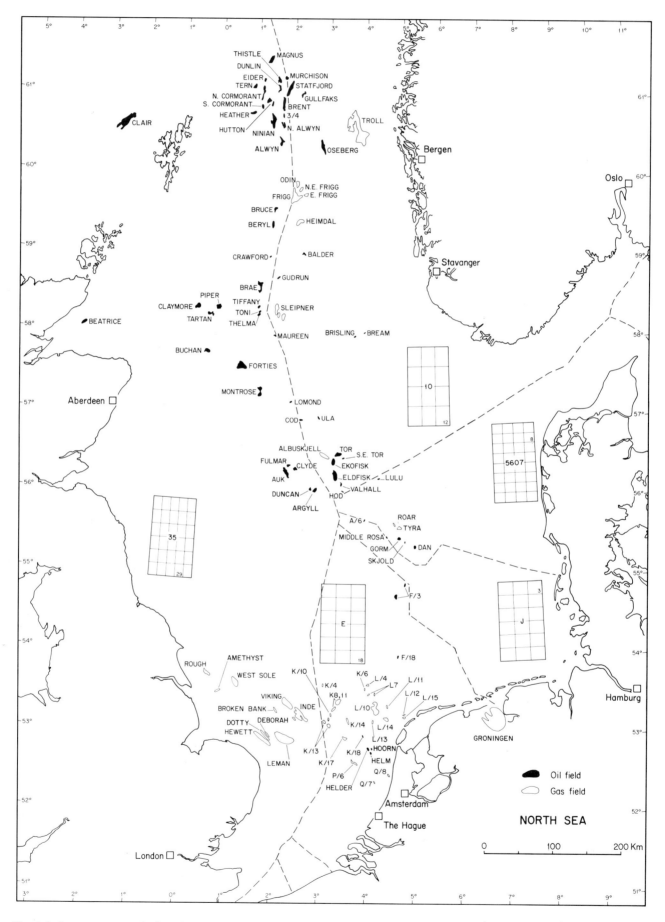

Fig. 1.1. Important named oil and gas fields of the North Sea area and an example of the number of blocks per quadrant for each of the national sectors of the North Sea.

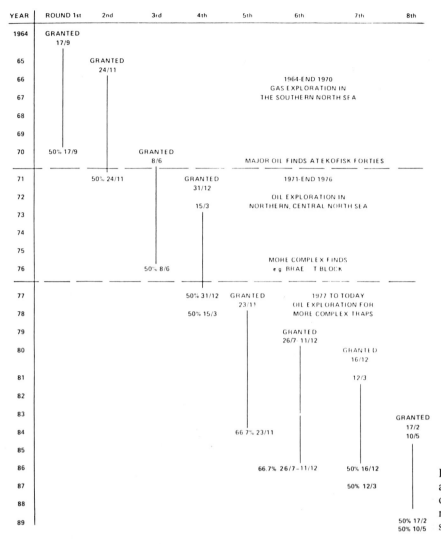

Fig. 1.2. Dates of award and 50% and 66.7% statutory relinquishment of U.K. North Sea Licence Rounds in relation to the major exploration scene.

affecting U.K. licences, and to Rønnevik *et al.* (1983) for the history of exploration licencing in Norwegian waters.

A few offshore wells had already been drilled as small outsteps into Dutch waters from fields onshore, beginning as early as 1961 (Kijkduin). Also in 1963, just before the size of the Groningen discovery was generally appreciated, the Danish Government had granted the exploration rights for the whole of their offshore area to one consortium. Other countries, however, eventually brought in legislation offering individual concession areas of a relatively small size, ranging from about 200 km² (U.K.) to 550 km² (Norway). Such legislation was enacted by Britain and Germany in 1964, by Norway in 1965, and by The Netherlands in 1968. Following the 50% acreage relinquishment by the Dansk Untergrunds Consortium (DUC) in January 1982, the Danish Government invited applications for the surrendered acreage in 1983.

1.3 General factors governing the growth of new knowledge

One of the most striking features of the discovery of the major oil and gas deposits offshore north-west

Europe has been the rapidity with which today's position was reached after the tentative beginnings in the mid sixties. In only twenty years, over fifty named fields of oil or gas have been discovered in the U.K. sector, and a similar number of the combined Norwegian, Danish and Dutch sectors, to which must be added at least as many small hydrocarbon accumulations that could possibly achieve economic maturity at a future date. Furthermore, while the most widely appreciated outcome of the enterprise was the discovery of large reserves of oil and gas, less appreciated has been the discovery of the hitherto unknown geology of a very large area of the north-west European Continental Shelf. Fig. 1.1 illustrates the distribution of many of the oil and gas fields that had been named by the end of 1983. Fig. 1.3 contrasts the relative absence of geological knowledge in the North Sea in the mid-sixties, immediately after the first phase of gas exploration in the southern North Sea, with the vastly improved position today.

To achieve this result required the shooting of many thousands of kilometers of seismic lines and the drilling of well over 1500 exploration wells including over nine hundred in U.K. waters alone (Fig. 1.4)—an effort that in other major hydrocarbon plays has historically

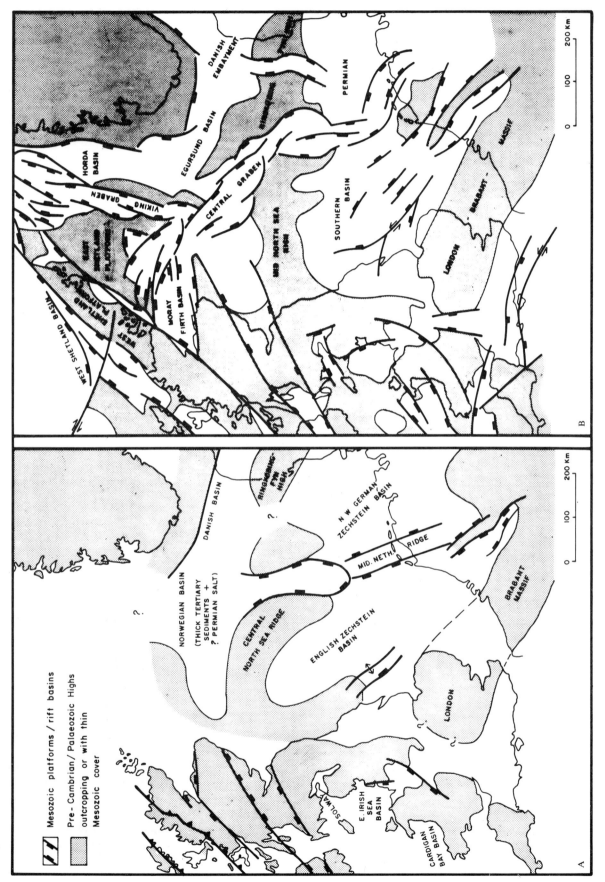

Fig. 1.3. The tentative tectonic framework of the North Sea area as understood immediately after the first phase of exploration in the southern North Sea (A), contrasted with the greatly improved position in 1983 (B). Adapted from Donovan *et al.* (1968) and Ziegler (1982a).

taken many decades. That it was achieved in this short time, was due to a combination of several favourable factors.

Like most mineral 'rushes' in the past, that of the North Sea was fired by initial discoveries of a size that attracted the attention of the whole industry. The huge gas find at Slochteren in Groningen in 1959, and the rapid understanding of the distribution of its Carboniferous source and Permian reservoir and capping formations, ensured that a major exploration campaign would ensue in the North Sea, and, significantly, in that part of the North Sea where it could most easily be undertaken—in the south. Ten years later, a billion barrel (160×10^6 m³) oil find at Ekofisk in the Norwegian sector provided exactly what was needed to spur on the tentative northward forays into deeper and more exposed waters. The Forties discovery a year later provided a similar stimulus in U.K. waters.

The pace of exploration was certainly aided by the fact that the play was offshore from the start. This enabled a large coverage of relatively inexpensive seismic surveys to be obtained in the shortest possible time. Moreover, the drilling that followed, though it had to overcome severe physical and technical difficulties, was not hampered by lengthy planning applications and site preparations as would have been the case onshore. The technology of drilling was fortunately developing at a rate which ensured that when exploration interest moved northward there was at least the beginnings of the capability to handle the problems there. While the physical conditions being met with in the central North Sea were among the most difficult yet faced by the drilling industry, from both the weather and supply points of view, the new generations of semi-submersible rigs eventually proved equal to the task. In the late 1960s it would take an average of 3.7 string months to drill a northern exploration well. By 1976 the average time was reduced to 2.1 string months, a figure that has remained fairly constant since then. In the early phases of drilling in the north, winter activity was not normally attempted. By 1972, this seasonal barrier had been overcome by a number of operators, and since then, year round mobile drilling has become commonplace.

In the vanguard of exploration, geophysical surveying technology also made great strides during the late 1960s. The switch to non-dynamite sources of seismic energy, which took place at this time, deprived the industry of an optimum signal, but vastly simplified their operations by eliminating the shooting boat—a major plus in the North Sea conditions. Very beneficial too, in terms of suppression of 'noise' and in increased clarity of reflectors on seismic lines, was the greater ease in obtaining higher multiplicities. In 1966, 4-fold coverage was usual, but by 1970, 24-fold shooting was the rule, increasing to 48-fold and 96 channels by 1980.

Also of key importance was the timely appearance of digital seismic recording techniques. At about the time of the discovery of Ekofisk, this new technique was becoming widespread, and the following decade witnessed a remarkable improvement in the quality of seismic records. Digital recording came about as a result of the increase in the power and cost effectiveness of digital computers. In the 1960s, deconvolution allowed much higher resolution to be achieved in time sections. This allowed seismic reflection sections to be tied with well-log data, thereby opening the way to detailed seismo-stratigraphy and the recognition of direct hydrocarbon indications. In the 1970s, a corresponding improvement in spatial resolution was achieved by advances in seismic migration.

In contrast to many hydrocarbon provinces in other parts of the world, the intervals of interest in the North Sea area have an unusually wide geological range. Hydrocarbons have now been proven in reservoirs ranging in age from the Devonian to the Tertiary, and the improving seismic technology played a crucial role throughout the seventies in clarifying what proved to be complex structural and stratigraphic geology. As in the early stages of most petroleum plays, the larger simpler fields were recognised early, and it is now typically becoming harder to locate the smaller or more subtle structures.

Continuing improvements in seismic resolution are therefore as necessary as ever. Improvements in both seismic and drilling techniques, therefore, were factors that speeded up the gathering of new information. This brought about accelerated activity, because the faster data was obtained, the faster could new activity be planned and justified.

A new petroleum play without an economic push, and without enough licence room within which to operate, would not get far however technically attractive. The rapid pace of North Sea development benefited from the start from mostly favourable conditions in both these areas.

As regards oil, the quadrupling of the price in 1974 provided a real incentive for non-OPEC countries to establish home-based sources of supply, and this boost for exploration came at a time when many of the larger U.K. discoveries had already been made. Thus, although remaining prospects tended to be smaller and less straightforward than those that had been found previously, and although companies were becoming more clearly aware of the huge costs of North Sea oil development, exploration was able to continue without a downturn.

It must be noted, however, that the decline of exploration for gas in the southern sector of the U.K. North Sea in the 1970s was due to a lack of economic incentive, and not to an exhaustion of prospective potential. In The Netherlands' offshore, where prospects are technically no better, exploration was maintained at a high level, undoubtedly as a result of the higher value accorded to gas there than was obtainable in the U.K. during the seventies (Fig. 1.4). Indeed, since 1982, a higher price for new gas has resulted in renewed activity in the U.K. southern North Sea.

The pace of North Sea exploration has been controlled to a considerable extent by the various rounds

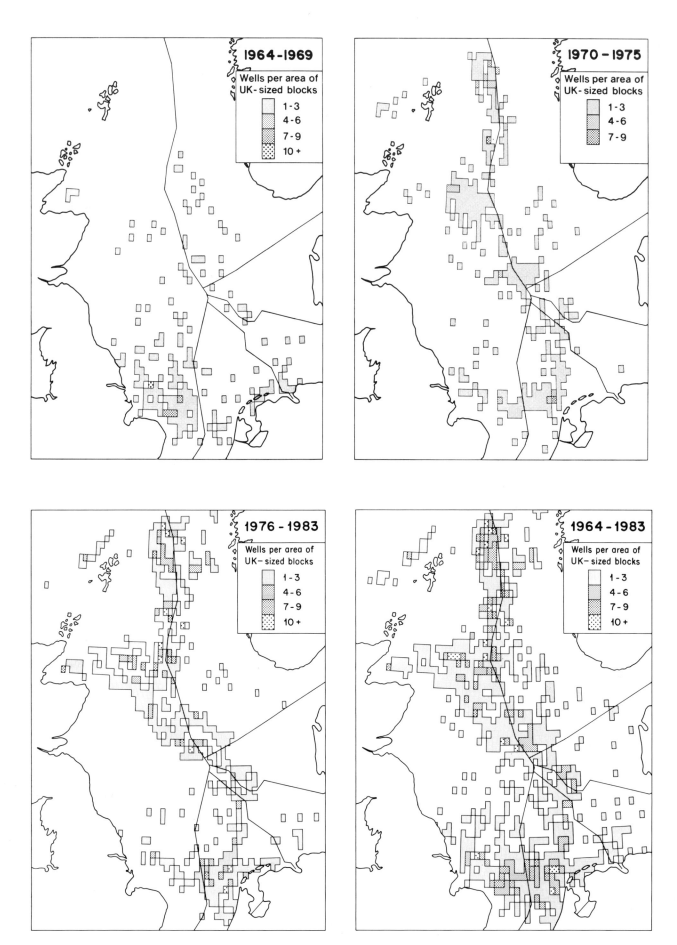

Fig. 1.4. North Sea mobile drilling effort for three periods between 1964 and 1983. Note that the area has been treated as if all blocks were the same size as those in British waters (i.e. 30 blocks per quadrant as opposed to 18 per quadrant in Dutch and German waters or 12 per quadrant in the Norwegian sector).

of licensing; for example, the general lag in Norwegian discoveries relative to those across the Median Line with U.K. is a direct result of Norwegian licensing policy. As regards the licensing of U.K. waters, the industry overall has consistently shown itself at home with the discretionary award system, as is witnessed by its generally eager response to individual rounds. This method of allocating acreage by the Department of Energy remains in very high regard after eight rounds in nineteen years, and this fact has also contributed greatly to the rate at which the continental shelf has been evaluated.

1.4 Main phases of North Sea exploration

1.4.1 The period 1964–1970

The growth of new geological insight was initially greater in the south, where early drilling was concentrated—and the dimensions of the Southern Permian Basin and of the Rotliegend dune belt were soon established (Glennie, 1972). Although at that time the Rotliegend objective still could not be seen on seismic records despite the above advances, the major part of the southern gas volume had by then already been discovered (Fig. 1.5). The larger structures at Hauptdolomite level proved to be indicative also of large structures at the top Rotliegend. As time progressed, the remaining structures still to be tested were of progressively slighter relief, and defining closure at Rotliegend level became more dependent upon accurate regional knowledge of the thickness to be expected of the Zechstein carbonates. Drilling had revealed that overlying the Rotliegend sands were thick shelf carbonates (Taylor *et al.*, 1975), which thinned northward into a basinal facies, and the velocity gradient that resulted imposed considerable effects on the depth conversion of seismic time sections.

By 1968 it was becoming clear that drilling a structure within the sand belt was no guarantee of success.

In the well 48/13-1, for instance, where the sands were the thickest so far found, they proved to be tightly cemented, even though gas bearing over hundreds of feet. Further drilling and detailed petrological work showed that certain areas of the dune belt had suffered deep burial during the Mesozoic with attendant porosity destruction by secondary silica and the growth of authigenic illite. Inversion of this structural situation occurred locally in the latest Cretaceous, giving rise to attractive looking structural highs, such as in the Sole Pit area. Other high areas also proved disappointing, such as those to the east of the Indefatigable field in block 49/19, where the sands were porous but unexpectedly devoid of gas. To explain this, it was necessary to invoke an unfavourable relationship between the timing of structure formation (probably Early Tertiary) and the onset of gas migration (Mesozoic).

The slow enactment of petroleum legislation for The Netherlands offshore prevented drilling from taking place until 1968. There were immediate successes with the discovery of Rotliegend gas in blocks K7, L2 and L12, which were followed by further discoveries in the K and L quadrants in the succeeding years (Fig. 1.6).

While the commercial limits of the Rotliegend sands appeared to be constrained to a relatively narrow east-west trending zone in the southern half of the Southern Permian basin, the Triassic Bunter sands, of excellent reservoir quality, were found to be far more widely distributed, and furthermore, they were extensively structured by the underlying Permian salts. High hopes for a major Bunter play were short-lived, however, because the Permian evaporites proved in most places to be a barrier to gas migration from the Carboniferous source, and the traps were either completely water-bearing or only partly filled with gas. Rare exceptions to this rule were where faulting or salt thinning had breached the seal. The only finds with Triassic reservoirs of immediate commercial interest were the Hewett field found by Arco in 1966 (Cumming *et al.*, 1975), and the Dotty field discovered by Phillips in

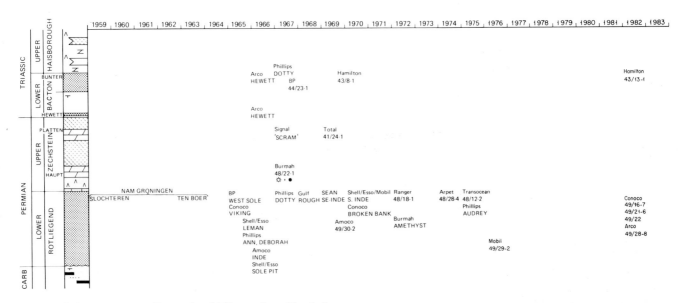

Fig. 1.5. Important gas discoveries, U.K. southern North Sea.

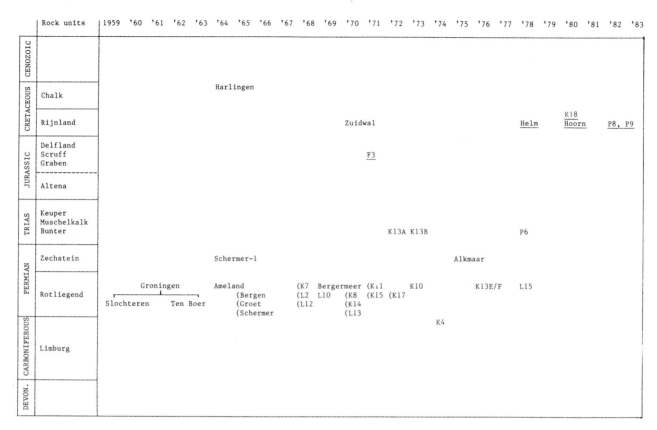

Fig. 1.6. Some important gas and *oil* discoveries in Netherlands land and marine areas since the discovery of the Groningen field in 1959. Marine drilling did not begin until 1968. The full stratigraphic range of hydrocarbon occurrences is from Upper Carboniferous to Lower Cenozoic.

1967. At Hewett, a basal Triassic sand unknown elsewhere was additionally gas-bearing.

The carbonates of the Permian Zechstein were rated as a secondary objective for gas following earlier experience in Germany and The Netherlands, and there were numerous instances where they were found to be gas-bearing during drilling for the deeper Rotliegend objective. So far, none of the offshore cases has proved to be of a size, or has indicated a sufficiently sustainable rate of production, that would warrant commercial development. Some wells obtained respectable initial production rates from the Haupt- and Plattendolomite, sometimes with sizeable associated condensate/oil production (e.g. 48/22-1, Fig. 1.5), but the high flow rates of both gas and liquids were typically short lived.

While the Zechstein and Triassic intervals yielded little developable gas, there was a rich yield of stratigraphic information. It was found that the lithostratigraphic correlation of individual carbonates, sands and evaporites was surprisingly precise far west of the German type areas. The new offshore evidence also provided the key to the integration of the Permian and Triassic basin-margin deposits of the U.K. with those of the continent. Regional work on these intervals provided the first published information from the offshore activities (Heybroek *et al.*, 1967; Geiger and Hopping, 1968).

Our stratigraphic understanding of the Triassic and older intervals was further enlarged by the recognition of palynological correlations with known areas of continental Europe and England. A firm calibration has now been established with Alpine ammonite-controlled sections. Work on planktonic and arenaceous foraminifera also was undertaken on the new Mesozoic and Tertiary sections encountered in wells.

North of the Mid North Sea High, exploration interest focussed immediately on the Lower Tertiary, where there was seismic evidence of 'growth-fault'-like features in the west, and where the first wells gave patchy encouragement for finding both sand and hydrocarbons. The 'growth-faults' proved illusory in so far as they might have indicated deltaic roll-over structural prospects. More significant, however, was the recognition, in 1967, of a second fault zone further east, which proved to be the western margin of the Central Graben.

In 1969, flagging enthusiasm for exploration in the north received a notable boost in the results of Amoco/Gas Council's 22/18-1 well (Fowler, 1975). After eight months with the Sea Quest to drill and test, encouraging oil flows were reported from Lower Tertiary sands. At about the same time, Phillips, in Norwegian waters, established major flows of oil, with gas, from their Ekofisk discovery well 2/4-1, elevating the Chalk at that time to the unexpected position of a major objective (Byrd, 1975). The Chalk's potential as a reservoir had already been indicated in Danish waters by the 1966 discovery of non-commercial oil and gas at Anne (Table 1.1) and by the discovery of gas in Roar and

Tyra (1968), and also in Norwegian waters by the Valhal oil discovery (1967; Fig. 1.7). These finds were too small, however, to consider for early development.

This major discovery at Ekofisk was quickly realised to be of a very large oil field, and it proved to be a most important milestone with regard to oil exploration in the North Sea. It effectively closed a decade dominated by exploration for non-associated gas. From here on, the hunt for northern oil was engaged in earnest, attracting most companies active in the international oil industry, and bringing back some American companies that had earlier left the area because of their poor success in finding gas.

Table 1.1. The year of discovery of some oil and gas finds in Chalk reservoirs of the Danish Central Graben.

Year of discovery	
1966	Anne
1968	Roar, Tyra
1969	Arne
1971	Dan, Gorm
1975	N. Arne
1977	Skjold, Adda
1978	Nils
1980	Lulu
1981	Middle Rosa

During 1970, drilling in the U.K. northern area concentrated on fulfilling commitments in 1st and 2nd Round licences, which were due for 50% statutory reduction in 1970 and 1971 (Fig. 1.2). The Lower Tertiary and Chalk now provided the main interest in the Central Graben area, and the Permian Zechstein and Rotliegend were still sought for in the west and on or near the Mid North Sea High. By the end of the year, the atmosphere for exploration, which to some extent had become frustrated by the erratic sand development in the Lower Tertiary, and by tantalising non or poorly producible oil occurrences in the Cretaceous Chalk (no Ekofisk having appeared in U.K. waters), dramatically improved with two clearly commercial discoveries late in the year: BP's 21/10-1, the Forties discovery well, found a gross interval of 118 m (386 ft) of oil in Lower Tertiary sands, which on test flowed at a rate of 750 m^3 (4720 bbl) per day, with almost negligible gas (Walmsley, 1975; Carmon and Young, 1981); and Shell/Esso's 30/16-1 well, the Auk field discovery, produced at a rate of 940 m^3 (5920 bbl) per day of light, low-sulphur crude from a thin zone of collapse-brecciated and vugular Permian dolomites. Underlying the dolomites were over 450 m (1500 ft) of water-bearing Rotliegend sands. The results at Forties and at Auk strengthened the hypothesis that generation of oil was likely to be optimum in the Central Graben area, and from Mesozoic or Tertiary source rocks. Migration could be expected into older reservoirs, provided they were proximal to the Graben and were in a high structural position with respect to source rocks.

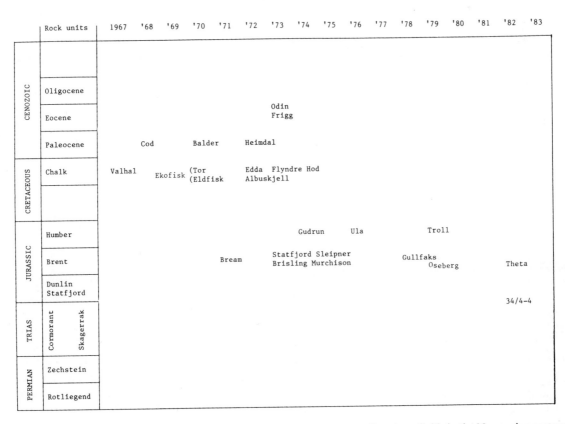

Fig. 1.7. Year of discovery and approximate age of reservoir of some important oil and gas fields in the Norwegian sector of the North Sea.

At Forties, the thick section of sands and the extensive core material obtained provided valuable data for a sedimentological understanding of the nature and potential distribution of the unpredictable Paleocene sands.

A third well in 1970 (Fig. 1.8) opened new and intriguing possibilities—that of deeper pre-Cretaceous Mesozoic sands within the Central Graben domain. Phillips' Josephine discovery, 30/13-1, was the first in the North Sea to test oil from sands beneath the Chalk. A flow rate of 128 m³ (800 bbl) per day was attained from a thin sand below 3600 m (12 000 ft). This moderate flow rate, and the thinness of the reservoir sand, did not arouse great commercial interest at the time.

By the end of 1970 the centre of interest in U.K. waters had swung in a decided fashion away from gas in the Southern Basin where, except perhaps for Conoco at Broken Bank, diminishing returns had set in for exploration. At that time there was no sign of an improvement for gas on the pricing front, and in the absence of an incentive to pursue what was becoming harder to find, exploration drilling in the south declined (Fig. 1.4). This trend was expressed in a relinquishment of 1st Round licences on 17 September 1970 that was massively in excess of statutory requirements (75% rather than 50%).

1.4.2 The period 1971–end 1976

The lack of exploration interest in the Rotliegend and Bunter gas sands was not matched in the Dutch sector where free market prices prevailed. There, a whole string of gas discoveries continued to be made through-out this period in the K and L quadrants (Fig. 1.6) and to the NW of Amsterdam both on land and straddling the coastline (van Lith, 1983). Gas was also found in Bunter sands in block K/13 (Roos and Smits, 1983). In German waters, on the other hand, with the exception of a small gas discovery in block A/6, the few Rotliegend and Bunter gas-bearing structures found since the mid 1960s were all too rich in nitrogen and carbon dioxide for their methane content to be marketed economically.

The Ekofisk discovery led to a series of oil finds in Chalk reservoirs of the Central Graben in southern Norwegian and Danish waters (Fig. 1.7, Table 1.1). While the U.K. 3rd Round of licensing came too early to catch the post-Ekofisk enthusiasm for northern oil exploration (the blocks being awarded in June 1970), the 4th Round, in 1971, fully reflected the intense industry interest aroused by the discoveries.

Following a period of very active seismic acquisition, which infilled the wide reconnaissance grids in the northern North Sea in the area south of the Shetland Platform and between the Forties field and the Scottish coastline, a complex variety of structural elements began to emerge in increasing but never sufficient detail. The 'landscape' of buried topography, the surface of which appeared to be pre-Lower Cretaceous in age, created numerous objectives of potential interest in the 4th Round of licensing. The shortcomings in seismic resolution beneath the ubiquitous unconformity, and the virtual absence of well penetrations below it, left the interpretation wide open to as to what geological material composed the 'highs'. Interpretations ranged from the very favourable possibility of a Juras-

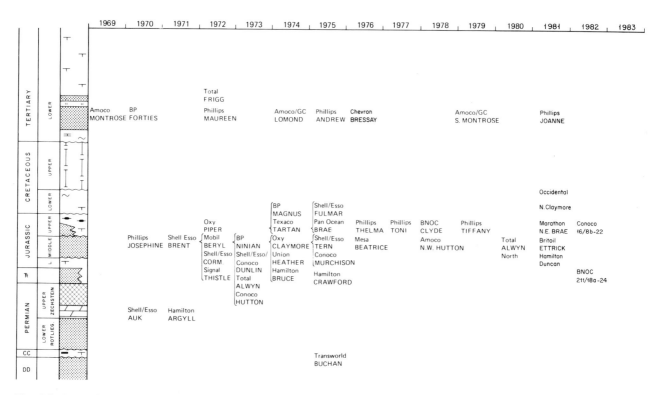

Fig. 1.8. Year of discovery and approximate age of reservoir of some important oil fields of the U.K. Central and Northern North Sea.

sic sequence, to those envisaging the less favourable Triassic, Devonian or older sequences.

The largest visible 'buried highs' were clearly in the northern North Sea. In the 3rd Round, Shell/Esso had acquired a block 211/29, covering what then had been a very loosely defined deep feature. The first well on block 211/29, which was completed without testing before the 4th Round closed, revealed that the pre-unconformity sequence contained good quality oil-bearing Jurassic sands of deltaic origin and of good porosity. The Brent field reservoirs were first tested in the second well on the structure the following year and were proved capable of 1040 m³ (6500 bbl) per day of 38°API oil with a GOR of 1550 scf/bbl. An important gas cap was also present (Bowen, 1975). The discovery of the giant Brent field (350×10^9 m³ recoverable oil) had a profound influence on exploration, and in succeeding years resulted in a spate of discoveries of similar type in the northern North Sea.

A second Auk-type field was also discovered in 1971 by Hamilton—the Argyll field in block 30/24, confirming the possibility of accumulations marginal to the Graben (Pennington, 1975). Here, both Permian Zechstein and Rotliegend reservoirs proved productive at attractive rates.

The subsequent four years witnessed an exciting succession of discoveries in the U.K. northern North Sea, where the Brent-type 'buried highs' proved commercially oil-bearing in eight major fields, with other structures such as Stratjford and Sleipner being discovered over the Median Line in the Norwegian sector. These were now resolved into tilted fault blocks draped with organic-rich Kimmeridge Clay, which were later onlapped by Lower Cretaceous muds and marls. In the U.K. 15 and 14 quadrants also, a shrewd 4th Round selection by Occidental yielded two prolific fields— Piper in 1971, and Claymore in 1974.

The major drilling effort these finds represented, produced a wealth of new regional information on the Jurassic succession in particular. Knowledge of the sequences in E. Yorkshire, Sutherland and E. Greenland had for some time indicated scope for mid-Jurassic clastics in the northern North Sea. A prophetic illustration of this possibility is contained in a small inset map of the Middle Jurassic in Will's Palaeogeographic Atlas (1951) where, in the Bathonian, clastics were visualised as infilling a narrow north-south basin open to the Boreal Sea.

The generally shallow-water nature of the Lower and Middle Jurassic Statfjord and Brent sands, and the gentle transition upwards of the Statfjord sands from the underlying Triassic, itself of very widespread uniformity, encouraged the view that the regressive episodes that they represented would also be of regional extent; thus the resulting sands could justifiably be sought over most of the northern North Sea where Middle Jurassic was present. This view has been borne out by extensive drilling, and the main limitation to the Middle to Lower Jurassic sand play is the depth to which these intervals have descended in the process

of Late Cimmerian (pre-Lower Cretaceous) faulting and Tertiary subsidence.

A new element in the Jurassic play appeared in 1972, when the results of Occidental's Piper discovery became known (Williams *et al.*, 1975). Here was the Upper Jurassic (Oxfordian) in an attractive coastal sand facies, and oil-bearing. Two years later, in BP's Magnus field in the far north, oil-bearing sands were found in a deep-water facies in the Kimmeridgian (De'Ath and Schuyleman, 1981). This stratigraphic unit, now widely recognised for its source rock characteristics, had not been foreseen as a potential significant reservoir-bearing formation.

From the beginning of interest in the northern North Sea, the coastal outcrops at Brora had been of particular interest. Here, the intervals of extremely coarse and ill-sorted debris of the Kimmeridgian sequence were thought to display neither the porosites and permeabilities needed for a productive reservoir, nor the persistence to suggest that such beds might be widespread. In 1975 further light was thrown on the commercial possibilities for the Upper Jurassic reservoirs by Pan Ocean's discovery of the Brae field in the Viking Graben. Here, a thick succession of conglomerates and sandstones were found oil-bearing to a maximum thickness of 450 m (1500 ft), in what was interpreted to be a suite of coalescing fans within which abrupt lateral lithological discontinuities occur (Harms *et al.*, 1981). Significantly, this development of Upper Jurassic clastics occurs at the level of the basal Cretaceous Late-Cimmerian unconformity at a point of high relative relief between the platform area to the west and the Graben.

In the same year, 1975, and far to the south, again in a location near to high pre-Lower Cretaceous relief, Shell/Esso, at Fulmar, found Upper Jurassic sands oil-bearing in well 30/16-6. As the main flush of discoveries related to the Middle Jurassic regressive sands dwindled in the mid-seventies, interest increased in the more-difficult-to-predict Upper Jurassic play.

Throughout this later period, a large amount of new data was also accumulated on the Tertiary. Every well drilled to the Jurassic added new Tertiary information, and many wells had combined Jurassic/Tertiary objectives. Discoveries of oil in the Tertiary continued to be made at a moderate pace relative to that of the Jurassic. From 1972 to 1975, finds of oil in the Lower Tertiary were reported by Phillips (Maureen and Andrew) and Amoco/Gas Council (Lomond). And in 1976, an accumulation of heavy oil was discovered by Chevron (Bressay) on the Shetland Platform margin. Lower Tertiary gas was found in 1973 by Total in the Frigg field (Heritier *et al.*, 1981), which straddles the U.K.-Norway median line, and by Esso in the Odin field.

During this period also, notable developments in seismic technology took place which enabled the Jurassic play to be pursued with no loss of momentum. An important growth in the practice of in-house post-stack processing enabled operators to concentrate upon the particular problems that applied in their areas. Advances in migration techniques and in velocity studies were aided by such in-house dedicated systems.

In bio-stratigraphy, the first half of the seventies witnessed an enormous increase in the volume of new information, and computer recording of data became mandatory. By the mid-seventies, calcareous nanno-plankton provided the key to dating the Cretaceous and Danian chalks, while in the Jurassic the value of arenaceous foraminifera as environment indicators became recognised. In palynology, improvements in sample preparation enabled higher concentrations of microflora to be obtained, thus accelerating the establishment of a palynozonation based on dinoflagellates in the Jurassic and in parts of the Tertiary. Important too was the integration of palyno- and palaeontological zonations in the Tertiary sequence.

In 1975 and 1976, the pressure was on for licensees to complete evaluation of 3rd Round blocks before the 50% surrender required by June 1976. Also looming was the more demanding task of preparing for the 4th Round relinquishment due at the end of 1977.

1.4.3 The period 1977-1983

Since 1977, activity has remained generally high, with companies pursuing a variety of objectives in the central and northern North Sea. Upper Jurassic targets have provided the main interest in both areas, but other objectives, Palaeozoic, Lower Cretaceous, Paleocene and Eocene have continued to generate interest, and considerable success has been met with in certain areas.

Activity in the U.K. sector of the southern gas basin remained depressed until 1982 when, in response to an economic improvement in the prospects for gas, and in anticipation of 8th Round offerings, a significant jack-up drilling campaign resumed. One result of this renewed drilling activity was the discovery of Bunter gas in block 43/13a (Esmond field). This was followed a year later by an announcement by Hamilton Brothers that they plan to have this field and two other earlier discoveries in the same quadrant, Forbes (43/8) and Gordon (43/15 and 43/20) in production by mid 1985.

The 7th Round of licensing, which closed in late 1980/early 1981, put on offer most of the prime and proven oil exploration areas in the U.K. North Sea, and this event and the succeeding year of activity has underlined the fact that exploration has reached a mature phase in much of this area. The relatively small sizes of the discoveries made, and the fact that many companies concentrated their exploration close to their existing facilities, are characteristic of this period (Fig. 1.9).

By 1982/83, the acreage available for the Department of Energy to put on offer in the 8th Round was largely undrilled and of a highly speculative nature with regard to its prospectivity. Exceptions were provided by selected auction blocks in the prime areas which had been left untaken in the 7th Round, and

some southern gas areas, the first significant offers in this area since the 4th Round in 1972.

The period between 1977 and the U.K. 8th Round has been characterised by exploration for increasingly difficult objectives. The seismic expression of Upper Jurassic traps is less straightforward than was the case of the horsts, tilted fault blocks, overlying drape structures and salt-induced diapirs that constituted the main objectives in earlier periods. Growing experience in recognising and drilling on the relatively modest seismic indications at the base-Cretaceous level have led, however, to a number of discoveries spread widely across the northern and central North Sea. In the southern Viking Graben, Phillips, continuing the Brae-type play, found Toni (1977) and Tiffany (1979) in their T-block 16/17. The Upper Jurassic has continued to reward exploration, discoveries being made in the Central North Sea by BNOC at Clyde (1978), Occidental in block 14/18 (1978), Marathon at N.E. Brae (1980), BNOC at Ettrick (1981) and Shell/Esso in the 21/19 block (1981).

Upper Jurassic fields have also been found in Norwegian waters by, for example, Shell at Troll (1979). But mid-Jurassic oil- and gas-bearing reservoirs continued to be discovered throughout the period with Statoil's Gulfaks (1978) and Oseberg (1979).

Near to the Argyll field in U.K. waters, Hamilton has successfully tested small fault traps involving Jurassic sands satellite to the main Permian trap (Duncan and East Duncan in 1981).

The Lower Cretaceous has proved to be hydrocarbon-bearing and productive in the relatively limited area favouring sand deposition in the Witch Ground Graben and Fisher Bank Basin areas of quadrants 14 and 16. Discoveries include Gulf's find in block 16/26 (1977) and Occidental's North Claymore Field (1980). Outside this region, with the exception of the Moray Firth area, the Lower Cretaceous has proven to be singularly poor in potential reservoir lithologies.

Oil-bearing Lower Cretaceous sands have long been known onshore in The Netherlands (e.g. NAM's Schoonebeek field near the German border, and the Rijswijk field (Bodenhausen and Ott, 1981) in the vicinity of The Hague, which were discovered respectively in 1942 and 1952). Offshore, Union discovered oil in their Helm (1976) and Hoorn (1980) fields, and Conoco in their K/18 field (1980).

Exploration in the Tertiary sequence has concentrated upon local distributions of deep-water sand fans and upon the Shetland Platform margin, where oil has been found of a gravity heavier than usual for the U.K. North Sea. The heavy oil accumulations include Conoco's find in block 9/3 and Chevron's in block 2/10. In the Central Graben a number of Tertiary discoveries have been made of which Phillips' Joanne find is one.

Deeper, older objectives in the Carboniferous and Devonian are relatively unexplored, and in several areas they are likely to claim increasing attention as the

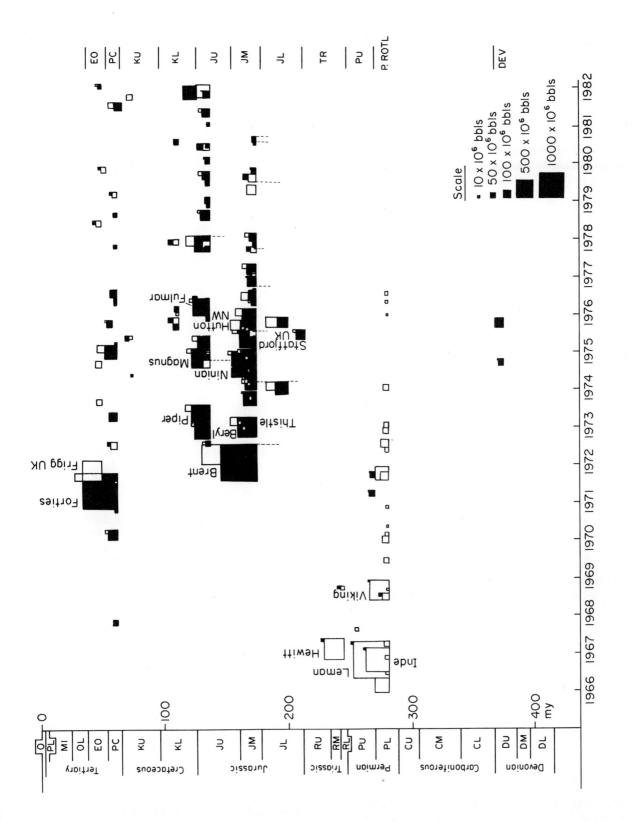

Fig. 1.9. Discovery history of plays in U.K. sector of the North Sea. Note that fields are displayed in size classes.

prime areas approach an ever maturer stage of exploration.

Advances in the application of the seismic method continue to be made both in the field of acquisition and of processing. On the acquisition side a trend toward three-dimensional coverage is noticeable, reflecting the greater concern for precise structural and stratigraphic control in appraisal and development drilling as the number of discoveries grows. Improvements in offshore position-fixing have made parallel progress.

Seismostratigraphical methods have an increasing part to play. In the balance of remaining prospects for oil a stratigraphic component of trapping is more commonly required than used to be the case. The recent successes mentioned above, for example, underline the importance of a rigorous examination of seismic data for the minor indications of small clastic fans and lobes adjacent to synchronous relief. The multidisciplinary approach that is called for in pursuing such objectives typifies the demands made in petroleum geology in the oil industry today.

1.5 Concluding remarks

The value of a proper petroleum geological understanding of the natural resources of the North Sea is crucial in judging what the policies should be regarding their future development. The debate concerning the eventual total recoverable hydrocarbons that took place in the mid-seventies between the optimistic statisticians on the one hand and the industry on the other would have been more valuable had a better petroleum geological understanding been available.

An encouraging feature of the discovery process we have reviewed has been the readiness of the industry to disseminate its newly won knowledge in conferences, such as the two in London, 'Petroleum and the Continental Shelf of North-west Europe'' in 1974 and 'Petroleum Geology of the Continental Shelf of Northwest Europe' in 1980, the 1977 'Mesozoic northern North Sea Symposium' in Oslo, and the 1982 conference in The Hague on 'Petroleum Geology of the south-eastern North Sea and the adjacent onshore areas'. The contents of this book are fully in line with this theme.

1.6 Acknowledgements

This contribution is published by permission of Shell International Petroleum Mij, The Hague, and Shell Exploration and Production, London. I am grateful to Ken Glennie for providing additional data on the non-U.K. parts of the North Sea.

1.7 References

1.7.1 Selected regional references

Bender, F. and Hedemann, H.A. (1983) Zwanzig Jahre erfolgreiche Rotliegend-Exploration im Nordwestdeutsch-land—weitere Aussichten auch im Präperm? *Erdoel-Erdgas-Zeitschrift* **99**, 39-49.

Birkelund, T. and Perch-Nielsen, K. (1976) Late Palaeozoic-Mesozoic evolution of central East Greenland. In Esher A. and Watt, W.S. (Ed.) q.v. p. 304-339.

Day, G.A., Cooper, B.A., Anderson, C., Burgers, W.F.J., Ronnevik, H.C. and Schoneich, H. (1981) Regional seismic structure maps of the North Sea. In: Illing, L.V. and Hobson, G.D. (Ed.) q.v. p. 76-84.

Donovan, D.T. (Ed.) (1968) *Geology of Shelf seas.* Oliver and Boyd.

Esher, A. and Watts, W.S. (1976) *Geology of Greenland.* Gronlands Geologiske Undersogelse, 603 p.

Heybroek, P., Haanstra, V. and Erdman, D.A. (1967) Observations on the geology of the North Sea area. *7th World Petroleum Cong. Proc.*, v. **2**, 905-16.

Plein, E. (1979) Das Deutsche Erdöl und Erdgas. *Jh. Ges. Naturkde. Württemburg* **134**, 5-33.

Rønnevik, H., Eggen, S. and Vollset, J. (1983) Exploration of the Norwegian Shelf. In: Brooks, J. (Ed.) *Petroleum geochemistry and exploration of Europe.* Geol. Soc. London. Special Publication No. 10. 71-93.

Sorgenfrei, T. (1969). A review of petroleum development in Scandinavia. In: Hepple, P. (Ed.) q.v. p. 191-208.

Stheeman, H.A. and Thiadens, A.A. (1969) A history of the exploration for hydrocarbons within the territorial boundaries of The Netherlands. In: Hepple, P. (Ed.) q.v. p. 259-269.

Wills, L.J. (1951) *A palaeogeographic atlas of the British Isles and adjacent parts of Europe.* Blackie & Sons, London.

Ziegler, P.A. (1975) North Sea basin history in the tectonic framework of N.W. Europe. In: Woodland, A.W. q.v. p. 131-149.

Ziegler, P.A. (1977) Geology and hydrocarbon provinces of the North Sea. *GeoJournal* 1/1.

Ziegler, P.A. (1981) Evolution of sedimentary basins in North-West Europe. In: Illing, L.V. and Hobson, G.D. (Ed.) 1981 q.v. p. 3-39.

Ziegler, P.A. (1982) *Geological atlas of Western and Central Europe.* Elsevier, Amsterdam. 130 p.

Ziegler, W.H. (1975) Outline of the geological history of the North Sea. In: Woodland, A.W. q.v. p. 165-190.

1.7.2 Publications of North Sea conferences

Finstad, K.G. and Selley, R.C. (Ed.) Mesozoic northern North Sea Symposium, Oslo, 1977. *Proc. Norwegian Petroleum Soc.* 6/1-6/26.

Hepple, P. (Ed.) (1969) *The exploration for petroleum in Europe and North Africa.* Inst. of Petroleum.

Illing, L.V. and Hobson, D.G. (1981) *The petroleum geology of the Continental Shelf of N.W. Europe.* Hayden, 521 p.

Kaasschieter, J.P.H. and Reijers, T.J.A. (Ed.) (1983) Petroleum Geology of the southeastern North Sea and the adjacent onshore areas. *Geol. Mijnbouw* **62** (1), 1-239.

Norwegian Pet. Soc. The sedimentation of North Sea reservoir rocks. *Proc. Geilo Conf.*, May 1980.

Woodland, A.W. (Ed.) (1975) *Petroleum and the Continental Shelf of N.W. Europe.* Vol. 1: Geology, Applied Science Publ. 501 p.

1.7.3. Selected papers on North Sea facies studies

Barnard, P.C. and Cooper, B.S. (1983) A review of geochemical data related to the Northwest European gas province. In: Brooks, J. (Ed.) *Petroleum geochemistry and exploration of Europe.* Geol. Soc. London. Special Publication No. 10. 19-33.

Brennand, T.P. (1975) The Triassic of the North Sea. In: Woodland, A.W. (Ed.) q.v. 295-311.

Geiger, M.E. and Hopping, C.A. (1968) Triassic stratigraphy of the southern North Sea basin. *Phil. Trans. R. Soc.* (B) **254**, p. 1-36.

Glennie, K.W. (1972) Permian Rotliegendes of Northwest Europe interpreted in light of modern desert sedimentation studies. *Bull. Am. Assoc. Pet. Geol.* V. 56, No. 6, p. 1048-1071.

Hancock, J.M. and Scholle, P.A. (1975) Chalk of the North Sea. In: Woodland, A.W. (Ed.) q.v. p. 413-427.

Lutz, M., Kaasschieter, J.P.H. and Wijhe, D.H. van (1974) Geological factors controlling gas accumulations in the Mid-European Basin. *Proc. 9th World Pet. Cong.* (*Tokyo*), V. **2**, p. 93-103.

Parker, J.R. (1975) Lower Tertiary sand development in the Central North Sea. In: Woodland, A.W. (Ed.) q.v. p. 447-453.

Surlyk, F. (1978) Submarine fan sedimentation along fault scarps on tilted fault blocks (Jurassic-Cretaceous boundary, East Greenland). *Gronlands Geologiske Undersogelse*, Bull. **128**, 108 p.

Taylor, J.C.M. and Colter, V.S. (1975) Zechstein of the English sector of the southern North Sea basin. In: Woodland, A.W. (Ed.) q.v. p. 249-263.

Taylor, J.C.M. (1981) Zechstein facies and petroleum prospects in the central and northern North Sea. In: Illing, L.V. and Hobson, G.D. (Ed.) q.v. p. 176-185.

1.7.4 North Sea oil and gas field papers

Bodenhausen, J.W.A. and Ott, W.F. (1981) Habitat of the Rijswijk Oil Province, The Netherlands. In: Illing, L.V. and Hobson, G.D. (Ed.) q.v. p. 301-309.

Bowen, J.M. (1975) The Brent oil field. In: Woodland, A.W. (Ed.) q.v. p. 353-363.

Brennand, T.P. and Veen, F.R. van (1975) The Auk oil field. In: Woodland, A.W. (Ed.) q.v. p. 275-285.

Butler, J.B. (1975) The West Sole gas field. In: Woodland, A.W. (Ed.) q.v. p. 213-223.

Byrd, W.D. (1975) Geology of the Ekofisk field, offshore Norway. In: Woodland, A.W. (Ed.) q.v. p. 439-447.

Carman, G.J. and Young, R. (1981) Reservoir geology of the Forties oil field. In: Illing, L.V. and Hobson, G.D. (Ed.) q.v. p. 371-379.

Cumming, A.D. and Wyndham, C.L. (1975) The geology and development of the Hewett gas field. In: Woodland, A.W. (Ed.) q.v. p. 313-327.

De'Ath, N.G. and Schuyleman, S.F. (1981) The geology of the Magnus oil field. In: Illing, L.V. and Hobson, G.D. (Ed.) q.v. p. 342-351.

Dept. of Energy (1975-1980) *Development of the oil and gas resources of the United Kingdom* (D of E 'Brown Book') H.M.S.O.

Fowler, C. (1975) The geology of the Montrose field. In: Woodland, A.W. (Ed.) q.v. p. 467-477.

France, D.S. (1975) The geology of the Indefatigable gas field. In: Woodland, A.W. (Ed.) q.v. p. 233-241.

Gray, I. (1975) Viking gas field. In: Woodland, A.W. (Ed.) q.v. p. 241-249.

Gray, W.D.T. and Barnes, G. (1981) The Heather oil field. In: Illing, L.V. and Hobson, G.D. (Ed.) q.v. p. 335-341.

Hallett, D. (1981) Refinement of the geological model of the Thistle field. In: Illing, L.V. and Hobson, G.D. (Ed.) q.v. p. 315-325.

Harms, J.C., Tackenberg, P., Pickles, E. and Pollock, R.E. (1981) The Brae oil field area. In: Illing, L.V. and Hobson, G.D. (Ed.) q.v. p. 352-357.

Heritier, F.E., Lossel, P. and Wathne, E. (1981) The Frigg gas field. In: Illing, L.V. and Hobson, G.D. (Ed.) q.v. p. 380-391.

Hurst, C. (1983) Petroleum geology of the Gorm field, Danish North Sea. In: Kaasschieter, J.P.H. and Reijers, T.J.A. (Ed.) q.v. p. 157-168.

Maher, C.E. (1981) The Piper oil field. In: Illing, L.V. and Hobson, G.D. (Ed.) q.v. p. 358-370.

Oele, J.A., Hol, A.C.P.J. and Tiemans, J. (1982) Some Rotliegend gas fields of the K and L blocks, Netherlands offshore (1968-1978)—A case history. In: Illing, L.V. and Hobson, G.D. (Ed.) q.v. p. 289-300.

Pennington, J.J. (1975) The geology of the Argyll Field. In: Woodland, A.W. (Ed.) q.v. p. 285-295.

Roos, B.M. and Smits, B.J. (1983) Rotliegend and main Bundstandstein gas fields in block K/13—A case history. In: Kaasschieter, J.P.H. and Reijers, T.J.A. (Ed.) q.v. p. 75-82.

Stauble, A.J. and Milius, G. (1970) Geology of Groningen gas field. *AAPG Memoir* **14**, p. 359-369.

Veen, F.R. van (1975) Geology of the Leman gas field. In: Woodland, A.W. (Ed.) q.v. p. 223-233.

Walmsley, P.J. (1975) The Forties field. In: Woodland, A.W. (Ed.) q.v. p. 477-487.

Williams, J.J., Conner, D.C. and Peterson, K.E. (1975) The Piper oil field, U.K. North Sea: A fault-block structure with Upper Jurassic beach-bar reservoir sands. In: Woodland, A.W. (Ed.) q.v. p. 363-379.

1.7.5 Other pertinent references

Walmsley, P.J. (1983) The role of the Department of Energy in petroleum exploration of the United Kingdom. In: Brooks, J. (Ed.) *Petroleum geochemistry and exploration of Europe*. Geological Society London. Special Publication No. 10, 3-10.

Chapter 2

The Structural Framework and the Pre-Permian History of the North Sea Area

K.W. GLENNIE

2.1 Introduction—a basic philosophy

The contours of the depth to the top of the Chalk beneath the North Sea (see Lovell, Fig. 8.3, this volume) indicate a simple pattern of subsidence centred along a north-south line roughly down the middle of the area. In contrast to this relative simplicity, anyone who has tried to construct a fault map of an older Mesozoic or late Palaeozoic seismic reflector knows that, at depth, the structure of some parts of the North Sea area is far from simple. This contrast is well illustrated by the seismically-derived cross-section in Figure 2.1, and, on the grand scale, the structural complexity of the older geological units is clearly seen on Figure 2.2, which is a map of the main Permian and pre-Cretaceous Mesozoic structural units of the region. Here, we see that northwest Europe has been broken into a large variety of basins and narrower grabens separated by areas of more positive relief.

As a part of the NW European continental shelf, the North Sea area has had a long and complex geological history. In much the same way that, in an evolutionary sense, the human race has inherited physiological features first developed by its zoological ancestors, so the later structural and stratigraphic development of the North Sea area seems to have been controlled to quite a considerable extent by relics of its earlier history.

In general, Precambrian cratonic blocks seem to have remained, or tried to remain, positive features throughout most of the Phanerozoic. Carrying this reasoning a little further, one could speculate with a fair degree of certainty that many of those areas in which, for instance, the Carboniferous sequence is thin, are probably underlain by blocks of relatively light and buoyant Precambrian or Early Palaeozoic continental crust. Similarly, where late Palaeozoic granites intrude sequences of older Palaeozoic deep-marine sediments, the granites probably reflect the mobilisation either of underlying silica and feldspar-rich continental craton, or of an accretionary wedge of quartz-rich sedimentary rocks overlying a subduction zone. Had oceanic crust been involved, then the resulting intrusions are much more likely to have been gabbroic in nature.

An outline of the regional geological history is given in the following pages, and will show that the existing structural and stratigraphical framework is largely the outcome of:

(a) a sequence of divergent, convergent and tangential plate movements,
(b) crustal subsidence and uplift resulting in basins and highs, and crustal shortening leading to orogenesis,
(c) erosion leading to sedimentation in both marine and continental environments,
(d) changes in the rate of heat flow through the crust commonly leading to uplift and igneous activity in areas of high heat flow, or to subsidence, commonly with no igneous activity in areas of low heat flow.

The simple structural outline and associated historical development given here is intended to set the scene for the more detailed discussions given in later chapters. As a scene setter, therefore, emphasis is given to the pre-Permian history, whereas the Mesozoic and Cenozoic tectonics are considered largely on the mega scale, reflecting events that took place mainly beyond the limits of the North Sea (Table 2.1).

2.2 Outline of the structural framework and its evolution

2.2.1 Sources of data

Our understanding of the geology of the North Sea area is derived from two main sources:
1. The surface (and subsurface) geology of the surrounding land areas, and
2. The subsurface geology of the North Sea itself.
This latter is based largely on the structural and stratigraphic interpretation of thousands of kilometres of seismic lines, which have been calibrated by the use of cores, drill cuttings and wireline logs from an ever-increasing number of exploration wells and, to a lesser extent, the production wells of oil and gas fields. Our knowledge is still patchy, however, the degree of available detail in any particular area being dependent on the Industry's assessment of its commercial prospectivity.

Line tracings of the key horizons on selected modern seismic lines (e.g. Fig. 2.1) illustrate the sort of regional geological interpretations that are now possible. For the benefit of those with only a limited knowledge of geology, a more extensive interpretative caption to the figure is given as an appendix to the chapter.

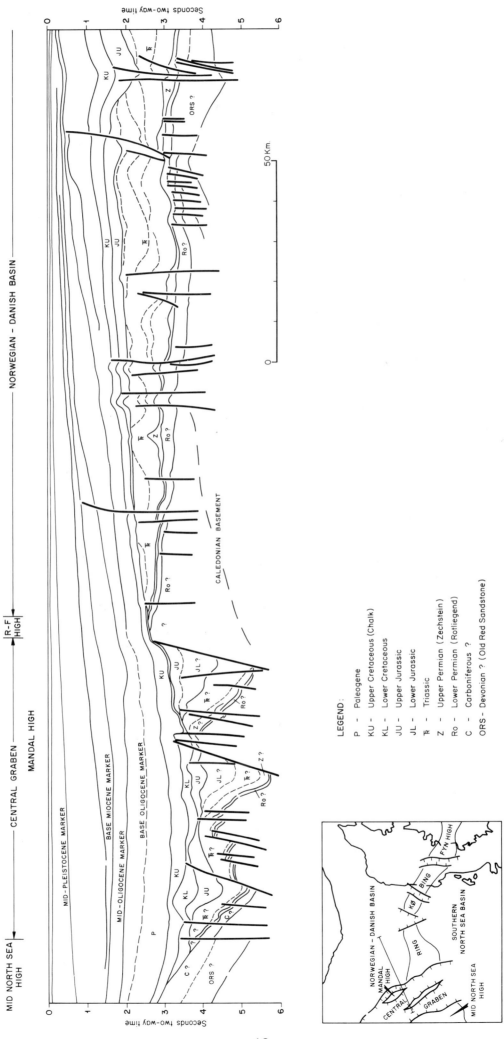

Fig. 2.1. Line drawing of a composite seismic line across the northern Central Graben and part of the Norwegian-Danish Basin. (A more extended caption is given as an appendix to this chapter).

18

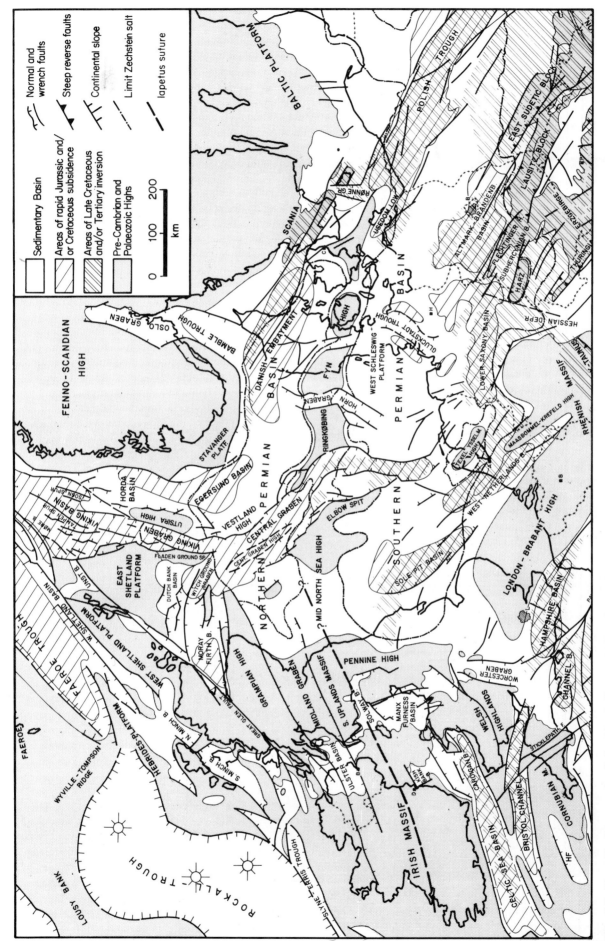

Fig. 2.2. Tectonic framework of the North Sea area: Permian and Mesozoic tectonic units (simplified from Ziegler, 1982a, and Iapetus suture added).

Table 2.1. Simplified evolution of the Tethys and Atlantic oceans tentatively related to some parochial North Sea post-Caledonian structural events.

Ma.	PERIODS		REGIONAL EVENTS		NORTH SEA			
			TETHYS-RELATED	ATLANTIC-RELATED	NORTHERN	MORAY FIRTH	CENTRAL	SOUTHERN
CENOZOIC	Miocene		ALPINE OROGENY	SPREADING OF PRESENT MID ATLANTIC RIDGE	REGIONAL SUBSIDENCE CENTRED OVER GRABEN SYSTEM			Zechstein diapirism
	Oligocene							
	Eocene		PLATE COLLISION	Plateau basalts				Inversion of NW-SE trending sub-basins
100	Paleocene		GRADUAL CLOSURE OF TETHYS — Rotation of Iberia	SPREADING OF ROCKALL TROUGH EXTENDING NORTH TO NORWAY-GREENLAND SEA	Doming of West Shetlands / LATE RIFTING PHASE	Uplift of Moray Firth & Scottish Highlands	Renewed faulting in Central Graben / Inversion of Danish Embayment	Zechstein diapirism
CRETACEOUS	Late			SEA-FLOOR SPREADING IBERIA – NEWFOUNDLAND		Rapid subsidence of inner Moray Firth	SUBSIDENCE in CENTRAL GRABEN / Extrusives in Danish Embayment	Indefatigable erosion / Rapid Sole Pit subsidence
	Early		SEA-FLOOR SPREADING IN TETHYS	ONSET OF SEA-FLOOR SPREADING IN CENTRAL ATLANTIC	Doming in N. Viking / ROTATIONAL FAULTING IN VIKING GRABEN		DOMAL COLLAPSE & MAIN PHASE OF GRABEN FORMATION	RIFT & WRENCH TECTONICS
JURASSIC	Late						DOMING AND LIMITED VOLCANIC ACTIVITY	
	Mid		RIFTING PHASE	RIFTING IN CENTRAL ATLANTIC	MAIN RIFTING PHASE	Volcanic activity in East		
200	Early							
TRIASSIC					DEVELOPMENT OF NW EUROPEAN BASIN/GRABEN SYSTEM		Earliest Zechstein diapirism	Zechstein diapirism / 4000m Triassic in Polish Trough
PERMIAN	Late		LATE HERCYNIAN WRENCH TECTONICS	Rifting in Norway-Greenland Sea	Zechstein flooding of sub-sealevel basins			
	Early		EARLY COLLAPSE OF VARISCAN FOLD BELT IN EUROPE		SUBSIDENCE OF SOUTHERN & NORTHERN PERMIAN BASINS BEGAN	SUBSIDENCE OF MORAY FIRTH	Extrusion of L. Rotliegend volcanics began / Right-lateral faulting: inversion of Sole Pit Basin	
300					E-W Scottish & Scanian dyke swarms: volcanics			
CARBONIFEROUS	Stephanian		VARISCAN OROGENY	INITIATION OF NORTH ATLANTIC FRACTURE PATTERN	RIFTING IN NORTH BRITISH ISLES		2500m U. Carboniferous	
	Westphal. D–A		PLATE COLLISION				VARISCAN FOREDEEP	
	Namurian				Renewed uplift of Scottish Highlands			
	Dinantian		STEP-WISE CLOSURE					BACK-ARC RIFTING
DEVONIAN	Late		OF PROTO TETHYS	Probable strike-slip movement of Great Glen Fault & extension in N. Atlantic		Volcanics in Orcadian Basin	Marine Limestones in Auk & Argyll	
	Mid				Granites in Scot. Highlands	Subsidence in Orcadian Basin	Volcanics in S. Scotland	Granites in Lake District
400	Early				CALEDONIAN OROGENY			

20

2.2.2 Major structural features

As we have already seen from the Top Chalk contour map (Fig. 8.3), the North Sea contains the site of an axis of considerable Tertiary subsidence, which is flanked by the positive areas of the British Isles to the west and Scandinavia (including the Danish peninsula) to the east. The southern margin is marked by the northern terrestrial limit of the Alpine foreland (The Netherlands, Germany), but, both in a physiographical and in a structural sense, the northern limit of the North Sea is marked by the NNE trending Atlantic continental margin just beyond the Shetland Isles (see, e.g. Fig. 2.2), at about latitude 62°N.

As mentioned earlier, by digging deeper below the surface we find a much more complex erosional and depositional pattern (Fig. 2.1, 2.2). The area seems to be dominated by two east-west trending basins, the larger Southern and the smaller Northern Permian Basins which, as their names imply, came into existence during the early Permian or perhaps even the latest Carboniferous; they are separated by the Mid North Sea-Ringkøbing-Fyn system of highs, and are surrounded by positive areas of older deformed rocks.

Cutting both basins and highs almost at right angles to the basins' axes is a system of grabens. The most important, in terms of both structural effect and its association with hydrocarbon accumulations, is the zigzag N-S trending Viking and Central graben system. The Viking Graben lies north of the Northern Permian Basin and separates the Shetland Platform from the Fenno-Scandian High. The Central Graben, however, cuts both the Permian basins and the intervening high.

Other areas of major subsidence include the Horn-Bamble-Oslo system of grabens, which came into existence during the Late Carboniferous, and the Moray Firth Basin, whose history of subsidence probably did not begin until the early Permian. Other basins, grabens and half grabens also developed at about this time beyond the limits of the North Sea (e.g. Manx-Furness, Solway and Ulster basins in the British Isles). Although following a different pattern of subsidence from that seen in the North Sea area, their origins must be causally related.

Mining, drilling and seismic data reveal the presence of an older sedimentary basin beneath the floor of the Southern Permian Basin (Fig. 2.3). This deeper basin is important to the oil industry because it contains abundant Carboniferous coal seams, which are the source rocks for most southern North Sea gas. With an ever decreasing cumulative thickness of coal, the surface of this basin extended as a broad plain northward towards the Caledonian Highlands of Scotland and Norway; to the south, the Carboniferous sediments became marine along the northern edge of the Variscan foredeep, which is now largely represented by the Variscan Mountains (Fig. 2.3). Relative uplift and erosion following the Late Carboniferous Variscan orogeny (e.g. Mid North Sea High) has reduced the area within which the Carboniferous Coal Measures are preserved.

Crossing southern Scotland obliquely and extending beneath the North Sea is the Midland Valley Graben, which virtually dies out in the Forth Approaches Basin before the Central Graben is reached; and the Southern Uplands, a positive area of strongly folded Lower Palaeozoic strata, which, as a structural unit, seems to be continuous with the north-western part of the Mid North Sea High (Fig. 2.2).

Further to the north, the folded and strongly metamorphosed Caledonian rocks of the Scottish Highlands and western Norway are wedged over and against the separate Precambrian cratons of the Hebrides Platform in the west and the Fennoscandian (Baltic) Shield to the east (Fig. 2.3). During the Devonian, the synorogenic and post-orogenic Old Red Sandstone filled all the areas of relative subsidence within and beyond the Highlands with the erosional products of the Caledonian Mountains.

2.2.3 Evolutionary outline

An outline of the most important events in the geological evolution of the North Sea area can be given very briefly, and are highlighted in historical order as a series of cartoons in Figure 2.4.

The Caledonian rocks beneath the northern North Sea were deformed into a major mountain range during the mid Palaeozoic (Fig. 2.4C), were transected by the Viking Graben in the Permian (?) and Mesozoic (Fig. 2.4F), and were virtually completely buried by the end of the Cretaceous. At the southern limit of the yet unborn North Sea Basin, the Late Carboniferous Variscan Orogeny (Fig. 2.4E) marked the closure of another important ocean known as Proto Tethys, and the creation of the super-continent Pangaea. After the initial subsidence of the Northern and Southern Basins had been completed, the Viking/Central Graben rift system began to play a dominant role in the structural history of the area. Following its earlier inception, it began to subside rapidly during the Triassic, and reached its maximum structural development in the late Jurassic and early Cretaceous. The duration of its active existence coincided with the slow break-up of Pangaea into Laurasia and Gondwana; the termination of North Sea graben development was linked to the separation of Laurasia into North America and Eurasia by the onset of sea-floor spreading in the North Atlantic Ocean (Table 2.1). Spreading was initially along the line of the Rockall Trough (Fig. 2.4G) and then shifted to its present axis (Fig. 2.4H) early in the Tertiary. Some idea of the tensional complexity involved in these movements can be gleaned from the map of the Triassic rift systems in the North Atlantic realm (Fig. 2.5).

Thus the structural geometry of the North Sea area can be considered as the result of plate movements involving tension or compression in different directions at different times in its developmental history. The first of the cartoons outlining this history (Fig. 2.4A)

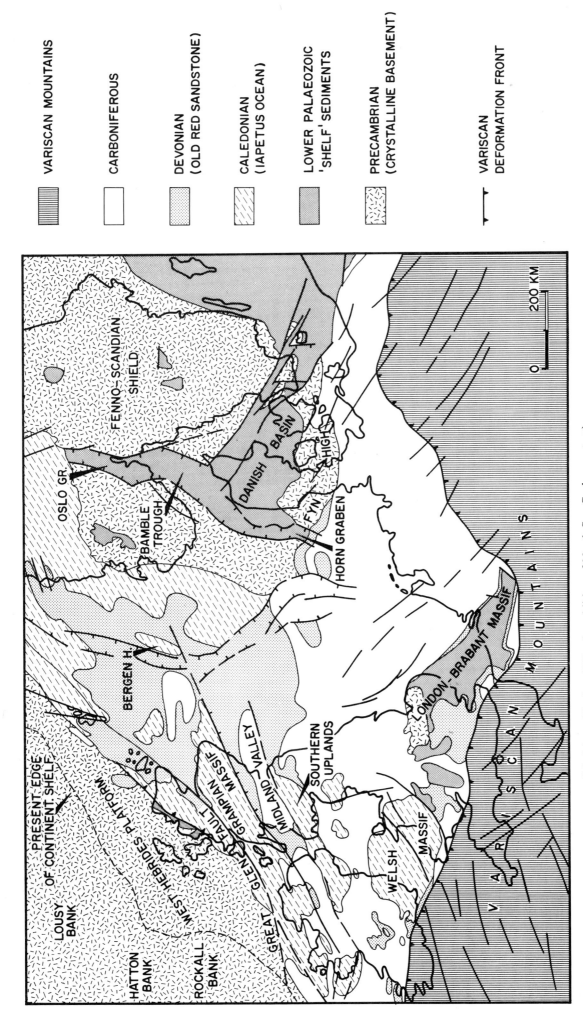

Fig. 2.3. Pre-Permian geological map. (Modified from Ziegler, 1982a, by addition of North Sea Graben system).

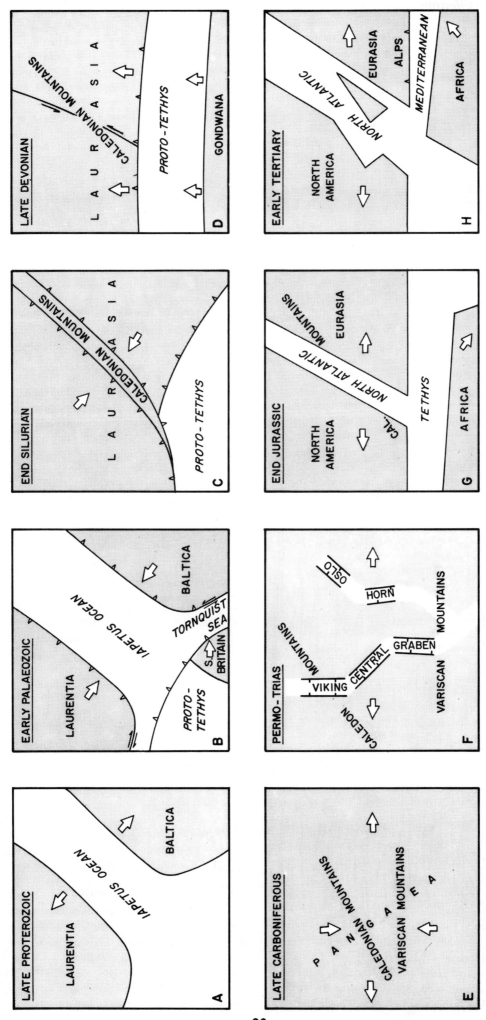

Fig. 2.4. Cartoons of relative plate movements that affected the North Sea area during the late Proterozoic and the Phanerozoic. (Caution: the scales are not constant).

23

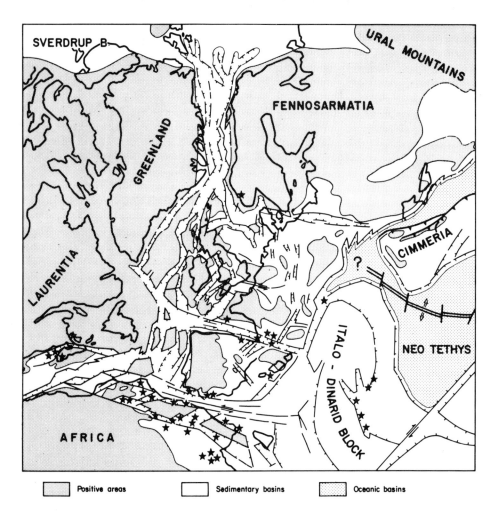

Fig. 2.5. Triassic rift systems of the North Atlantic area; largely reactivated relics of an earlier history. Stars indicate Triassic volcanic activity. From Ziegler (1982a).

Positive areas	Sedimentary basins	Oceanic basins

indicates that we must return to the Precambrian to follow it fully.

2.3 Pre-Permian history

2.3.1 Iapetus Ocean and Caledonian Orogeny

The history of the North Sea area prior to the Late Silurian-Early Devonian Caledonian Orogeny is known only in the simplest terms. The results of studies in Britain, Scandinavia, Greenland and Canada, however, are beginning to give a logical, if still tentative, history of events that have direct implications for the North Sea area. From our point of view, perhaps the best starting point for discerning this history is in the British Isles.

Precambrian beginnings

Apart from the Archaean Lewisian Gneiss, so characteristic of the Outer Hebrides (Fig. 2.6), the metamorphic rocks of the Scottish Highlands can be divided into two major sequences, the Moines and the Dalradians. The Moines are confined to the more northwesterly exposures (NW Highlands and Grampians), and are overlain by the Dalradians (dominating the area adjacent to the Highland Boundary Fault, Fig. 2.6), which include basic igneous rocks possibly representative of former oceanic crust (e.g. basic rocks of

Unst, Shetland, and Morven, Aberdeenshire).

The Caledonian Orogeny was the outcome of geological events that, so far as we are concerned, perhaps had their origins almost 1000 Ma ago. Following the Grenvillian Orogeny (∿ 1040 Ma), tensional movements were responsible for graben formation within the Lewisian crystalline crust of an Archaean continent, and for deposition of the earliest terrestrial sediments of the Torridonian Group (Stewart, 1982). West of the Moine Thrust (MT on Fig. 2.6), these ancient sediments are not metamorphosed. To the east of the thrust, however, they were first metamorphosed together with the underlying 'Grenville' basement around 740 Ma (Morarian event) and then again during the Latest Cambrian and Early Ordovician together with the laterally adjacent and partly overlying Dalradian Series (Johnson *et al.*, 1979; Watson and Dunning, 1979).

Palaeozoic development

Much of the marine sediment involved in the Caledonian Orogeny was deposited in the Iapetus Ocean (Harland and Gayer, 1972), which separated the Baltic and Laurentian shields throughout the early Palaeozoic. In Scotland, the site of the rupture which gave rise to that ocean possibly lay south of the present Midland Valley. Anderton (1982) suggests that the opening was signalled by the eruption of the Tayvallich

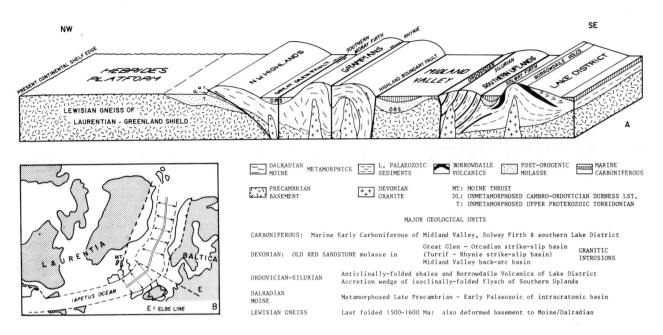

Fig. 2.6. A. Block diagram of the main Caledonian structures between the edge of the Hebrides Platform and southern edge of the Lake District. B. Tentative reconstruction of the Iapetus Ocean showing the Elbe-line wrench fault (E) as the possible line of closure of the Tornquist Sea.

Volcanics around 600 Ma (latest Precambrian, earliest Cambrian) in an ensialic basin north of the Midland Valley. This narrow basin along the margin of the Laurentian Continent became the site of early Palaeozoic marine sediments of the Highland Border Complex, which were later metamorphosed as part of the Dalradian Series (Henderson and Robertson, 1982); associated serpentinites, gabbros and spilitic lavas suggest that the basin was floored by crust of oceanic type. As yet, however, it is still uncertain whether the rift originated about the same time as the opening of the Iapetus Ocean, or as a back-arc basin during later ocean closure.

An increase in the volume of the world's mid-ocean ridges is thought to have caused a Cambrian transgression over the older Moine and Dalradian terrestrial sequences. This transgression is also recognised in the Baltic area, for instance, where the still-undeformed Alum oil-shale sequence was deposited (see Cornford, this volume). Unfortunately for Denmark, these rocks seem rapidly to become post-mature for oil further west (Thomsen *et al.*, 1983).

The Laurentian Plate, of which Scotland formed a part, occupied a near equatorial location (Cocks and Fortey, 1982) and became the site of carbonate-platform deposition, and sediment starvation of more basinal areas, from the mid-Cambrian onwards (Fig. 2.7).

It has long been known that the faunas of the Scottish equivalent of this equatorial Cambro-Ordovician carbonate shelf (Durness Limestone, Fig. 2.6) have a much greater affinity to those of the Beekmantown Limestone of Pennsylvania (see, e.g. Phemister, 1960) than to the physically now much closer sequences of Wales. Cocks and Fortey (1982), used the climatic effects on assemblages of shallow-marine, planktonic and deep-water benthic faunas to reconstruct probable

changes in the width of the Iapetus Ocean with time (Fig. 2.7). In addition to the east coast of Laurentia, which included northern Scotland, they found that the two highest nappes in the Trondheim area of Norway must also have lain close to the Equator during the early Ordovician. The rest of Scandinavia, Wales and England, however, lay some 40-60° to its south (Fig. 2.7A). Furthermore, using the provinciality of faunal assemblages as a guide, they recognised that Scandinavia and England must have been separated by a deeper-marine area, which they called the Tornquist Sea, corresponding to the suture marked by the North German-Polish Caledonides (Ziegler, 1982). The Tornquist Sea had closed by the end of the Ordovician (Fig. 2.7B).

Caledonide orogenic events

The late Proterozoic and early Palaeozoic sedimentary and volcanic rocks of the Caledonides were deformed and metamorphosed at several different times during the 600 million years of its developmental history. We have already noted the metamorphism of equivalents of the Torridonian sediments of the Scottish Highlands around 740 Ma (Morarian event) to form the Moine Series. The Highlands were also involved in the Grampian Orogeny, which had its peak in the Late Cambrian to Early Ordovician and resulted in the strong folding and metamorphism of both the Dalradian and the already deformed Moine Series, and in the emplacement of granites within them (e.g. the older granites of the N.E. Highlands: Johnstone, 1966; Bradbury *et al.*, 1976). Associated uplift of the Highlands resulted in the southward spread of turbidite fans across the Iapetus Ocean. Northward subduction of these fans during the later Ordovician and Silurian gave rise to

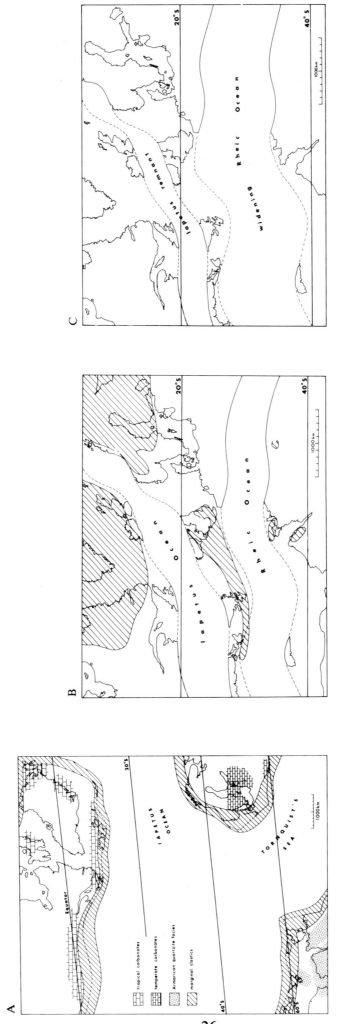

Fig. 2.7. Three stages in the oceanic separation of Britain during the early Palaeozoic. A: Early Ordovician (Arenig), B: Early Silurian (Llandovery), C: Late Silurian (Ludlovian), based on faunal evidence. The Tornquist Sea probably had closed by the Late Ordovician, uniting most of Baltica with Gondwana (including Wales and much of England). By the Late Silurian, the Iapetus Ocean had almost closed, uniting all Britain except Cornubia, which was about to be deposited in the widening Late Palaeozoic Rheic (Proto-Tethys) Ocean. Adapted from Cocks and Fortey, 1982.

the construction of the imbricate (McKerrow *et al.*, 1977; Webb, 1983) accretionary wedge of the Southern Uplands.

The Grampian Orogeny was probably caused by a violent collision between the Scottish Highlands and a Midland Valley landmass. The stresses generated by this collision were possibly transmitted through the Midland Valley 'massif' as the result of a collision on its southern side with the sialic basement of the Southern Uplands (Watson and Dunning, 1979) the existence of which is deduced from the occurrence of younger granitic intrusions within its sedimentary overburden. In keeping with the above interpretation, the Southern Uplands basement may have been an underthrust southern continuation of the Midland Valley massif. On the basis of seismic refraction data, however, Leggett *et al.* (1983) suggest that the basement is of English origin and was partly subducted beneath the Southern Uplands accretionary wedge (Fig. 2.6) during the late Silurian-early Devonian final closure of the Iapetus Ocean. If correct, then the basement of the Southern Uplands can have played no part in the Grampian Orogeny.

Closure of the Iapetus Ocean seems, at different times, to have been achieved by both NW-directed and SE-directed subduction (see e.g. Phillips *et al.*, 1976, Fig. 4). Within the British Isles, the line of closure is marked by a suture that can be traced from the Shannon estuary in western Ireland, via the Solway Firth (Fig. 2.2, 2.6) into the northeast-trending Northumberland Trough. Its straightness and lack of major disruption implies a strike-slip origin with relatively slow convergence of the two sides of the ocean. This suture has not been traced with certainty beneath the North Sea.

South of the Iapetus suture, the Lake District of northern England and the Wicklow Mountains of southeast Ireland were the sites of extensive mid-Ordovician andesitic volcanism (e.g. Borrowdaile Volcanics, Fig. 2.6) typical of a subduction-related island arc. The current proximity of the Lake District portion of this arc to the line of suture can be accepted if the missing part of the accretionary wedge is assumed to have been removed by oblique strike-slip subduction; such a wedge is still preserved north of the Wicklow Mountains.

With its asymmetrical flanks, the broad structure of the Lake District (Fig. 2.6) suggests that a zone of décollement exists between the pre-volcanic sedimentary sequence and the crystalline basement that is interpreted to underlie it. Such a zone would be in line with the partial subduction of this basement beneath the Southern Uplands accretionary wedge during the late Silurian (Fig. 2.6; see also Leggett *et al.*, 1983, Fig. 3). This basement possibly represents the northern edge of a microcontinent that stretches south to the Midlands of England, with the Welsh Basin occupying a site of strong subsidence on its western flank (Fig. 2.6B). The extent of the microcontinent beneath the North Sea is uncertain.

Closure of the Tornquist Sea did not give rise to a major fold belt. It is suggested here that the subduction geometry of the British and Scandinavian portions of the Iapetus Ocean was such that the Tornquist Sea closed as a result of the oblique convergence of the Baltic and English plates, and of other microcontinents enclosed in the Caledonides of central and eastern Europe.

2.3.2 Devonian Old Red Sandstone

Closure of the Iapetus Ocean resulted in the creation of the supercontinent Laurasia (Figs. 2.4, 2.7, 2.8) and in uplift of a major mountain range that stretched from the southern United States to eastern Canada (Appalachians) through northern Britain to the northern end of the united Greenland—Scandinavia (Fig. 2.8A). Widespread early Devonian granitic intrusions (e.g. the newer granites of NE Scotland; Johnstone, 1966) and associated volcanic activity testify to the anatectic remobilisation of subducted crustal material of continental type during orogenesis. Some of these intrusions were probably responsible for the destruction of older Palaeozoic source rocks for oil and for their conversion to graphite, as at Seathwaite in the Lake District (Parnell, 1982a).

Erosion in an almost vegetation-free continental climate that seems to have had a seasonal rainfall (Barrell, 1916) resulted in deposition of widespread fluvial sequences, which in some cases terminated in intramontane lacustrine basins (e.g. Lake Orcadie, Geikie, 1879; Fig. 2.9). Along the southern margin of the newly formed Caledonian Range, Early Old Red Sandstones drainage was parallel to the mountain front and, over a considerable distance, converged towards the present North Channel between Scotland and Ireland. A changing basin configuration resulted in Late Old Red Sandstone sediment transport in the Midland Valley and Northumberland Trough being directed towards the North Sea area (Simon and Bluck, 1982). The Devonian sea lay across southern Britain (Fig. 2.9).

The Orcadian Basin (Fig. 2.9), within which the Middle Old Red Sandstone is up to 5 km thick, seems to have developed on both sides of the major Great Glen wrench fault (Fig. 2.2, 2.8B), across which both the sense and amount of horizontal displacement is in dispute. Originally given a post-Devonian sinistral offset of 110 km by Kennedy (1946), an offset of up to 2000 km in the same sense (see Fig. 2.7B) has been proposed by Van der Voo and Scotese (1981) on the basis of palaeomagnetic data. This proposal has been hotly disputed by Donovan and Meyerhoff (1982), Parnell (1982b) and by Smith and Watson (1983) on the grounds that the unique facies patterns and biostratigraphy present on either side of the fault have not been recognised elsewhere. These latter writers therefore suggest that if there has been any movement, it has been in a dextral sense and, especially since the onset of Middle Devonian time, of small amount.

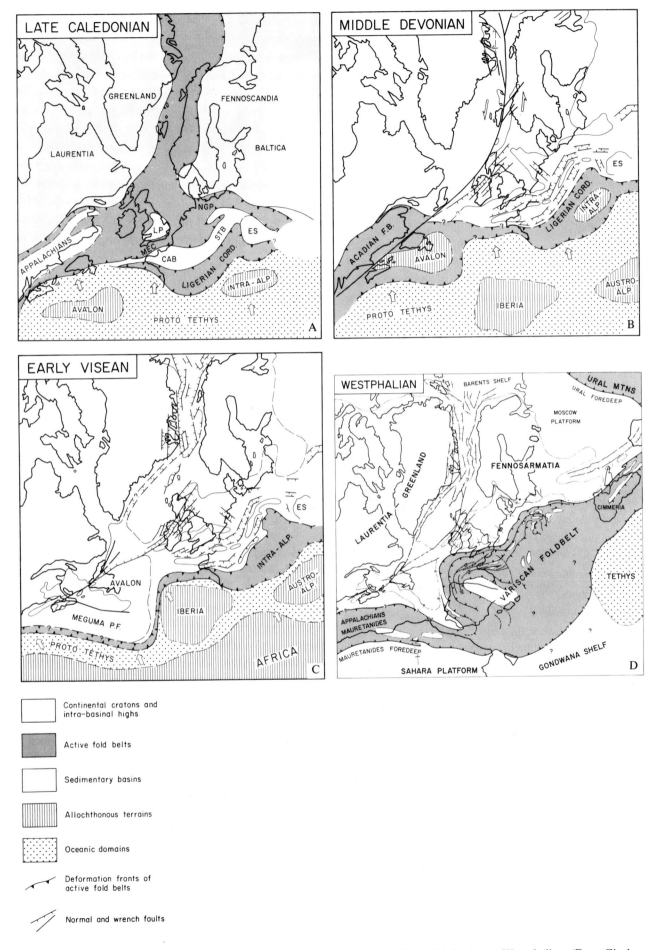

Fig. 2.8. The tentative tectonic framework of the North Atlantic realm from Late Caledonian to Westphalian. (From Ziegler, 1984).

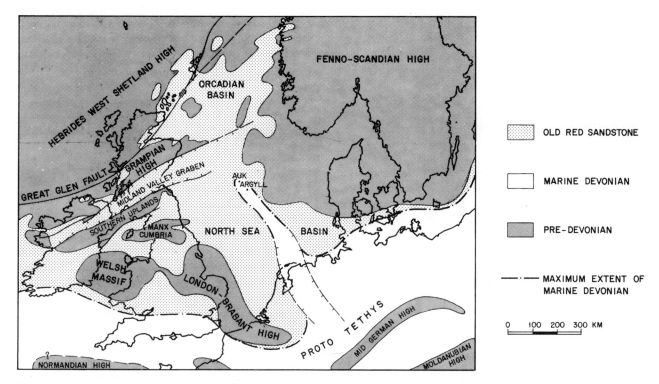

Fig. 2.9. Maximum extent of continental Old Red Sandstone, and marine Devonian sediments of Proto-Tethys. (Modified from P.A. Ziegler, 1982a).

Nevertheless, as P.A. Ziegler (1982b) points out, the deposition and deformation of the thick Old Red Sandstone sediments in the Orcadian Basin are probably the result of an interplay between tensional and compressional stresses associated with horizontal displacements. This seems to be supported in particular on Shetland, where the lithological contrast seen across the Walls Boundary Fault (the probable northern extension of the Great Glen Fault) and the late Devonian deformation along approximately E-W fold axes of Mid Devonian sandstones on the Walls Peninsula are strongly suggestive of, respectively, N-S offsets and associated compression in the same sense (see e.g. Mykura, 1976; Fig. 2, Pl. VII). After folding, the sandstones just west of the Walls Fault were intruded by dykes and sill-like sheets of diorite and granite, suggesting deep crustal tension in the latest Devonian, again probably related to strike-slip movement. Both Devonian and post-Devonian offsets could well have been relatively small, however, and, as suggested by Smith and Watson (1983), possibly of no more than 100-200 km. In this respect, it is pertinent that volcanic activity in the Orcadian Basin was decreasing during the Mid and Late Devonian. Also pertinent is that the offshore basins of the Mid-Norway area seemed to have been dominated by shearing during the Devonian and Early Carboniferous (Bukovics, *et al.*, 1984; see also Fig. 7).

Fish beds in the lacustrine sediments of the Orkney Basin are famed for the richness and diversity of their contained species, and have been considered as a potential source of oil. Indeed, a little oil can be seen to have been generated locally but, in general, the fish beds are barely mature at outcrop. They conceivably could have contributed to accumulations in the West Shetland and Inner Moray Firth basins, however, where they are more deeply buried. Potential reservoirs for oil can be found in the abundant Mid and Late Devonian fluvial and aeolian sandstones now exposed in Caithness, Orkney and Shetland (Mykura, 1976; Allen and Marshall, 1981), which are locally impregnated with dead oil; and on the flank of the Central Graben, Old Red sandstones form the reservoir for commercial oil of Jurassic origin in the Buchan Field.

To the south of the Old Red Sandstone continent, the Proto Tethys (Fig. 2.4C, D and 2.8A, B, C) had been in existence since the beginning of the early Palaeozoic, and would remain so until its closure led to the Late Carboniferous Variscan Orogeny (Fig. 2.4E). Following a structural trend that was possibly initiated during closure of the Tornquist Sea, an arm of the Mid Devonian sea spread as far north as the Auk and Argyll oil fields, where shallow-marine carbonates and evaporites were deposited (Fig. 2.9). This arm of the sea appears to have closely followed the line later adopted by the Central Graben; repeated inheritance of an old line of weakness seems most likely.

In the Late Devonian, Old Red Sandstone continental sediments occupied a broad NW-SE trending depression in the North Sea area between the Scandinavian Highlands in the east and, to the west, a combination of the Scottish Highlands and isolated positive areas of erosion comprising the Southern Uplands, Manx-Cumbria, and a SE Ireland-Wales-London-Brabant High (Fig. 2.9). Although ill defined, the size and shape of this Devonian basin bears some similarity to the Cenozoic North Sea Basin (cf. Fig. 8.3, this volume).

2.3.3 Carboniferous

The Carboniferous of NW Europe is very important to the hydrocarbon industry because it was during this period that the carbonaceous source rocks for the whole of the southern North Sea and Dutch-German-Polish gas belt were deposited. The main period of coal deposition was during the Late Carboniferous Westphalian stage, when over 2000 m of strata were deposited locally with a cumulative thickness of coal well in excess of 50 m.

The generalised outline of preserved Carboniferous strata north of the Variscan deformation front is seen on Figure 2.3, and the distribution of some of the major rock units of the Carboniferous is displayed on Figure 2.10. Figure 2.11 depicts the relationship between the presence of Westphalian A and B coal measures and the accumulations of gas in the overlying Lower Permian (Rotliegend) sandstone reservoirs of the southern North Sea.

Dinantian

With the slow northerly drift of Laurasia, Early Carboniferous sedimentation represents a transition from the relatively arid Old Red Sandstone conditions of the southern hemisphere tropics to the more humid equa-

torial conditions of coal-measure deposition (cf. Habicht, 1979).

Erosion of the Caledonian Highlands and deposition of non-fossiliferous continental clastics in basinal areas continued to dominate the region north of the Highland Boundary Fault throughout the early Carboniferous (Fig. 2.10). South of the fault, and in continuation of the scene already set during the Late Devonian, a broad relatively flat plain occupied much of the southern half of the North Sea area and extended south to the axial parts of the Variscan foredeep basin. Marine conditions were established across the southern part of this plain, but further north, sun-cracked shales and argillaceous dolomites of the basal Carboniferous Cementstone Group suggest deposition beneath inland sheets of water subjected to periodic desiccation (McGregor and McGregor, 1948). In the Midland Valley of Scotland, the succeeding oil shales were also deposited under lacustrine conditions.

Shallow-marine carbonates dominated early Carboniferous deposition in the western British Isles, and it is probably from the west that the Midland Valley of Scotland was flooded by incursions of the sea with the resulting deposition of shallow-marine limestones. The pre-Permian strata forming the floor of the Danish Embayment (cf. Figs. 2.1, 2.2, 2.9) are mostly at depths that are too great (5000 m; Ziegler, 1982, Encl. 34) to form an economic target for the drill. It is still uncertain, therefore, whether or not they include a Carboniferous sequence. The reported presence of marine strata of Carboniferous age in the Oslo Graben

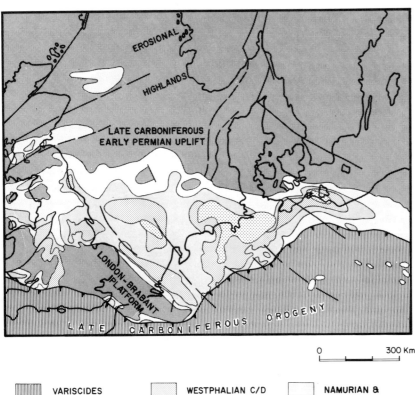

▦ VARISCIDES	▨ WESTPHALIAN C/D	☐ NAMURIAN & DINANTIAN
▦ STEPHANIAN	☐ WESTPHALIAN A/B	▨ PRE-CARBONIFEROUS

Fig. 2.10. N.W. European Carboniferous basin. Simplified from Ziegler, 1982a).

(Olaussen *et al.*, 1982) poses the question of their former connection with the open ocean. This might have been from the west along an extension of the Midland Valley of Scotland or, perhaps more likely, from the south via either an incipient Horn Graben (see Olsen, 1983) or along that old line of crustal weakness marked by the Tornquist-Teisseyre fault system. This problem is unlikely to be resolved until some very deep wells are drilled in the Danish Embayment, or a well by chance exposes evidence in the vicinity of Denmark.

Cyclic alternations of marine and terrestrial sedimentation (including the formation of coal seams) during the later Dinantian may reflect eustatic changes in sea level resulting from the earliest of the episodic Permo-Carboniferous glaciations of Gondwana. Such glacially-induced fluctuations in sea level probably continued to affect the pattern of sedimentation throughout the Carboniferous, and interglacial rises in sea level were possibly responsible for some of the horizons of marine fossils, especially those of great regional extent, that are so characteristic of the Westphalian Coal Measures. In NE England, late Visean cyclic sedimentation involving both marine limestone and coal deposition is known as the *Yoredale* facies. Poorly dated strata of this type and approximate age are known from wells drilled over parts of the Mid North Sea High.

A number of grabens subsided during the Dinantian, including the Midland Valley of Scotland and the line of the old Iapetus Suture along the Dublin-Solway-Northumberland Trough (Fig. 2.2). In some areas such as the Midland Valley, subsidence was accompanied by igneous activity. Both graben development and igneous activity largely ceased during the Namurian and early Westphalian, presumably in response to N-S compression related to the Variscan Orogeny.

The morphologically smooth and almost horizontal depositional area of the southern North Sea formed a marked contrast to the early Carboniferous structural relief that must have been present between the Scottish border and North Wales (Fig. 2.11). Visean limestones locally reach a thickness of 3000 m in the Solway-Northumberland Trough and yet are less than 500 m thick over the Alston Block to the south (Taylor *et al.*, 1971, fig. 17). Much of this relief was smoothed out by Visean sedimentation (see also Fig. 2.6).

Namurian

Renewed uplift of the Scottish Highlands in the late Visean was probably responsible for the southerly progradation during the Namurian of a fluvio-deltaic sequence of sandstones, coals and marine shales known generally as the Millstone Grit. The best development of the Millstone Grit seems to have been

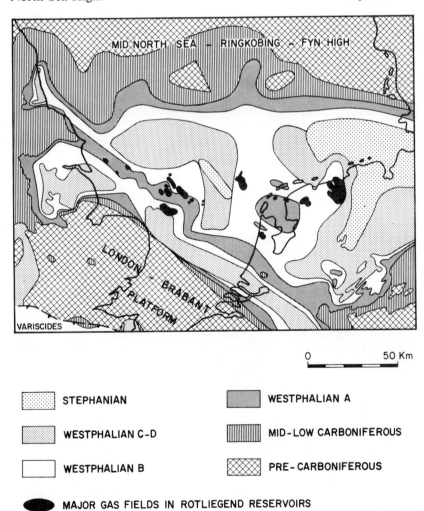

STEPHANIAN

WESTPHALIAN C–D

WESTPHALIAN B

MAJOR GAS FIELDS IN ROTLIEGEND RESERVOIRS

WESTPHALIAN A

MID–LOW CARBONIFEROUS

PRE–CARBONIFEROUS

Fig. 2.11. Pre-Permian of the Southern North Sea area. Modified from Eames, 1975.

confined to a N-S trending zone whose axis coincided with the present Pennine uplift; the sequence thins markedly to both east and west.

Off the southern end of the Millstone Grit delta, a turbidite sequence was deposited in a deeper-marine environment, which also seems to have been the site of deposition of the source rocks from which the Eakring oil was derived. Separate developments of local troughs have been recognised in the Grantham, Edale and Widmerpool areas. The first two follow a NW-SE Precambrian (Charnoid) structural trend whereas the third aligns with the locally E-W orientation of the northern edge of the London-Brabant Platform (Fig. 2.11).

Westphalian and Stephanian

The main development of thick coal seams was during the Westphalian, when coastal-plain paralic sediments prograded southward around (and only possibly across) the London-Brabant Platform. Coals were deposited on the south flank of the Platform in Kent, Belgium and northern France, however. To the east, Westphalian coals were deposited in The Netherlands, Germany and Poland (Fig. 2.10).

The long history of the London-Brabant Platform as a positive area can be inferred from some southern North Sea wells such as 47/29a-1, where Namurian strata overlie Ordovician rocks, and in Kent, where Westphalian Coal Measures onlap Silurian strata. The Platform was probably not transgressed by the sea until the Cretaceous, when it received a relatively thin cover of sediments (see Hancock, this volume, Figs. 7.5 and 7.8).

Coals form between 3 and 4% of the total thickness of the Westphalian, with the greatest concentration of seams in the older Westphalian A and B stages, but gas-prone carbonaceous shales must add considerably to the importance of the sequence as a source of gas (see also Cornford, this volume).

In the U.K. sector of the North Sea, the Westphalian sequence reaches a maximum thickness of some 1200 m in the vicinity of the Sole Pit axis of Late Westphalian inversion (see next section), whereas in The Netherlands (see Fig. 3.13) and Germany the total coal-bearing sequence (including Namurian) reaches over twice that thickness (Ziegler, 1977). Control on the thickness of the Westphalian succession is far from complete, however. In many North Sea wells, the drill penetrated no more than 10 to 20 m into the Carboniferous, much of which had been oxidised to a purple-brown colour prior to deposition of the overlying Permian desert sediments. Because of this oxidation, the preservation of age-indicative spores in the top 50 m of the Carboniferous is a rarity. With insufficient stratigraphic control, oxidised early Westphalian strata can be confused with red bed sequences of latest Westphalian and Stephanian ages—the so-called Barren Red Measures.

Overlying the main development of coal seams, the younger Westphalian C is found locally in a red-bed facies that already heralds the approaching aridity of the latest Carboniferous and early Permian. The areas of coal-bearing Westphalian C presumably result from deposition close to rivers or in the centres of sub-basins where the supply of water was adequate for plant growth. Similarly, the Westphalian D also has a well-developed coal-bearing sequence locally, and is in a red-bed facies elsewhere.

Although strata of Stephanian age occur in Dutch and German waters beneath the desert-lake sediments of the early Permian Rotliegend (see following chapter) none have been proven in U.K. waters. The writer has suggested (Glennie, 1983) that, away from the influence of permanent water, strong winds, associated with the effects of the Permo-Carboniferous glaciation in Gondwana, were possibly already having a marked deflationary effect upon late Carboniferous sediments during the Stephanian. Such winds are presumed to have resulted from an enlarged area of polar high barometric pressure, which caused all other global air-pressure belts to be squeezed towards the equator (see also Fig. 3.9).

It was also during the Westphalian to Stephanian time span that southern parts of the North Sea area and the Palaeozoic grabens of the British Isles were marginally affected by compressional deformations associated with the Variscan Orogeny. One of the effects of the Orogeny recognised in the area of the southern North Sea is a change in provenance of some of the sands of the Westphalian C sequence. All older Westphalian sands were derived from the north, but the more southerly Westphalian C sands came from the south.

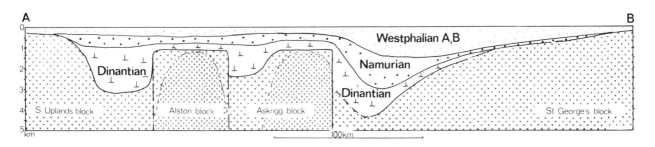

Fig. 2.12. Schematic N-S section from the Southern Uplands, along the Pennine axis to central Wales, to show the variable thicknesses of Dinantian, Namurian and Westphalian A/B sequences. From Leeder, 1982.

2.3.4 Variscan Orogeny

During the Devonian and early Carboniferous, northerly subduction of the oceanic crust of Proto Tethys caused Laurasia and Gondwana to move closer together. Continent-to-continent collision probably began in the late Visean. Orogenic activity, resulting in the creation of the Variscan Mountains, persisted until the Late Westphalian to Early Stephanian (Table 2.1). This mountain range, which lay in the heart of the newly formed supercontinent Pangaea (Fig. 2.4E), can be traced from Morocco, through Spain, France and Germany to southern Poland and Romania (Fig. 2.8D).

Following the North-South compression associated with the Variscan Orogeny, relative movement between Laurasia and Gondwana became E-W oriented (Ziegler, 1982b). The resulting system of NW-SE trending right-lateral wrench faults brought about the collapse of the Variscan fold belt during the Stephanian and Autunian. This fault system became largely inactive by the onset of the Saxonian, and over the Variscan foreland, the Northern and Southern Permian basins began to subside. In the U.K. southern North Sea, many of these wrench faults are aligned parallel to the Precambrian Charnoid structures of the East Midlands of England. N-S compression is invoked to account for late Carboniferous inversion in a NW-SE zone of crustal weakness in the Sole Pit area (Figs. 2.11, 2.13; Glennie and Boegner, 1981). Prior to the Mid Permian Zechstein transgression of the area, this same zone had developed as an important depocentre. In the Late Carboniferous inversion phase, right-lateral movement resulted from N-S compression, and in the Permian subsidence phase, E-W tension is presumed to have resulted in further right-lateral movement along approximately the same faults.

Similar tensional stresses acting fairly uniformly across the whole of the southern North Sea are believed to have resulted in swarms of NW-SE oriented faults that cut the Carboniferous surface. (Note, for example, the variable density of pre-Zechstein faults in Fig. 2.13, Section 2 and in Fig. 3.16). Many of these faults were probably reactivated during the Cretaceous inversion of the area, but were transmitted through the overlying Zechstein salt only along major zones of shear. It seems, therefore, that east-west tension already played an active role in the North Sea area during the Variscan Orogeny, a role that was to be

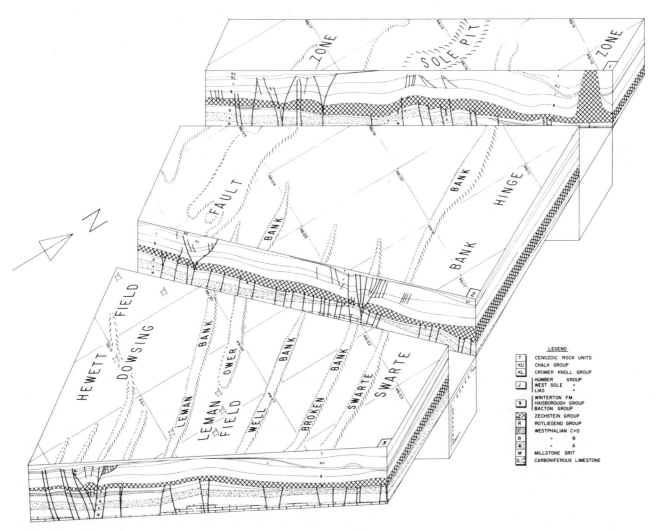

Fig. 2.13. Isometric block diagram of the Sole Pit area, U.K. Southern North Sea. Note that the area of Late Cretaceous inversion is underlain by an axis of Late Carboniferous inversion, and is flanked by zones of probable strike-slip faults ('flower structures'). Elsewhere, sub-Zechstein faults do not cut the post-Zechstein sequence.

repeated at various times during the succeeding 200 million years.

2.4 Post-Variscan basin development

2.4.1 Basin distribution

It has just been shown that within the northern orogenic foreland in NW Europe, two apparently conflicting structural processes took place either at the same time or followed each other in fairly rapid succession; the one involved N-S compression and the other involved E-W tension.

1. As can be seen from Figure 2.2, the area between the Caledonian fold belt of Scotland and Scandinavia in the north and the Hercynian fold belt to the south is dominated by two E-W trending basins, the Northern and the Southern Permian Basins. These basins are separated by the Mid North Sea-Ringkøbing-Fyn system of highs.

2. Cutting these basins and highs almost at right angles are the Central and Horn Grabens, which have northerly extensions in, respectively, the Viking and Oslo Grabens. East of the Horn Graben is a series of smaller grabens that also cut the Ringkøbing-Fyn High. In addition, to the west of the Pennine High, an arcuate series of half grabens stretches from the Paris Basin in the south to the Minch Basins in the north.

Furthermore, within the zone of strong Variscan folding in the south, another system of E-W and SW-NE trending horsts and grabens (e.g. English Channel, SW Approaches, Bristol Channel and Celtic Sea basins) possibly all had their origins within the latest Carboniferous to earliest Permian time span.

2.4.2 Wrench faults and Early Permian volcanism

Late Carboniferous to Early Permian E-W wrench movements were associated with the development of complex patterns of conjugate shear faults and related pull-apart structures, which resulted in local rapid subsidence and in the extrusion of locally thick Lower Rotliegend volcanics (see Glennie, 'Early Permian', and Fig. 3.3). Volcanic activity was especially marked in northern Germany and Poland, where it was probably associated with wrench movement along the Tornquist-Teisseyre fault system (Ziegler, 1982b). Similar volcanic activity at about the NW limit of the Tornquist-Teisseyre fault system was involved in the development of the Oslo-Bamble-Horn graben, which began to form in the latest Westphalian to earliest Stephanian probably by exploiting an early Palaeozoic or even Precambrian structural weakness (Russel and Smythe, 1983). This volcanic activity was preceded in the Oslo Graben by the deposition of shallow-marine, alluvial and aeolian sediments (Olaussen, 1982). Similar thick volcanics occur in the Horn Graben (Olsen,

1983), on the flanks of the Mid North Sea-Ringkøbing High and within the Central Graben (Skjerven, *et al.*, 1983), and E-W dyke swarms cut the Midland Valley of Scotland and the Northumberland Trough.

Although no Rotliegend-type volcanics have been found in association with the Viking Graben, the writer believes that its earliest origins could have been contemporaneous with those of the Central, Horn and Oslo grabens. Perhaps the Late Carboniferous-Early Permian volcanics of the Midland Valley of Scotland (McGregor and McGregor, 1948) and Sunnhørdland area of SW Norway (Dixon *et al.*, 1981) mark the northern limit of the Stephanian-Autunian fracture system. North of the Highland Boundary Fault and its extension across the North Sea lay more rigid Caledonian crust.

This interpretation differs from that of Ziegler (1982a, b), who believes that there is no evidence to support the opening of the Viking and Central grabens any sooner than the early Triassic or perhaps the latest Permian. Glennie ('Early Permian', this volume) is impressed by the close proximity of Lower Rotliegend volcanics with the Central Graben, and by the presence of Late Permian Zechstein salt within the southern 200 km of the Viking Graben (Fig. 4.9, see also Fagerland, 1983) that is sufficiently thick to act diapirically. Furthermore, he reasons that the rapid marine flooding of the Rotliegend desert basin by the Zechstein Sea must have involved the flow of water from the vicinity of the Arctic Circle via a pre-existing North Atlantic fracture system (Glennie and Buller, 1983). This fracture system was initiated in the far north during the late Dinantian (Table 2.1), but whether it had propagated southward to the North Sea area by the end of the Carboniferous or early Permian is still open to question. The thick Zechstein salt of the southern Viking Graben could also be explained by pre-Zechstein crustal sag followed by Early Triassic uplift and erosion of the Zechstein on the flanks of a young rift, an interpretation that is preferred by Ziegler (1982a); Triassic strata now overlie Old Red Sandstone on these flanks (Fig. 2.14).

2.4.3 Permian basin fill

Following the Variscan Orogeny, much of Europe north of those mountains was an area of arid desert. With strong deflation and little rainfall, the rate of subsidence in the Northern and Southern Permian basins exceeded that of sedimentation throughout the early Permian. By the time of the mid-Permian Zechstein marine transgression, the deepest parts of these basins were occupied by desert lakes whose surfaces were probably some 200-300 m below sea level.

After flooding, the Zechstein basins were still areas of relative sediment starvation. Especially during the early part of Zechstein deposition, the shallow-water basin margins were the sites of prolific manufacture of organic carbonate, but the floors of the basin centres were too deep, and therefore too dark, for the rapid growth of such carbonates and, with low rainfall and little river supply, the rate of sediment influx was very low. Because of high temperatures and rates of evaporation, and limited access to supplies of both marine and fresh water, the enclosed Zechstein basins became giant evaporating pans with deposition first of gypsum (now anhydrite), followed by thick sequences of halite and, finally, highly soluble potassic salts. Thus, although the Zechstein basins lacked a good supply of sediment, they were eventually filled largely by chemical precipitates of which halite was by far the most important (see Taylor, this volume).

The Zechstein sequence is the product of five depositional cycles, each of which shows the effects of increasing salinity with time. Each cycle began with a supply of normal sea water of relatively low salinity, and ended with evaporation to probable dryness. The water was probably derived from the open ocean somewhere between Norway and Greenland, via a barred channel, the rate of supply varying with a global sea level that fluctuated in concert with the waxing and waning of the last of the Gondwana ice caps.

2.5 Highlights of the post-Permian structural evolution

So far as the limited area of the North Sea is concerned, the details of its post-Variscan sedimentary and structural development can be found in succeeding chapters. This geological evolution was dependent to a considerable extent, however, on events that took place beyond the limits of the North Sea, and it is to the simple outline of their inter-relationships that this section is devoted. Some of these correlations can be found in Table 2.1. The eventual understanding of these events has depended to a very large extent on interpretation of the mass of seismic data of ever-improving quality that has been shot over most of the North Sea. In most cases, major post-Variscan seismic events can now be interpreted regionally with a remarkable degree of ease and accuracy (e.g. Fig. 2.1). Because of a lack of clear, continuous reflectors, however, this ease in interpretation can rarely be extended to the seismic expression of older rock units in the area.

There were three over-riding events that affected the post-Permian history of NW Europe. The first, and possibly the least of the three in its influence on the structure of the North Sea, was the Late Triassic-Jurassic opening of Tethys (Table 2.1) and the resulting splitting of Pangaea back into Laurasia and Gondwana. The second, which had been struggling to reach fulfilment since the mid-Carboniferous, was the creation of the North Atlantic Ocean and the separation of Laurasia into North America and Eurasia (Fig. 2.4G). A seaway between the Arctic and Central Atlantic had already existed in the early Jurassic (Hallam, 1977). Sea-floor spreading between the two continents was only partially achieved in the middle Jurassic, at which time it was confined to the Central Atlantic. Complete crustal separation between North America

and Europe did not occur until the Late Paleocene-Early Eocene. Supporting evidence can be derived from the Mid-Norway Continental Margin (see, e.g. Bukovics *et al.*, 1984). The third event was the Cretaceous to earliest Tertiary closure of the Tethys Ocean that had separated Africa and Eurasia, and the creation of the Alpine fold-chain (Fig. 2.4H).

Superimposed upon the changing pattern of crustal fragmentation and re-unification was an overall slow northward passive drift of the continents. This drift took the North Sea area from south of the equator during the Devonian to its present latitude over half way from the Equator to the northern pole (Habicht, 1979). It had a latitudinal climatic effect on fauna and on sedimentation but cannot be considered of structural importance without relating this drift to other plate movements.

The E-W tensional regime that gave rise to the North Sea graben system was to dominate basinal development from the late Palaeozoic until the complete crustal separation of Laurasia by activation of the modern North Atlantic spreading axis during the Eocene. The effects of this tension and resulting crustal attenuation was to manifest itself in three main ways (see also Table 2.1).

1. Further graben development by crustal extension and subsidence of rotational fault blocks; this became important during the Late Jurassic, for instance, in the Viking Graben (Fig. 2.14) and adjacent to the Rockall-Faeroe Trough.

2. Adjustments along strike-slip faults. Inherited from an earlier history, these faults were present throughout the area. Activity along such faults on a more regional scale resulted in major basin development. When movement was concentrated along a few isolated zones at any one time, only local subsidence ensued. A small-scale example in the Viking Graben is illustrated with seismic data by Fagerland (1983). On a somewhat larger scale, the Broad Fourteens Basin (Fig. 6.1, this volume) subsided locally during the Jurassic and Early Cretaceous by over 4000 m (see also Figs. 3.13, 3.14), whereas just over the median line in British waters, the later part of that time interval corresponds to a major erosional phase that cut deep into Triassic strata (Fig. 3.16). A corollary to this strike-slip activity is that by destabilising the overburden, it was possibly responsible for activating the diapiric movement of Zechstein salt in those areas where the salt was more than about 150 m thick and overlain by 1000 m or more of overburden (see Taylor, this volume, Section 4.8). Since North Sea diapirism probably was not spontaneous, the converse should also apply, and the stratigraphic dating of the onset of diapiric activity probably indicates the approximate time of local fault movements (Fig. 2.1, 2.13, 4.13, 4.14, 4.15, 7.2). This seems to have reached a peak of activity in the Central and Southern North Sea

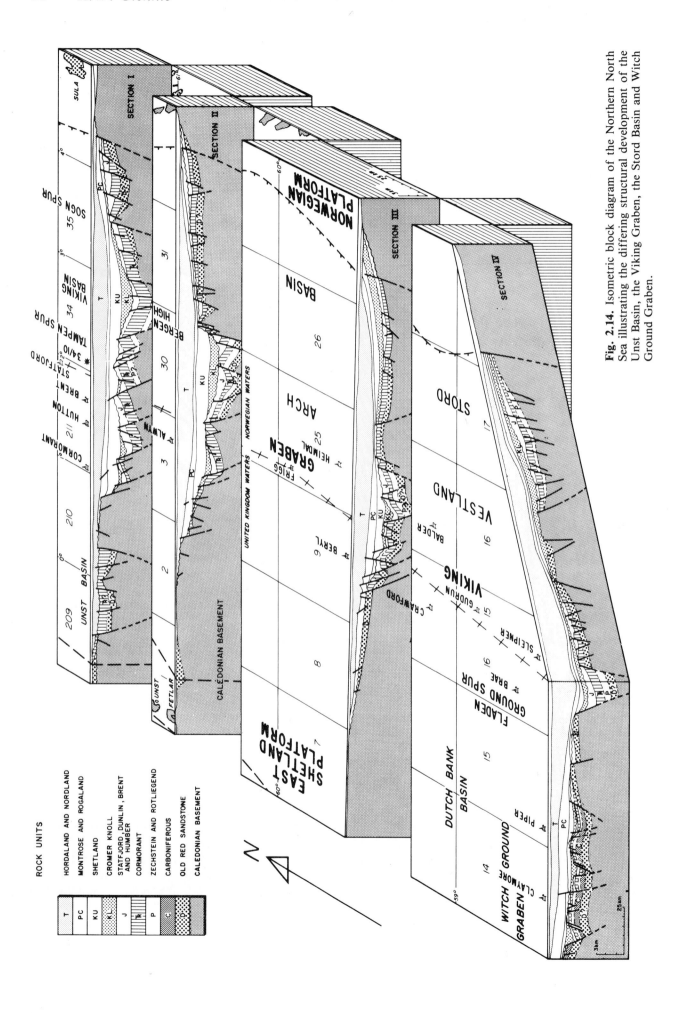

Fig. 2.14. Isometric block diagram of the Northern North Sea illustrating the differing structural development of the Unst Basin, the Viking Graben, the Stord Basin and Witch Ground Graben.

ROCK UNITS

HORDALAND AND NORDLAND
MONTROSE AND ROGALAND
SHETLAND
CROMER KNOLL
STATFJORD, DUNLIN, BRENT AND HUMBER
CORMORANT
ZECHSTEIN AND ROTLIEGEND
CARBONIFEROUS
OLD RED SANDSTONE
CALEDONIAN BASEMENT

during the Late Jurassic and Early Cretaceous, the so-called Late Cimmerian movements. In several places, however, it can be clearly demonstrated that diapirism had started locally by the mid Triassic (e.g. Best *et al.*, 1983, Fig. 4) and was probably triggered by movements related to the main development phase of the North Sea graben system (Table 2.1).

3. Doming is inferred to have occurred where crustal attenuation coincided with an increased thermal gradient, which in the extreme case resulted in volcanic activity. Such doming has been recorded at the intersection of the Moray Firth fault system and the Viking and Central Grabens, and was associated with Middle Jurassic volcanic activity (see Brown, this volume, Section 6.3). A similar doming of the West Shetland Platform during the Paleocene (Table 2.1) resulted in renewed sedimentation of coarser clastics within the Viking Graben (W.H. Ziegler, 1975; Fig. 8.13, this volume). This latter doming preceeded the widespread volcanic activity of the continental shelf bordering the Rockall-Faeroe Trough, the extrusion of plateau basalts in Northern Ireland and Western Scotland, and the emplacement of a swarm of dykes along pre-existing fractures, some of which reached the margins of the North Sea. The cooling of such centres of igneous activity is thought to result in crustal subsidence and basin formation (e.g. domal collapse of Central Graben in Late Jurassic; Table 2.1).

Superimposed upon the above effects of tension was that of isostatic subsidence of the crust in response to loading with water or sediment. Erosion, of course, is followed by isostatic uplift. The effects of eustatic changes in sea level on crustal movement are commonly masked by other, stronger movements. Thus a global highstand in sea level during the early Kimmeridgian (Vail *et al.*, 1977) was superimposed upon subsidence within the Viking and Central Grabens that was related to source-rock formation; and the succeeding lowstand during the Valanginian was accompanied by widespread Early Cretaceous erosion, a little of which possibly resulted from uplift related to the removal of an overburden of water (see also Rawson and Riley, 1982; and Hancock, this volume, Section 7.2).

The North Sea Basin, as we now know it, started to subside in the Late Cretaceous and continued throughout the Cenozoic. The axis of subsidence closely follows the line of the Central and Viking Grabens where Cretaceous subsidence alone had locally exceeded 2 km (Fig. 2.14). Today, the Late Jurassic Kimmeridge Shale has reached depths in excess of 3 km within much of the Central and Viking Grabens (Fig. 9.18) and is undoubtedly the reason why so much oil and gas was generated from its oil-source rock and accumulated in adjacent structural highs. And by comparison with Figure 10.10, it is clear that the greater the depth of burial of the source rock, the higher is the API

gravity of the oil generated until, at the greatest depths, the source rock is post-mature for oil generation and expels only gaseous hydrocarbons.

The pattern of Cenozoic subsidence has been ascribed to cooling and contraction of the underlying lithosphere, and to isostatic adjustment to its increasing load of sediment and water. The crust beneath the North Sea appears to thin from values of 30 to 35 km beneath Norway and the Shetland Isles to about 10-15 km below the Viking Graben. Ziegler (1983) argues that not all this thinning can be accounted for by mechanical stretching, and suggests that the balance might be the result of subcrustal erosion, a poorly understood process that is active during the rifting stage and apparently is irreversible.

There was virtually no volcanic activity associated with the Central and Viking Grabens following the Mid Jurassic eruptions in the vicinity of their junction. Thus the Cenozoic basinal subsidence beneath the North Sea is possibly related to the decay of a thermal anomaly beneath a zone of considerable crustal attenuation, and this decay may already have begun during the Cretaceous.

In retrospect, it seems that the structural history of the North Sea has involved the repeated utilisation of a zone of crustal weakness that was possibly generated in the Ordovician during the closure of the Tornquist Sea. Recognised as an axis of subsidence during the Devonian, it later became the site of the Central Graben. The history of this zone of weakness also reflects the superimposed effects of tensional, compressional and strike-slip movements, and changes in the heat-flow regime.

2.6 Appendix on seismic cross-section in Fig. 2.1

This composite seismic cross-section can be divided into four major units, individual components of which have been dated regionally by the use of cores, cuttings and wireline-log correlations.

1 A pre-Permian basement unit, the deepest parts of which contain only weak indications of internal structure. It has a surface relief of some 3 seconds two-way-time, or around 3000-4000 m depending on the acoustic velocity of the overlying strata. The strongest relief occurs on the flanks of the Central Graben. Regional evidence suggests that one basement reflector possibly represents the erosional contact between strongly-folded Caledonian basement rocks and the overlying Old Red Sandstone. A character change within the basement sedimentary sequence of the Mid North Sea High suggests the possible presence of Carboniferous strata roughly conformable with the underlying Old Red Sandstone, an interpretation that is supported regionally by well data.

2 Within the Norwegian-Danish Basin, the basement rocks are overlain by a truncated wedge of sediments that, on grounds of seismic and structural character, are believed to range in age from the Early Permian

(Rotliegend) to the Late Jurassic (confirmed by well data). The Rotliegend has been penetrated by the drill along the northern edge of the Ringkøbing High, but not in the basin centre. The Late Permian Zechstein is recognised by its characteristic salt diapirs (many of which are now relics of collapse following salt dissolution at a later date), and its base forms a regionally correlatable seismic marker. The pre-Zechstein strata presumably represent the Early Permian Rotliegend, which here may be mostly in a desert-lake facies.

The post-Zechstein part of the sequence is largely Triassic in age, reaching an acoustic 'thickness' of about 1½ seconds two-way-time. The geometry of internal reflectors indicates that Zechstein salt diapirism was already active locally during the middle and possibly late-Early Triassic, and Zechstein salts still disrupt Triassic strata at a few localities.

The Triassic is strongly and uniformly truncated towards the Central Graben boundary fault, and is overlain by a sedimentary sequence that has been dated by well correlation as Late Jurassic in age. And the Upper Jurassic itself wedges out before the margin of the Central Graben is reached, and is overlain by the Upper Cretaceous Chalk, which locally is separated from the basement by only a thin sliver of Lower Cretaceous.

3 Within the Central Graben, the basement is covered by a sedimentary sequence that ranges in age from Early Permian (Rotliegend) to Lower Cretaceous. Here, however, the pre-Late Jurassic sequences have been rotated into half grabens, and both grabens and highs have been draped by Upper Jurassic and Lower Cretaceous sedimentary sequences of irregular thickness. There has been limited Zechstein diapirism.

4 Units two and three are overlain unconformably by the Upper Cretaceous Chalk, which attains its greatest thickness within the Central Graben. Most faults do not extend above the Early Cretaceous surface, thus implying a much greater degree of tectonic calm after that event than before it. The few faults that show slight post-Cretaceous activity are all away from the flanks of the Central Graben. The overlying sequence records the relatively calm conditions of Cenozoic subsidence centred over the Central Graben. The prograding sequences of Early to Mid Cenozoic strata indicate that a source of sediment must have been present to the east throughout that time span, with a major phase of outbuilding possibly coinciding with the Oligocene global low-stand of sea level.

It is clear from the cross-section that this part of the western flank of the Norwegian-Danish Basin underwent considerable uplift during the Mid Jurassic at approximately the same time as major fault-block rotation was taking place within the Central Graben. Together with the Central Graben, this same basin flank has been subsiding steadily since the early Cenozoic.

2.7 Acknowledgements

This contribution is published by permission of Shell Internationale Petroleum Mij., The Hague. I am indebted to Peter Ziegler not only for the use of many of his figures, but also for discussion and advice. Dan Griffin, Bob Hartstra and Paul Veeken all contributed to the interpretation of Fig. 2.1.

2.8 References

Allen, P.A. and Marshall, J.E.A. (1981) Depositional environments and palynology of the Devonian southeast Shetland Basin. *Scott. J. Geol.* 17 (4), 257-273.

Anderton, R. (1982) Dalradian deposition and the late Precambrian-Cambrian history of the N. Atlantic region: a review of the early evolution of the Iapetus Ocean. *J. geol. Soc. London* 139 (4), 423-431.

Barrell, J. (1916) The dominantly fluviatile origin under seasonal rainfall of the Old Red Sandstone. *Bull Geol. Soc. Amer.* 27, 345-386.

Best, G., Kockel, F. and Schöneich, H. (1983) Geological history of the Southern horn Graben. *Geol. Mijnbouw* 62, 25-33.

Bradbury, H.J., Smith, R.A. and Harris, A.L. (1976) Older granites as time markers in Dalradian evolution. *J. geol. Soc. London* 132 (6), 677-684.

Bukovics, C., Cartier, E.G., Shaw, N.D. and Ziegler, P.A. (1984) *Structure and development of the Mid-Norway Continental Margin.* Proc. North European Margin Symposium, Trondheim, 1983.

Cocks, L.R.M. and Fortey, R.A. (1982) Faunal evidence for oceanic separations in the Palaeozoic of Britain. *J. geol. Soc. London* 139 (4), 465-478.

Dewey, J.F. (1982) Plate tectonics and the evolution of the British Isles. *J. geol. Soc. London* 139, (4), 371-412.

Dixon, J.E., Fitton, J.G. and Frost, R.T.C. (1981) The tectonic significance of post-Carboniferous igneous activity in the North Sea Basin. In: Illing, L.V. and Hobson, G.D. (Eds.) *Petroleum geology of the Continental Shelf of North-West Europe.* Heyden, London, 521 p.

Donovan, R.N. and Meyerhoff, A.A. (1982) Comment on 'Paleomagnetic evidence for a large (∼2000 km) sinistral offset along the Great Glen fault during Carboniferous time'. *Geology* 10, 604-605.

Eames, T.D. (1975) Coal rank and gas source relationships— Rotliegendes reservoirs. In: A.W. Woodland (Ed.) *Petroleum and the Continental Shelf of North-West Europe.* Applied Sci. Pub., Barking, 191-201.

Fagerland, N. (1983) Tectonic analysis of a Viking Graben Border Fault. *Am. Assoc. Petroleum Geol. Bull.* 67 (11), 2125-2136.

Geikie, Sir A. (1879) *The Old Red Sandstone of Western Europe.* Trans. Roy. Soc. Edin., vol. XXVIII, p. 345.

Glennie, K.W. (1983) Lower Permian Rotliegend desert sedimentation in the North Sea area. In: M.E. Brookfield and T.S. Ahlbrandt (Eds.) *Eolian sediments and processes.* Elsevier, Amsterdam, 521-541.

Glennie, K.W. and Boegner, P. (1981) Sole Pit inversion tectonics. In: Illing, L.V. and Hobson, G.D. (Eds.) *The petroleum geology of the Continental Shelf of NW Europe.* Heyden, London, 521 p.

Glennie, K.W. and Buller, A.T. (1983) The Permian Weissliegend of N.W. Europe: the partial deformation of aeolian dune sands caused by the Zechstein transgression. *Sedimentary Geology* 35, 43-81.

Habicht, J.K.A. (1979) Palaeoclimate Palaeomagnetism and continental drift. *Am. Assoc. Pet. Geol. Stud. Geol.* 1-31.

Hallam, A. (1977) Biogeographic evidence bearing on the creation of Atlantic seaways in the Jurassic. *Milwaukee Publ. Mus. Spec. Publ. Biol. Geol.* **2**, 23-34.

Harland, W.D. and Gayer, R.A. (1972) The Arctic Caledonides and earlier oceans. *Geol. Mag.* **109**, 289-314.

Hospers, J. and Holte, J. (1980) Salt tectonics in block 8/8 of the Norwegian sector of the North Sea. *Tectonophysics* **68** (3/4), 257-282.

House, M.R., Richardson, J.B., Chaloner, W.G., Allen, J.R.L., Holland, C.H. and Westoll, T.S. (1977) *A correlation of Devonian rocks of the British Isles.* Geol. Soc. London. Spec Report No. 8, 110 p.

Johnson, M.R.W., Sanderson, D.J. and Soper, N.J. (1979) Deformation in the Caledonides of England, Ireland and Scotland. In: Harris, A.L., Holland, C.H. and Leake B.E. (Eds.) *The Caledonides of the British Isles—reviewed.* Geol. Soc. London, 165-186.

Johnstone, G.S. (1966) *The Grampian Highlands.* British Regional Geology, Inst. Geol. Sci. H.M.S.O., Edinburgh. 107 p.

Kennedy, W.Q. (1946) The Great Glen Fault. *Quart. J. Geol. Soc.* **102**, 41-76.

Leeder, M.R. (1982) Upper Palaeozoic basins of the British Isles—Caledonide inheritance versus Hercynian plate margin processes. *J. geol. Soc. London* **139** (4), 479-491.

Leggett, J.K., McKerrow, W.S. and Soper, N.J. (1983) A model for the crustal evolution of Southern Scotland. *Tectonics* **2** (2), 187-210.

Martin, H. (1981) The late Palaeozoic Gondwana glaciation. *Geol. Rundschau* **70** (2), 480-496.

McGregor, M. and McGregor, A.G. (1948) *The Midland Valley of Scotland.* British Regional Geology, Inst. Geol. Sci. H.M.S.O., Edinburgh, 95 p.

McKerrow, W.S., Leggett, J.K. and Eales, M.H. (1977) An Imbricate thrust model for the Southern Uplands of Scotland. *Nature* **267**, 237-239.

Mykura, W. (1976) *Orkney and Shetland.* British Regional Geology. Inst. Geol. Sci. H.M.S.O., Edinburgh, 149 p.

Olaussen, S., Larsen, B.T., Midtkandal, P.A. and Steel, R. (1982) *Sedimentation in Upper Palaeozoic Oslo Graben.* Abstract for KNGMG Conference: Petroleum Geology of the Southeastern North Sea and adjacent onshore areas. The Hague, November 1982.

Olsen, J.C. (1983) A structural outline of the Horn Graben area. *Geol. Mijnbouw* **62**, 47-50.

Parnell, J. (1982a) Genesis of the graphite deposit at Seathwaite in Borrowdale, Cumbria. *Geol. Mag.* **119** (5), 511-512.

Parnell, J. (1982b) Comment on "Palaeomagnetic evidence for a large (∿ 2000 km) sinistral offset along the Great Glen fault during Carboniferous time". *Geology* **10**, 605.

Phemister, J. (1960) *Scotland: the Northern Highlands*, 3rd edn. British Regional Geology. Inst. Geol. Sci. H.M.S.O.; Edinburgh. 104 p.

Phillips, W.E.A., Stillman, C.J. and Murphy, T. (1976) A Caledonian Plate Tectonic model. *J. geol. Soc. London* **132** (6), 579-609.

Ramsbottom, W.H.C., Calver, M.A., Eagar, R.M.C., Hodson, F., Holliday, D.W., Stubblefield, C.J. and Wilson, R.B. (1978) *A correlation of Silesian rocks in the British Isles.* Geol. Soc. London. Spec. Report 10. 81 p.

Russel, M.J. and Smythe, D.K. (1983) Origin of the Oslo

Graben in relation to the Hercynian-Alleghenian orogeny and lithospheric rifting in the North Atlantic. *Tectonophysics* **94**, 457-472.

Simon, J.B. and Bluck, B.J. (1982) Palaeodrainage of the southern margin of the Caledonian Mountain Chain in the Northern British Isles. *Trans. Roy. Soc. Edinburgh: Earth Sci.* **73** (1), 11-15.

Skerven, J., Riis, F. and Kalheim, J.E. (1983) Late Palaeozoic to Early Cenozoic structural development of south-southeastern Norwegian North Sea. *Geol. Mijnbouw* **62**, 35-45.

Smith, D.I. and Watson, J. (1983) Scale and timing on the Great Glen Fault, Scotland. *Geology* **11**, 523-526.

Steel, R.J. (1976) Devonian basins of Western Norway: sedimentary response to tectonism and varying tectonic context. *Tectonophysics* **36**, 207-224.

Stewart, A.D. (1982) Late Proterozoic rifting in NW Scotland: the genesis of the Torridonian. *J. geol. Soc. London* **139** (4), 413-420.

Taylor, B.J., Burgess, I.C., Land, D.H., Mills, D.A.C., Smith, D.B. and Warren, P.T. (1971) *Northern England.* British Regional Geology, Inst. Geol. Sci. H.M.S.O., London. 121 p.

Thomsen, E., Lindgreen, H. and Wrang, P. (1983) Investigation of the source rock potential of Denmark. *Geol. Mijnbouw* **62**, 221-239.

Vail, P.R., Mitchum, R.M. and Thompson, S. (1977) Seismic stratigraphy and global changes in sea level, Pt. 4: Global cycles of relative changes in sea level. In: Payton, C.E. (Ed.) *Seismic stratigraphy—applications to hydrocarbon exploration.* Am. Assoc. Pet. Geol. Memoir 26. p. 83-97.

Van der Voo, R. and Scotese, C. (1981) Palaeomagnetic evidence for a large (∿ 2000 km) sinistral offset along the Great Glen fault during Carboniferous time. *Geology* V. **9**, 583-589.

Watson, J. and Dunning, F.W. (1979) Basement-cover relations in the British Caledonides. In: Harris, A.L., Holland, C.H. and Leake, B.E. (Eds.) *The Caledonides of the British Isles—reviewed.* Geol. Soc. London, 67-91.

Webb, B. (1983) Imbricate structure of the Ettrick area, Southern Uplands. *Scott. J. Geol.* **19** (3), 387-400.

Ziegler, P.A. (1977) Geology and hydrocarbon provinces of the North Sea. *Geojournal* **1**, 7-32.

Ziegler, P.A. (1981) Evolution of sedimentary basins in North-West Europe. In: Illing, L.V. and Hobson, G.D. (Eds.) *Petroleum geology of the Continental Shelf of North-West Europe.* Heyden, London. 3-39.

Ziegler, P.A. (1982a) *Geological atlas of Western and Central Europe.* Shell Int. Pet. Maat. Dist. by Elsevier, Amsterdam. 130 p.

Ziegler, P.A. (1982b) Faulting and graben formation in Western and Central Europe. *Phil. Trans. R. Soc. Lond.* A305, 113-143.

Ziegler, P.A. (1983) Crustal thinning and subsidence in the North Sea. *Nature* **304** (5926), 561.

Ziegler, P.A. (1984) Caledonian and Hercynian crustal consolidation of Western and Central Europe—A working hypothesis. *Geol. Mijnbouw* **63** (1) (in press).

Ziegler, W.H. (1975) Outline of the geological history of the North Sea. In: Woodland, A.W. (Ed.) *Petroleum and the Continental Shelf of North-West Europe.* Applied Sci. Pub. Barking. 165-187.

Chapter 3   Early Permian—Rotliegend

K.W. GLENNIE

3.1 Introduction

The 'Rotliegendes'* is an old German miner's term for the red beds that underlie the Zechstein. The classical Rotliegend sedimentary sequence was deposited in a post-Variscan basin that extended some 1500 km from eastern England to the Russo-Polish border and has been referred to as the 'Southern Permian Basin' (Figs. 3.1, 3.2). Seismic surveys and offshore drilling have shown that another, much smaller, Rotliegend basin occurs between the fragmented Mid North Sea-Ringkøbing Fyn High and the Shetland and Egersund platforms, and is known as the 'Northern Permian Basin'. A third area of Rotliegend deposition is limited to the Moray Firth Basin. Of the same approximate age of creation is a series of small half grabens, which stretch from SW England to SW and W Scotland; their fill of Permian sediment has been correlated with North Sea sequences by Smith et al. (1974), and Lovell (1983); see also Smith (1972).

In the Southern Permian Basin, the Rotliegend can be divided into two distinct units, the Upper and the Lower Rotliegend. The Lower Rotliegend is characterised by the presence of volcanic rocks, which are not found in the Upper Rotliegend; the two units are possibly partly coeval (Fig. 3.4).

Sandstones of the Upper Rotliegend form a most important reservoir rock for gas in the Southern Permian Basin. They contain some 4.1×10^{12} m³ (145×10^{12} ft³) of proven recoverable reserves, of which 1×10^{12} m³ (35×10^{12} ft³) are in offshore fields of the Southern North Sea (Ziegler, 1980a) and 2.4×10^{12} m³ (86×10^{12} ft³) are in the giant Groningen gas field in The Netherlands (Fig. 3.12). The source for all this gas is the Coal Measures of the underlying Carboniferous, which, depending on the temperature gradient, gave up its gas when buried at depths of between 4000 and 6000 m (Lutz et al., 1975; Van Wijhe et al., 1980). The seal is provided by the overlying Zechstein sequence. The Zechstein cycle II (Stassfurt) halite (see the following chapter by Taylor) is the most important individual seal because it is regionally thick and is able to flow and thus heal any fault-induced fracture.

In the Northern Permian Basin, Rotliegend sandstones are oil bearing in both the Auk and Argyll fields, the source rock being the Upper Jurassic Kimmeridge Clay, which matured deep in the adjacent Central Graben.

The colour of the Rotliegend sedimentary rocks resulted from post-depositional diagenetic reddening

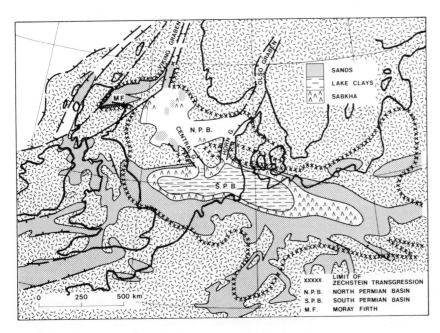

Fig. 3.1. Upper Rotliegend facies and palaeogeography, and limit of Zechstein transgression. Facies distribution poorly known in Northern Permian Basin. Modified from Glennie (1972) and Ziegler (1978).

* Rotliegendes: I am informed by my more linguistic colleagues that the term 'Rotliegend' is both simpler and more correct in English

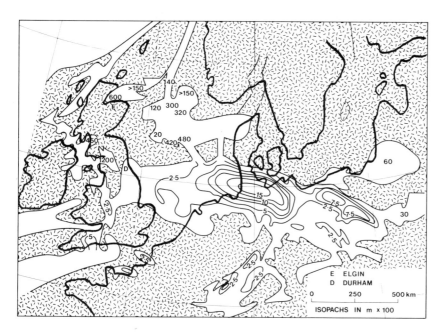

Fig. 3.2. Upper Rotliegend isopachs and isolated thicknesses. Modified from Ziegler (1980).

when ferrous ions in the ground water were oxidised to the ferric state. As shown by Walker (1967), a diagenetic environment conducive to such oxidation is commonly present beneath the surfaces of tropical deserts. The palaeogeographic, climatic and diagenetic significance of red beds formed in desert and other depositional environments is discussed in considerable detail by Turner (1980).

The uppermost part of the Rotliegend sandstone sequence is commonly grey or white in colour, which has given rise to the German name 'Weissliegend'. It is thought that the Weissliegend sands were above the water table at the time of the mid Permian Zechstein marine transgression and thus were never in a diagenetic environment in which they could become reddened (Glennie and Buller, 1983).

3.2. Lower Rotliegend

The Lower Rotliegend comprises an association of rocks that is predominantly volcanic in character but includes some sedimentary sequences, especially in Germany (Falke, 1972; Plein, 1978), that were deposited largely in fluvial and lacustrine environments under a climate that alternated between humid and arid. Aeolian sandstones also occur, and locally have good reservoir potential (Schneverdingen Sandstone; Drong *et al.*, 1982). The distribution of the Lower Rotliegend is very limited in comparison to the Upper Rotliegend (cf. Figs. 3.1, 3.3). The Lower Rotliegend volcanics are best developed in northern Germany and the Oslo Graben-Bamble Trough areas. Similar associations of sediments and volcanic rocks of about the same age (Late Stephanian-Early Autunian) but covering much smaller areas, are known in France, S.W. England, S.W. Scotland and some of the flank areas of the Mid North Sea-Ringkøbing-Fyn Highs.

The range from basic to intermediate volcanic rocks in the Lower Rotliegend, and their distribution adjacent to known or inferred faults, suggests that their origin was possibly related to the earliest tensional movements connected with the creation not only of the Permian basins, but also with the graben systems of the North Sea (Viking-Central grabens, Oslo-Horn grabens) and Germany (e.g. Schneverdingen Graben). If this interpretation is correct, then the North Sea graben system could have started to form during or shortly after the final stages of the Variscan Orogeny; this, in turn, coincides with a Late Westphalian age for uplift of the London-Brabant Platform and a Stephanian date for the earliest positive movements of the Mid North Sea High.

Some idea of the scale of Latest Carboniferous to Early Permian differential vertical movements can be gained when one considers that, prior to the Zechstein transgression, erosion removed all the previously deposited Carboniferous strata over large parts of the Mid North Sea High; and adjacent to the Central Graben, Zechstein erosion additionally cut deep into the Devonian Old Red Sandstone and even older strata, thus implying even greater uplift. Within the Southern Permian Basin, Late Carboniferous to Early Permian inversion along the Sole Pit axis (Fig. 2.13) resulted in erosion, locally, of the complete Westphalian sequence prior to deposition of the Upper Rotliegend.

3.3 Upper Rotliegend

3.3.1 Southern Permian Basin

The Upper Rotliegend is made up of four distinctive facies associations, which have been interpreted as the products of deposition in fluvial (wadi), aeolian, sabkha and lacustrine environments (Figs. 3.4, 3.6), cores of three of which have been illustrated by Glennie (1972). All four facies are widespread in the asymmetric Southern Permian Basin, whose floor sloped from south to north over much of its area (see Fig. 3.11). In the Southern North Sea area, the sandy facies is referred to as the Leman Sandstone in U.K. waters and

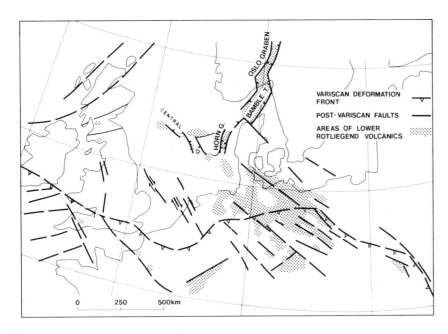

Fig. 3.3. Stephanian-Autunian fault patterns and distribution of Lower Rotliegend volcanics.

as the Slochteren Sandstone in The Netherlands (Figs. 3.4, 3.5). The clayey lacustrine facies of both areas is named the Silverpit Formation (Rhys, 1974; NAM, 1980). The characteristic lithologies that make up these formational units have been studied in cores and can be deduced from the wireline logs, for instance, of the type and reference sections depicted in figure 3.5 for different parts of the North Sea area.

Fluvial facies

The fluvial sequences are characterised by the occurrence of curled clay flakes, indicating frequent subaerial exposure and desiccation. Some thicker clays seem to have had their cracks infilled with sand from above; others were injected with a slurry of sand and water from below to form sandstone dykes (Glennie, 1970). The sandstones directly below clay beds display centimetre to decimetre foresets with low-dip and commonly discontinuous laminae, which are interpreted as having been deposited by flowing water. These sandstones are locally conglomeratic, with some of the contained pebbles consisting of red clay similar in character to the bedded claystones. Laminated sandstones locally grade up into apparently homogenous

sandstones that lack sedimentary structures. Some homogenous sands may contain large pebbles and are obviously of fluvial origin; others, however, grade down into sandstones with well-defined laminae typical of the aeolian sands that are commonly interbedded with those of fluvial origin. The origin of these structureless sands will be discussed below (Section 3.3.2). Fluvial sands tend to be well cemented with dolomite (originally calcite?) and only locally form good reservoirs for hydrocarbons (e.g. in the Groningen Field).

Many of the above features indicate that these essentially fluvial sequences should be interpreted as having been deposited by ephemeral streams in an arid or semi-arid environment; and may thus be referred to as wadi deposits.

Wadi sandstones are common along the southern margin of the Rotliegend basin (Fig. 3.6) and especially in Holland, Germany and Poland (see Ziegler, 1982, Encl. 13). In the U.K. sector of the North Sea, wadi sandstones are generally poorly developed, the Rotliegend sequence being dominated by sandstones of aeolian origin (see e.g. Nagtegaal, 1979). Such wadi sandstones are better displayed in the East Netherlands (Bungener, 1969) and in other basins with areas of greater relief, as in Arran, Scotland (Clemmensen and

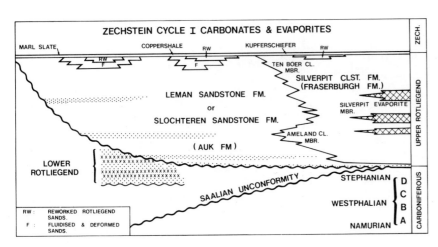

Fig. 3.4. Rock-stratigraphic diagram of the Permian Upper and Lower Rotliegend groups. Names in brackets refer to formations of the Northern Permian Basin (cf. Fig. 3.5). Dotted areas represent fluvial or mixed fluvial and aeolian sands.

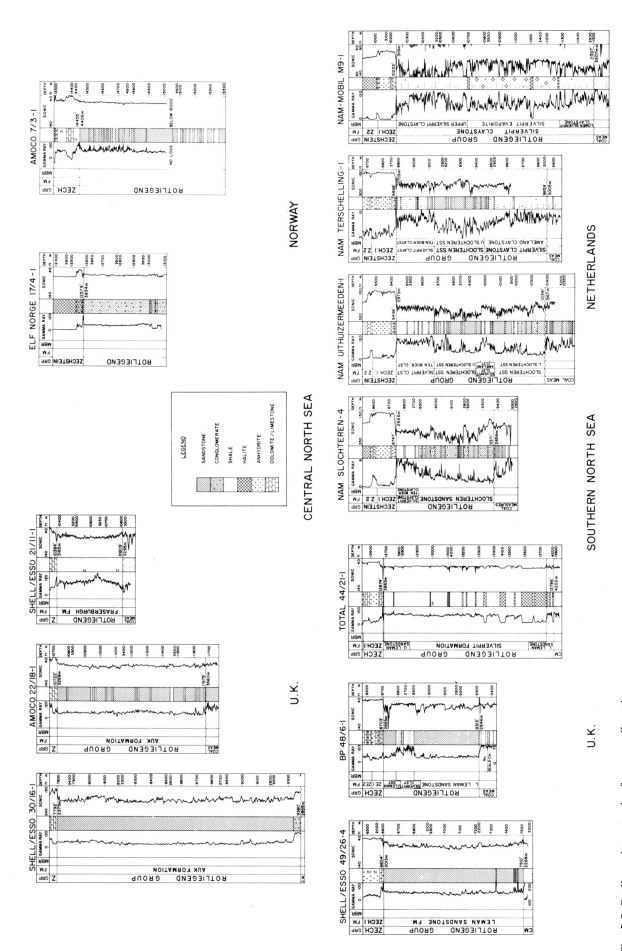

Fig. 3.5. Rotliegend type and reference well sections.

Abrahamsen, 1983), SW Scotland (Brookfield, 1980), and Devon, England (Laming, 1965). A sequence of cores from the southern North Sea that include wadi sands and conglomerates are illustrated by Glennie (1972, Fig. 10).

Aeolian facies

The aeolian sands are most readily recognised when they conform to a series of criteria which include: (a) well defined planar or trough-bedded strata in which adjacent laminae commonly show sharp grain-size differences between that of very fine sand and some 1 or 2 mm; (b) many intra-formational unconformities, above which the laminae are commonly horizontal; there is an upwards increase in the inclination of the laminae to an angle of some 20° to 25° before being terminated by the next low-angle truncation, and (c) a lack of mica flakes. Sequences of almost continuous aeolian bedding of this type locally reach thicknesses of one or two hundred metres or more. Analysis of the orientation of the aeolian bedding both in outcrop

(e.g. Durham, England) and in wells (dip-meter logs) indicate that both transverse and seif dunes were formed in the Southern Permian Basin (Fig. 3.11).

Within the Southern Permian Basin, the Early Permian winds blew from roughly east to west (Glennie, 1972, 1983a; Van Wijhe *et al.*, 1980), which, after correcting for the rotation of N.W. Europe since the Permian, suggests that the Rotliegend was deposited in a 'Trade Wind' desert of the Northern Hemisphere similar to the Sahara of today (Figs. 3.6, 3.8, 3.9).

In some North Sea wells, typical dune bedding grades into intervals of irregular wavy laminations, which are interpreted to be adhesion ripples, formed when wind-blown sand adheres to the damp surface of a sabkha. These adhesion ripples are believed to have been deposited over interdune areas whose surfaces coincided with the water table. The presence of small blebs of anhydrite within the adhesion ripples testifies to the hot and arid climate which caused gypsum to crystallise within the sands.

Dune sands form the main reservoir rock for gas in the Southern Permian Basin. At the time of deposition their

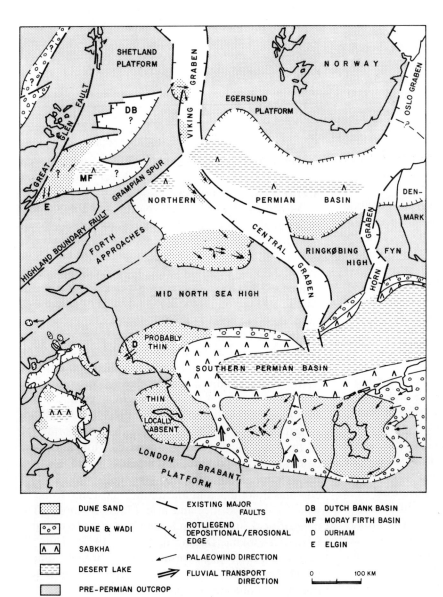

DUNE SAND	EXISTING MAJOR FAULTS	DB DUTCH BANK BASIN
DUNE & WADI	ROTLIEGEND DEPOSITIONAL/EROSIONAL EDGE	MF MORAY FIRTH BASIN
SABKHA	PALAEOWIND DIRECTION	D DURHAM
DESERT LAKE	FLUVIAL TRANSPORT DIRECTION	E ELGIN
PRE-PERMIAN OUTCROP	0 100 KM	

Fig. 3.6. Facies distribution and pattern of Early Permian (Rotliegend) winds in the North Sea area. Areas of dune sand are stippled.

porosities probably averaged around 42% (Hunter, 1977). With increasing depths of burial the porosity was progressively reduced, first by compaction and pressure solution and then by the growth of authigenic minerals including chlorite and illite (Glennie *et al.*, 1978; also Fig. 3.15). At greater depths, especially after the porosity has already been reduced to around 10%, illite develops in a fibrous form, which has a very deleterious effect upon permeability without significantly reducing porosity.

Lacustrine facies

The lacustrine facies consists primarily of red-brown mudstone with minor siltstone. Several halite horizons occur in the lower half of the sequence in U.K. waters (Figs. 3.4, 3.5, 3.11), but in North Germany, halite is best developed in the middle of the Upper Rotliegend sequence, thus implying a longer local history of sedimentation. The salts attain a sufficient thickness to react diapirically in northern Germany (see e.g. Plein, 1978, Fig. 1), and the deformed lacustrine sequence is then known as the Haselgebirge facies. In the Glückstadt Graben at the southern end of the Danish peninsula, Rotliegend halite occurs in the same diapiric structures as halite of Zechstein age (Best *et al.*, 1983, Fig. 4), mobilisation probably being triggered in the Triassic by earth movements associated with the Hardegsen disconformity (see Fisher, Section 5.2.1). Although the lacustrine facies achieves a thickness of some 1500 m in northern Germany, it seems to be devoid of fossils apart from the top metre or so (the fossils have strong Zechstein affinities and are probably attributed more correctly to the Zechstein marine transgression).

The basin-centre parts of this sequence also contain no known sedimentary structures indicative of sub-aerial desiccation or erosion, although this may merely reflect the lack of coring in commercially unproductive rocks. The desert lake is therefore believed to have been a constant feature throughout the early Permian. At its fullest extent, it must have covered an area of some 1200 km E-W by over 200 km N-S (Fig. 3.1).

Sabkha facies

Between the deposits of the desert lake and the more southerly depositional areas of the wadi and aeolian sands (Figs. 3.1, 3.6, 3.11), is a broad band of poorly bedded clays, silts and sands that display many features indicative of a sabkha (e.g. mud cracks, sandstone dykes, adhesion ripples, anhydrite nodules—see Glennie, 1970; Nagtegaal, 1973). These features collectively indicate a largely aquatic depositional area that was subject to limited aeolian deposition and to sub-aerial desiccation in an arid climate. The sabkha sediments represent the area that was covered by water only during the maximum extensions of the desert lake (Fig. 3.1).

The lacustrine and sabkha facies lack reservoir developments. To the contrary, they are much more likely to act as a seal for underlying hydrocarbon accumulations.

In the Southern North Sea and The Netherlands, the Rotliegend sequence is developed, from south to north and from base to top, in the following generalised facies (Figs. 3.4, 3.11): (a) mixed wadi and aeolian, (b) mainly aeolian, (c) sabkha, (d) desert lake. The whole sequence is capped by the Kupferschiefer with only a few centimetres to a maximum of three or four metres of marine-reworked sands between (Fig. 3.4).

The 'Weissliegend'

At many localities in the belt of dune sands, and for a distance beneath the Kupferschiefer of up to 50 m (150 ft) or more, a sequence of uncoloured ('Weissliegend') structureless sands alternate with, and grade into, highly deformed strata as well as into beds in which the original aeolian bedding is only weakly preserved. These non-depositional features are collectively believed to have resulted from a very rapid Zechstein transgression, and the escape of air trapped beneath the wetted surface of aeolian sand dunes during the rapid rise of water level (Glennie and Buller, 1983). Such a rapid transgression has already been invoked by Smith (1979) to explain other features in the overlying Zechstein of NE England.

3.3.2 Moray Firth and Northern Permian Basins

The facies distribution of the Rotliegend in these two basins is still incompletely known (Fig. 3.6). In the deeper parts of the basins the Rotliegend is rarely reached by the drill and, except near Elgin on the Moray coast, it is not seen in outcrop.

On present evidence, both basins probably contain rocks of the same general sedimentary facies already recognised in the southern basin (fluvial, aeolian and sabkha, although the presence of a bedded lacustrine halite has yet to be demonstrated). The sand-dominated sequences in U.K. waters have been designated the Auk Formation and the shaly sequence the Fraserburgh Formation (Deegan and Scull, 1975).

Much of the Auk sequence in the type locality has been interpreted as aeolian dunes. Dip-meter data indicate that here the winds blew in a direction opposed to that of the Southern Permian Basin (i.e. towards the E and SE; see Fig. 3.6); a barometric high must therefore have existed in the vicinity of the Mid North Sea structural high. The basal 14 m of the Auk Formation at the type locality (Fig. 3.5) is conglomeratic and contains clasts of quartz and schist. The shales of the Fraserburgh Formation in Shell/Esso well 21/11-1 contains dolomitic and micaceous sandstone stringers, which are anhydritic, and also adhesion ripples. The depositional environment is here interpreted as a dune-bordered sabkha (Deegan and Scull, 1975). Thus the Auk and Fraserburgh formations are broadly similar to the Leman Sandstone and Silverpit Claystone formational sequences of the Southern Permian Basin

(Fig. 3.4). Similar sequences have been recognised in released Norwegian wells (Fig. 3.5) but as yet are unnamed.

In the U.K. part of the Northern Permian Basin the thickness of the Rotliegend changes rapidly from place to place, making correlation between wells very difficult. The differences may reflect deposition in small rotated half grabens such as formed on the flank of the developing Central Graben (cf. Fig. 3.11). Such an interpretation could explain the derivation from adjacent fault scarps of the clasts of quartz and schist in the Shell/Esso well 30/16-1 mentioned by Deegan and Scull (1975).

It seems likely that in the Moray Firth Basin, Rotliegend sedimentation kept up with subsidence, as the overlying Zechstein is entirely in a shallow-marine facies. In the centre of the Northern Permian Basin, on the other hand, subsidence probably greatly exceeded sedimentation and the succeeding Zechstein is dominantly in a basinal halite facies (see Taylor, Chapter 4, Fig. 4.9; Taylor, 1981). Like the central parts of the Southern Permian Basin, the halite reached thicknesses over much of the northern basin that were great enough to permit diapirism (Taylor, Fig. 4.11).

On the Moray coast of Scotland, north of Elgin, around 200 feet (60 m) of dune sands, known locally as the Hopeman Sandstone (Peacock *et al.*, 1968), contain some reptile foot prints, which have been tentatively dated as Early Triassic or Late Permian. Bedding attitudes indicate that the contemporary winds blew in both southerly and south-westerly directions. Intermittently along 10 km of this coastline, and coinciding roughly with the deepest sands exposed, the dune bedding grades both laterally and upwards into structureless and distorted sequences up to 20 m or more thick (Fig. 3.7). As with the Southern North Sea

examples, the origin of these deformation structures also is ascribed to the escape of air through the wet surface of aeolian sand dunes prior to deposition of the basal Zechstein Kupferschiefer (Glennie and Buller, 1983). The structureless sands may result from the rapid upward replacement of escaping air by water, which causes slight differential grain movement that obliterates the bedding. The entrapment of sufficient air to deform these sands on such a large scale must have resulted from the differential capillary penetration of water into the dunes related to a very rapid rise in water level. The widespread occurrence of these deformation structures at the same sub-Kupferschiefer stratigraphic level in all three North Sea Permian basins implies that it must have been the Zechstein transgression that caused the rise in water level. For such a rapid rise in water level to be possible, the surface of the Rotliegend desert must have been below global sea level.

Although Zechstein strata are not recognised on land in the Moray Firth Basin, apart from the above described effect, they are known, from both bore hole and seismic data, to be present in the subsurface only a few kilometres offshore (Fig. 3.7). Thus these basin-margin Moray dune sands probably cover a greater time span than formerly believed, from Early Permian (Rotliegend) to Early Triassic. Offshore, Rotliegend sandstones alone attain a thickness of up to 600 m (Fig. 3.2).

Unlike the 'Weissliegend' sandstones of the North Sea basins, the Hopeman Sandstones of the Moray coast are stained red. This probably reflects a subsurface diagenetic environment induced by overlying desert conditions during the later Permian or early Triassic. The Hopeman sands were probably at the limit of the Zechstein Sea and were only temporarily covered by its waters.

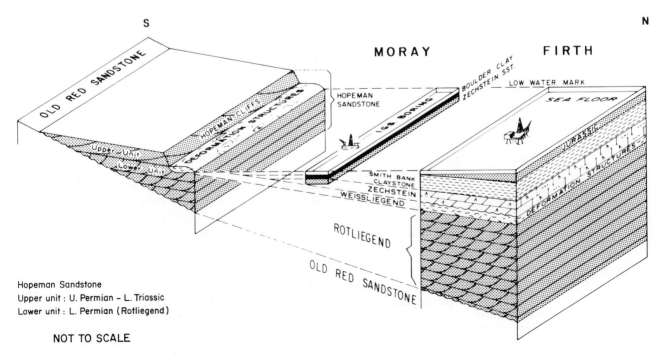

Hopeman Sandstone
Upper unit : U. Permian – L. Triassic
Lower unit : L. Permian (Rotliegend)

NOT TO SCALE

Fig. 3.7. Dune sand relationship—southern Moray Firth.

3.4 Historical development

3.4.1 Climate

The change from the humid equatorial conditions under which the Carboniferous Coal Measures were deposited to the arid climate of Rotliegend deposition, was probably mostly an effect of the passive northerly drift of Laurasia. The Southern Permian Rotliegend Basin came to occupy a latitudinal position north of the equator similar to that of the present North African-Arabian deserts (Fig. 3.8). Lower Rotliegend aeolian sandstones in Germany (Drong *et al.*, 1982) and evidence of strong pre-Saxonian deflation in the southern North Sea area (Glennie, 1983b), indicates that the Southern Permian Basin had already entered the Trade Wind desert belt early in the Permian.

The newly created E-W trending Variscan Highlands occupied a near equatorial location and will have had at least some tropical rainfall to judge from the Stephanian to Autunian coals in Saarland and Central France. A time-related reduction in the volume of fluvial sediments within the Southern Permian Basin may indicate that during the earlier part of the Rotliegend depositional history, the sources of fluvial activity were still just within the equatorial zone of higher rainfall; later, these source areas may also have entered the region of desert climate, resulting in an extension of the area of dune sands and some reduction in the size of the desert lake.

Well data, coupled with excellent quarry exposures of aeolian sands in Durham, England, indicate that many of the Rotliegend dunes attained a height of up to 50 m or more.

Fig. 3.8. Permian latitudes of N.W. Europe with superimposed wind directions.

Today's winds are capable of constructing only small seif dunes with an average maximum height of 5 to 10 m. The large modern seif dunes of Arabia and North Africa, possessing heights of some 100 m and wave lengths of 1 or 2 kilometres, were probably constructed during the Pleistocene. The most likely reason for these size differences is that the large areas of high barometric pressure associated with major glaciations (Permo-Carboniferous as well as Pleistocene) will have caused a concentration of the world's air pressure belts towards the equator and thus have created a shorter distance between the zones of high and low pressure than is now the case (Fig. 3.9). The resulting wind systems probably had higher average velocities than now, and were also colder than now, interpretations that are gaining support among many workers (e.g. Galloway, 1965; Bowler, 1976; Krinsley and Smith, 1981; Rea and Janecek, 1982).

These Pleistocene winds possibly persisted for a much greater part of the year instead of blowing for only a few hours or days at a time as is now the case. With ice caps the size of those found in Gondwana, the Permian dunes, like those of the Pleistocene glaciations, are likely to have been built on a scale that is impossible today (see also Glennie, 1983a, b).

If these deductions are correct, the Early Permian desert winds are likely to have been strong and, because of their long continental route before reaching the western part of the Southern Permian Basin, also very dry. Thus the area of dune activity will have extended during Gondwana glaciations, and strong evaporation will have reduced the size of the desert lake to a minimum. The dying stages of the Early Permian glaciations, on the other hand, should have coincided with a generally weaker wind system with a resulting combination of a higher convection-induced rainfall and less evaporation. This may explain the lateral extension of the lacustrine and sabkha facies for a considerable period prior to the Zechstein transgression. These points are suggested schematically in Figures 3.4 and 3.11.

3.4.2 Basin formation

Late Westphalian N-S compression brought about the creation of the Variscan Highlands to the south of Britain, and associated right-lateral shearing in the U.K. part of the Southern North Sea seems to have resulted in inversion of the Sole Pit area (Glennie and Boegner, 1981). The early collapse of the Variscan Highlands in the west, and the development of a horst and graben system that was possibly allied to the creation of a proto-Atlantic fracture system, gave rise to E-W tension and to right-lateral extension movements in the Southern North Sea area (Fig. 3.10; Glennie and Boegner, 1981) and in Germany (Drong, *et al.*, 1982). These movements resulted in subsidence of the Southern and Northern Permian Basins, leaving the Mid North Sea structural high as a relic of relative stability. It also resulted in the en-echelon development of NW-SE

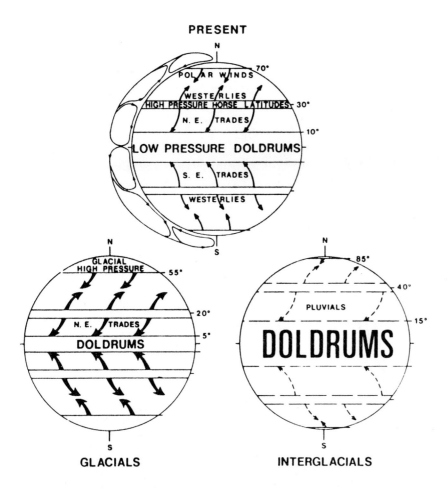

PRESENT

GLACIALS

INTERGLACIALS

VERY STRONG CONSTANT
WINDS BUILT MAJOR
DUNE SYSTEMS.

WEAK WIND SYSTEMS.
STRONG CONVECTION
INFLUENCE NEAR COASTS
& MOUNTAINS.
DESERT "PLUVIALS".

Fig. 3.9. Conceptual differences in width and location of the Earth's air-pressure belts in relation to the size of Polar ice caps.

trending sub-basins, such as the Sole Pit and Broad Fourteens basins, which continued to subside until the late Mesozoic, and possibly of other sub-basins as far east as Poland.

The same E-W regional tension is suspected of causing the development of the N-S and NW-SE oriented fracture systems of the North Sea (e.g. Oslo and Horn Grabens, Viking and Central Grabens) already during the Early Permian and perhaps even in the latest Carboniferous (Figs. 3.3, 3.10). Lower Rotliegend volcanics occupy the Horn Graben (Figs. 3.3, 3.11) and the Mid North Sea flanks of the Central Graben. And some of these subsiding grabens were deep enough by Zechstein time to allow the accumulation of salt that was sufficiently thick to move later diapirically (Southern Viking Graben and Central Graben). This contrasts with the areas flanking the grabens where early Zechstein strata are locally absent and the later Zechstein was deposited entirely in a shallow-marine carbonate/anhydrite facies (Fig. 3.11; see also Taylor, 1981).

3.4.3 Zechstein transgression

As shown earlier, the 'Weissliegend' sediments of the uppermost Rotliegend contain important evidence

concerning the rapidity of the Zechstein transgression, which is why the transgression is discussed in this rather than in the succeeding chapter.

It is along a combination of the proto-Atlantic and North Sea fracture systems that the waters of the Zechstein transgression are presumed to have been transported from the Permian open ocean somewhere between the northern coasts of Greenland and Norway (Fig. 3.10).

Future evidence from the offshore areas between East Greenland and Norway may indicate whether the mid-Permian proto-Atlantic rift was a narrow graben, as implied in Figure 3.10, or was a broader arm of the ocean as suggested by Callomon *et al.* (1972), and depicted by Taylor in Figure 4.1.

The Zechstein transgression probably started because a world wide rise in sea level, coinciding with the end of a phase of Permian glaciation, permitted oceanic water to flow along a pre-existing tensional fracture system. The surfaces of both the northern and southern Permian basins were probably well below the level of the open ocean, so that once the water began to flow south along the fracture the transgression continued until the level of the Zechstein Sea matched that of the ocean (Figs. 3.10, 3.11).

If the surface of the Rotliegend desert lake lay about

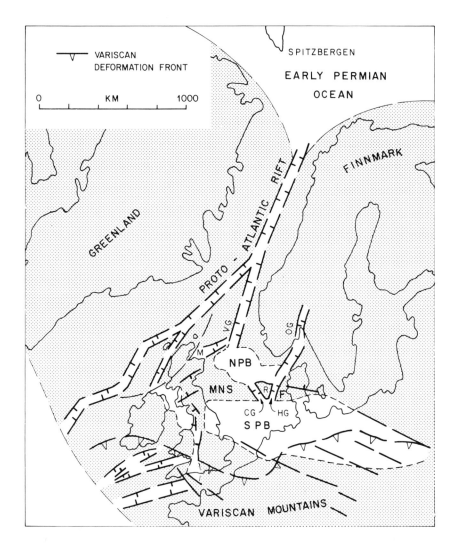

Fig. 3.10. Post-Variscan fault patterns. Modified after Russell (1976) and Ziegler (1978). CG Central Graben; HG Horn Graben; M Moray Firth; MNS Mid North Sea High; NPB Northern Permian Basin; OG Oslo Graben; RF Ringkøbing-Fyn High; SPB Southern Permian Basin; VG Viking Graben.

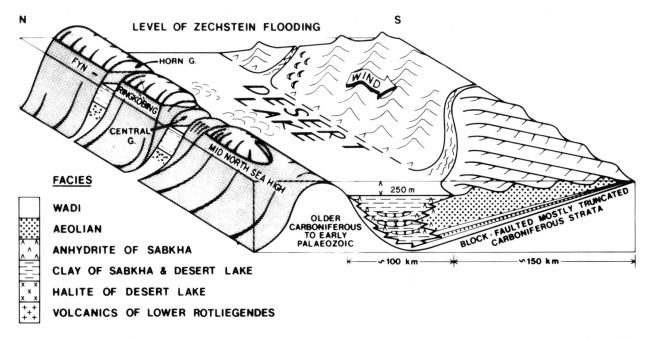

Fig. 3.11. Conceptual block diagram of the Southern Permian Basin and Central North Sea system of highs at the time of the Zechstein transgression. Zechstein Sea is presumed to have flowed into basin via Central and/or Horn Graben. Note suggested change in dune style from transverse in centre of basin to seif in western basin-margin location. From Glennie and Buller (1983).

250 m below the level of the open ocean, it would have required some 75000 km³ of water to fill the Southern Permian Basin (Fig. 3.1), and another 35000 km³ of water to fill the Northern Permian Basin. Ignoring seepage and evaporation, these basins could have been filled in about 6 years if they were jointly flooded at the rate of say 50 km³/day (e.g. channel of water 10 km wide, 20 m deep and average velocity of 3 m/sec).

Such a flood of water into the Southern Permian Basin may have caused some initial scouring of the desert lake sediments where it debouched into the basin, but on the other side of the lake, 100 km or more away (Fig. 3.11), erosion will have been minimal. If the proposed flow rates are correct, then initially the rise in water level will have been around 30 cm/day, and lakeside dunes, 50 m high, will have been covered with water in just over 150 days. It is this relatively rapid and continuous rise in water level which is thought to have been responsible for the in-situ deformation of the 'Weissliegend' upper part of the Rotliegend sedimentary sequence. The surfaces of these dunes were reworked by wave action only to a limited extent. Thus the original shapes of the dunes were only slightly modified and considerable relief was preserved, as can be seen from exposures in NE England. The succeeding Kupferschiefer draped this dune relief. It is unrealistic, therefore, to attempt a detailed correlation of Rotliegend reservoir sequences in the belief that the Kupferschiefer formed a horizontal datum plane.

3.5 Hydrocarbon occurrences

3.5.1 Gas fields

Exploration in the hostile environment of the North Sea was triggered by the realisation that there were sufficient reserves in the Rotliegend reservoirs of the giant Dutch Groningen field to alter the fuel economy of much of NW Europe from a reliance on coal and oil to one based more extensively on gas. Although discovered in 1959, it was not realised until several more wells had been drilled that the continued discoveries all belonged to the same field (Te Groen and Steenken, 1968). With a surface area of almost 800 km², and a porosity that ranged from 10-25% (permeability 0.1-1000 mD), the field was conservatively estimated in 1968 to have ultimate recoverable reserves of some 1650 × 10⁹ m³ (58 × 10¹² ft³) of gas (now believed to be nearer 2425 × 10⁹ m³ or 86 × 10¹² ft³); the unexplored southern North Sea lay down the palaeowind, roughly due west from Groningen (Figs. 3.1, 3.6, 3.12). The basic details of this giant are given in a series of articles by Te Groen and Steenken (1968), van der Laan (1968), Bungener (1969) and Stäuble and Milius (1970).

All the producing Rotliegend gas fields of Germany, The Netherlands and the Southern North Sea are underlain by Westphalian Coal Measures and overlain by Zechstein salt. The desert lake and sabkha facies contain no reservoir rocks and, apart from the Groningen

area and the small Rough field offshore Yorkshire (see Robertson, 1981, Fig. 6), the fluvial sands are generally too well cemented to form good reservoirs. The commercially productive reservoirs are therefore largely confined to the aeolian facies. These factors result in the southern North Sea Rotliegend gas fields being limited effectively to an E-W band some 100 km wide stretching from the North German Plain, through Groningen, to the east coast of England (Fig. 3.12).

Gas generation, migration and entrapment

Superimposed on the three basic requirements for an oil or gas field of source, reservoir and cap rocks, is the general need for structural deformation to create a trap.

It is axiomatic that a trap must be formed before migrating gas can be retained in a reservoir. In the Southern Permian Basin, some early traps formed in areas of slower subsidence, whereas others resulted from subsidence followed by inversion. In some of the latter cases, subsidence was continuous from the Early Permian until the Late Cretaceous, when uplift in the order of 1 to 4 km took place (Figs. 3.13, 3.14, 3.15, 3.16); this was followed by further differential subsidence during the Tertiary. In other areas, uplift began in the Mid Jurassic or Early Cretaceous, to be followed by more subsidence in the later Cretaceous. All these movements were probably coincident with widespread but minor transcurrent faulting related first to the development, and then to the final demise, of the major North Sea graben system (see Chapter 2.5).

Gas generation has been shown by Van Wijhe *et al.* (1980) to result from burial of the Coal Measures to depths of some 4000 m or more (Fig. 3.13). In the areas around Hamburg, Germany, Westphalian coals have reached the rank of anthracite (Bartenstein, 1979), and must by now have given up most, if not all, of their gas. This was an area of rapid subsidence throughout the Permian (e.g. 1500 m of Rotliegend desert-lake sediments) and Triassic (almost 9 km thick in the Glückstadt Graben—Best *et al.*, 1983, Fig. 4). Much of the succeeding Jurassic was removed during the 'Late Cimmerian' phase of erosion around the Jurassic-Cretaceous time boundary, preserved sequences being confined largely to the rim synclines of Zechstein diapirs, which were already active during the later Triassic (Best *et al.*, 1983, Fig. 4). Under these conditions of burial, gas generation in the Glückstadt Graben (Fig. 2.2) probably began already during the Triassic. With impervious desert-lake sediments immediately above, this gas must have migrated to the south to reach porous Rotliegend sandstones; to the north, the lack of a Rotliegend reservoir and the general absence of Mesozoic seals over the Ringkøbing-Fyn High will have resulted in the escape of much gas to the surface.

Gas generation in the Broad Fourteens and Sole Pit Basins probably took place during the later Jurassic and early Cretaceous, so that the Rotliegend reservoirs in contemporary highs flanking the basins were

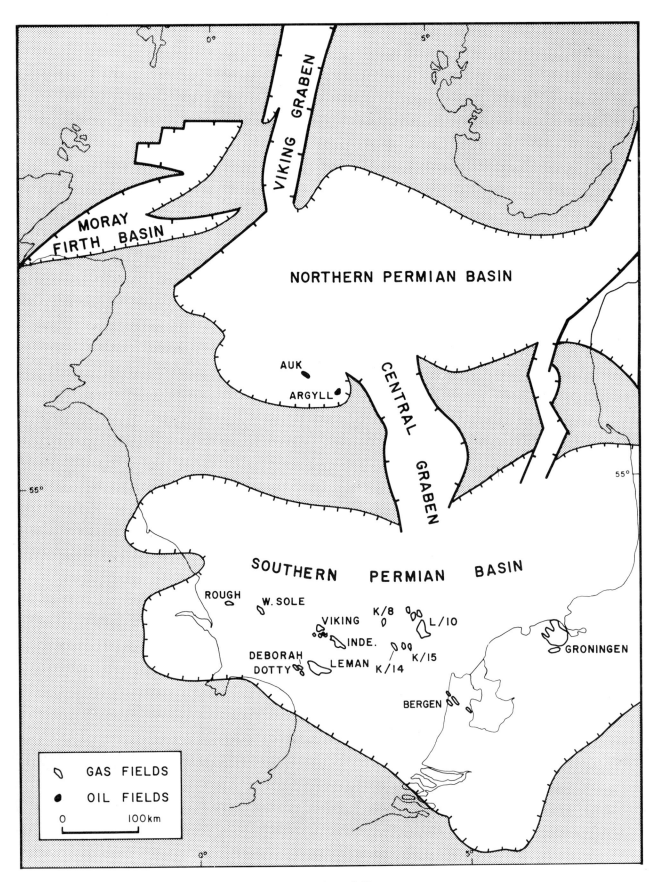

Fig. 3.12. Distribution of some Rotliegend producing oil and gas fields.

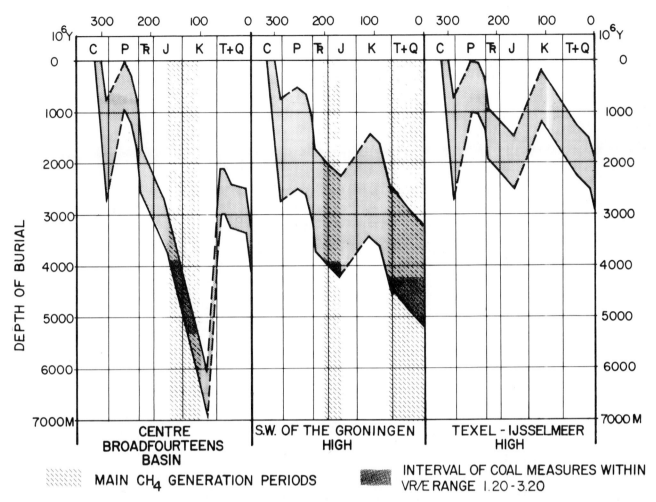

Fig. 3.13. Burial history of the Coal Measures in The Netherlands. From van Wijhe, Lutz and Kaasschieter (1980).

probably charged with gas at that time. After inversion of the basinal areas, however, gas remigrated into a reservoir that was already partly damaged by the growth of authigenic illite during the earlier deep burial (see Glennie *et al.*, 1978 and Fig. 3.15). In the Groningen area, however, Early Cretaceous uplift separated two periods of gas generation (Fig. 3.13) and it is probable that the Groningen field is still being charged with gas.

With such a long history of vertical and horizontal movement, it is not surprising that the Rotliegend reservoirs of the southern North Sea are highly faulted (Fig. 3.16). These fault movements undoubtedly triggered diapirism in the overlying Zechstein salt; in turn, some idea of the times of fault movement can be deduced from the erosional and depositional history of the rim synclines flanking the diapirs (see Taylor, Chapter 4, Figs. 4.13-4.15). Because of the relatively high acoustic velocity of halite, the rapid changes in its thickness resulting from halokinesis considerably distorts the shape of the underlying seismic reflectors, and must be taken into account when converting time maps of the Rotliegend to depth (see e.g. Butler, 1975; Christian, 1969).

The search for Rotliegend gas below the North Sea began in U.K. waters in 1964, and soon resulted in the discovery of a series of important fields, West Sole—

Viking—Leman—Indefatigable (Figs. 3.12, 1.5). of which the last two account for some 425×10^9 m³ $(15 \times 10^{12}$ ft³) of recoverable gas; these four fields have been described, respectively, by Butler, Gray, van Veen and France, all of which can be found in Woodland (1975).

Some gas fields of the Netherlands offshore that were strongly influenced by inversion movements have been described by Oele *et al.* (1981) for quadrants K and L (see Fig. 3.17) and by Roos and Smits (1983) for block K/13. The latter authors consider that Late Cretaceous inversion enabled the Triassic Bunter Sandstone in block K/13 to be charged from a gas-filled Rotliegend reservoir; the preservation of gas in two small Rotliegend fields (structures E and F in Figure 5.7) supports this interpretation. Similar fault movements probably caused the breakdown of the intervening Zechstein seal and permitted gas to transfer from the Rotliegend sandstone to both Zechstein (Plattendolomite) and Triassic reservoirs in the Bergen area of The Netherlands (van Lith, 1983) and to the Triassic reservoirs of the Hewett field on the margin of the Sole Pit Basin (see Fig. 5.5; and Cumming and Wyndham, 1975); gas was still retained in the Rotliegend reservoir of the adjacent small fields, Deborah and Dotty (Fig. 3.12).

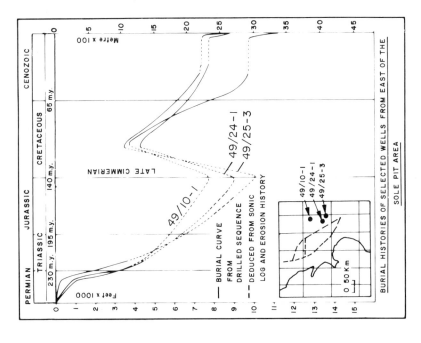

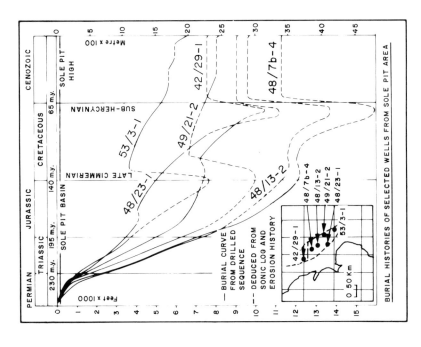

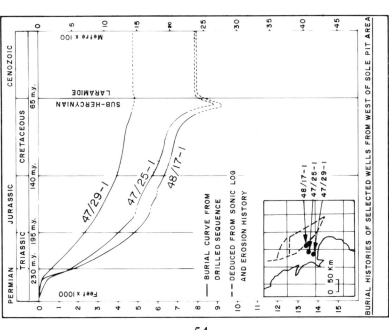

Fig. 3.14. Burial histories of the base Rotliegend at selected well locations, Sole Pit Basin, southern North Sea. From Glennie and Boegner (1981).

54

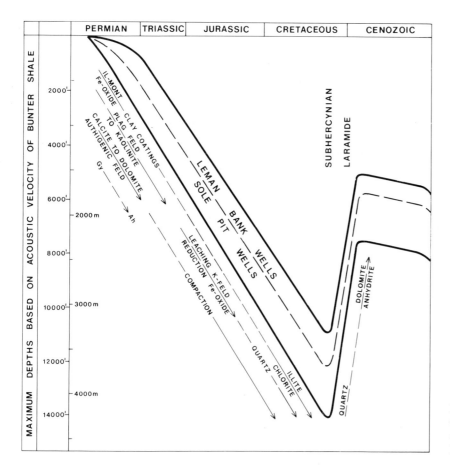

Fig. 3.15. Depth-related diagenesis in the Leman Bank and Sole Pit areas of the U.K. Southern North Sea. From Glennie *et al.* (1978).

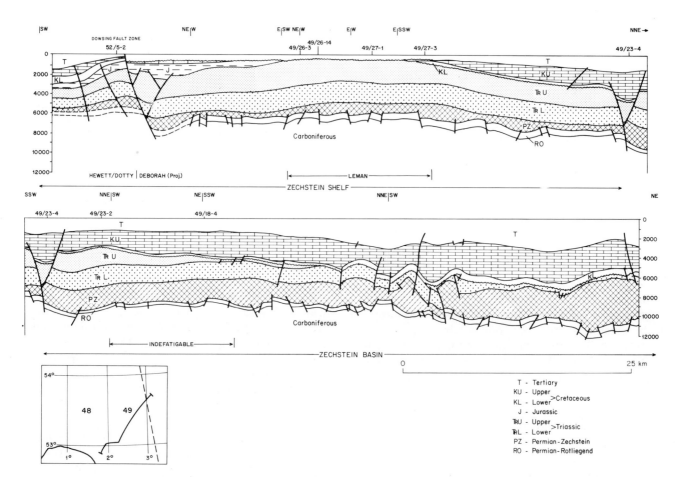

T - Tertiary
KU - Upper ⟩Cretaceous
KL - Lower ⟩Cretaceous
J - Jurassic
TrU - Upper ⟩Triassic
TrL - Lower ⟩Triassic
PZ - Permian - Zechstein
RO - Permian - Rotliegend

Fig. 3.16. Geological cross-section, U.K. Southern North Sea. Note fault-bounded relief of Rotliegend sequence, thick cover of Zechstein salt over Indefatigable, and much thinner salt over Leman and adjacent to the Dowsing Fault.

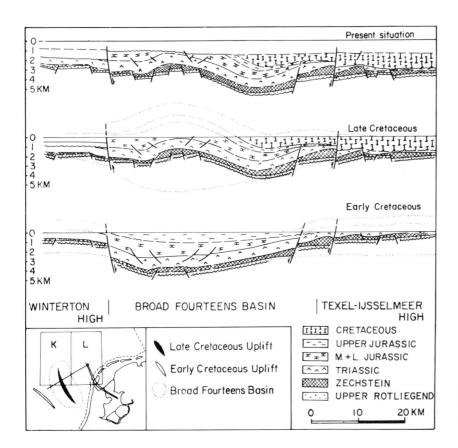

Fig. 3.17. Reconstruction of the burial history of the Rotliegend in The Netherlands offshore K and L blocks. From Oele *et al.* (1981).

Reservoir quality and diagenesis

In general, the best reservoir sands occur in the dominantly aeolian middle part of the Rotliegend sequence, with porosities in some fields ranging up to about 25%, and air permeabilities with values in excess of 100 mD where not damaged diagenetically. The porosity of the aeolian sands directly beneath the Kupferschiefer is commonly reduced for two different reasons: the original grain packing became tighter in the Weissliegend sands (the top 0-65 m) because of deformation associated with the Zechstein transgression; and the porosity of the upper 10 to 15 m of sands, whether deformed or not, is drastically reduced because of a dolomite cement believed to result from proximity to the overlying Zechstein carbonates. Robinson (1981) finds that the ratio between horizontal and vertical air permeability of most aeolian and wadi sands is between 1 and 100. This distinction is not seen in the Weissliegend sands because of either a lack of distinct bedding, or the presence of deformed bedding. Both these points are reflected by dip-meter logs, which generally show no dip data or, alternatively, only random dips.

The thick Rotliegend sequence found in the Leman Bank gas field (Figs. 3.12, 3.16) comprises mostly foresetted aeolian sandstones similar to those illustrated by Glennie (1972, Fig. 12) which are interpreted as having been deposited on the avalanche slopes of transverse dunes; these undercompacted sands had a naturally high primary porosity. In the Indefatigable area (Fig. 3.12, 3.16), on the other hand, the Rotliegend aeolian sequence is much thinner and has many horizons of silt and anhydrite-rich (former gypsum crystals) adhesion-ripple sands, which had an inherently poorer

porosity and permeability than the foresetted dune sands. The burial-related growth of authigenic minerals, and especially of illite, in areas of former deep burial such as the Leman Bank field, has resulted in the porosity and permeability of the dune sandstones now being much lower on average than those of the Indefatigable field, which was never so deeply buried (compare well 49/24-1 in Fig. 3.14 with Fig. 3.15).

The capillary effects associated with smaller pore connections in the diagenetically more damaged reservoir of Leman Bank probably account for the overall difference in recovery factors between the two fields, from 80% of the gas in place for Indefatigable to 75% for Leman Bank. Similarly, Robinson (1981) attributes a relatively high water saturation in the Amoco well 47/15-2 to the reduction in the size of pore throats related to the precipitation of authigenic minerals.

As we have seen, because of the effects of diagenesis, the quality of most gas reservoirs becomes poorer with increased depth of burial. It comes as a pleasant surprise, therefore, to find an example where Lower Rotliegend aeolian sandstones have better porosity and permeability at a depth of over 5000 m than the dune sands of the overlying Upper Rotliegend. This rather unusual situation is found in the Schneverdingen Graben in West Germany, about 50 km east of Bremen (Drong *et al.*, 1982). The almost 400 m thick Schneverdingen Sandstone was deposited in an actively subsiding narrow graben (Fig. 3.18) and has preserved porosities of up to 15% and permeabilities in the range of 1-10 mD, in contrast to the 1-5% porosity and < 0.1 mD permeability of the Upper Rotliegend sandstones. Drong and his colleagues suggest that the differences in reservoir quality might reflect differences in climate between

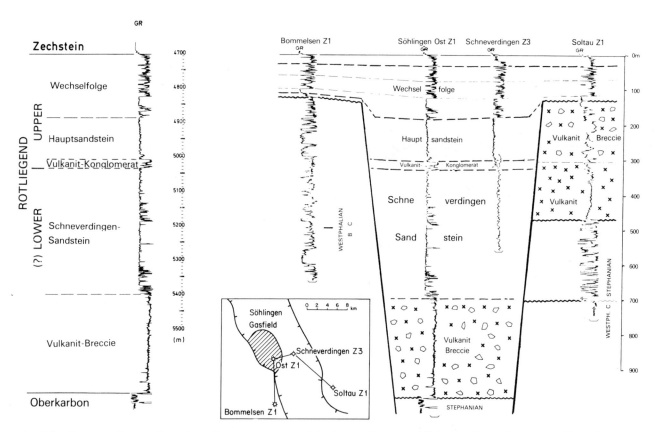

Fig. 3.18. Stratigraphic relationship and tectonic setting of the Lower Rotliegend Schneverdingen Sandstone, West Germany. (Slightly modified from Drong *et al.*, 1982).

the times of deposition of the two Rotliegend dune sequences (less calcite and anhydrite cement in the Schneverdingen Sandstone). It seems just as likely, however, that the porosity differences could also be related to differences in grain size, with stronger capillary retention of formation water (and hence stronger diagenesis on burial) in the finer grained (0.1-0.25 mm) Hauptsandstein than in the coarser (0.25-0.5 mm) deeper sands of the Schneverdingen Sandstein.

In the Northern Permian Basin, the Zechstein salt cap rock is present over the greater part of the area (Fig. 4.9), but the all important source rock for gas, the Carboniferous Coal Measures, is conspicuously absent (see Figs. 2.2, 2.8). Thus in this area, those Rotliegend reservoirs that have been penetrated by the drill are devoid of gas. Where the Rotliegend sandstones of this basin form reservoirs for hydrocarbons, they have been brought into the correct geometric relationship with a mature source rock for oil.

3.5.2 Oil fields

Although minor amounts of condensate and Natural Gas Liquids (NGL) are known from some Rotliegend

structures in the Southern Permian Basin, they occur as the heavy fractions of gas of Carboniferous origin.

The Auk and Argyll oil fields in the Central North Sea produce much of their oil from the basal Zechstein carbonates. In the shallower parts of the fields, however, Rotliegend sandstones are also saturated with oil. These fields are situated close to the western flank of the Central Graben. Pennington (1975) suggests that the oil found in the Argyll field is derived from structurally adjacent Paleocene shales. These shales, however, probably have not been buried sufficiently deeply to be mature. It seems much more likely that the oil-source rock for both the Auk and Argyll fields, like most of the fields of the Central and Northern North Sea, is the Kimmeridge Shale, which matured deep in the Central Graben sometime during the mid to late Tertiary. The oil probably migrated up graben-flank faults and became trapped beneath the general cover of a relatively impervious Upper Cretaceous chalk (Fig. 3.19).

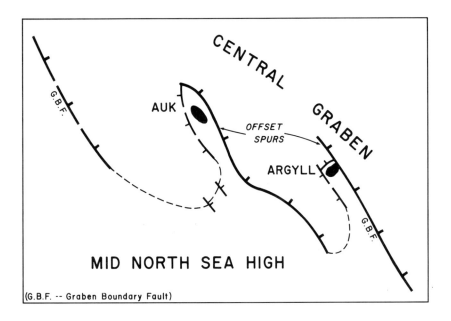

AUK

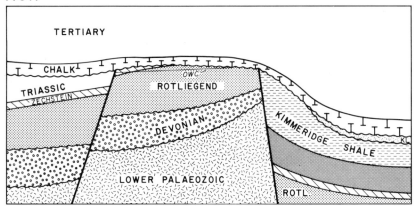

ARGYLL

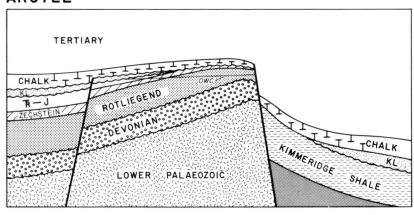

Fig. 3.19. Schematic structural setting and cross-sections of the Auk and Argyll oil fields. Source Rock—probably Upper Jurassic Kimmeridge Shales. Reservoir Rock—Zechstein dolomites and Rotliegend sandstones. Modified from Brennand and van Veen (1975), and Pennington (1975).

3.6 Acknowledgements

This contribution is published by permission of Shell U.K. Ltd. and Esso Petroleum Co. Ltd.

3.7 Selected references

Bartenstein, H. (1979) Essay on the coalification and hydrocarbon potential of the Northwest European Palaeozoic. *Geol. Mijnbouw* **58**, 57-64.

Best, G., Kockel, F. and Schöneich, H. (1983) Geological history of the southern Horn Graben. *Geol. Mijnbouw* **62**, 25-33.

Bowler, J.M. (1976) Aridity in Australia: age, origins and expression in aeolian landforms and sediments. *Earth Sci. Rev.* **12**, 279-310.

Brennand, T.P. and Van Veen, F.R. (1975) The Auk field. In: Woodland, A.W. (Ed.) q.v. 275-285.

Brookfield, M.E. (1980) Permian intermontane basin sedimentation in southern Scotland. *Sediment. Geol.* **27**, 167-194.

Bungener, M.J.A. (1969) Le Champ de gaz de Groningen. *Revue de l'Assoc. Franc. des Techniciens du Pétrole* **196**, 19-32.

Butler, J.B. (1975) The West Sole gas field. In: Woodland, A.W. (Ed.) q.v. 213-223.

Callomon, J.H., Donovan, D.T. and Trümpy, R. (1972) An annotated map of the Permian and Mesozoic formations of East Greenland. *Medd. Groenl.*, **168**, 1-35.

Christian, H.E. (1969) Some observations on the initiation of salt structures of the Southern British North Sea. In: Hepple, P. (Ed.). *The exploration for petroleum in Europe and North Africa.* Inst. Petroleum, p. 231-248.

Clemmensen, L.B. and Abrahamsen, K. (1983) Aeolian stratification and facies association in desert sediments, Arran basin (Permian) Scotland. *Sedimentology* **30**, 311-339.

Cumming, A.D. and Wyndham, C.L. (1975) The geology and development of the Hewett gas field. In: Woodland, A.W. (Ed.) q.v. 313-325.

Deegan, C.E. and Scull, B.J. (1975) *A standard lithostratigraphic nomenclature for the Central and Northern North Sea.* Report No. 77/25 Inst. Geol. Sci. H.M.S.O.

Drong, H.J., Plein, E., Sannemann, D., Schuepbach, M.A. and Zimdars, J. (1982) Der Schneverdingen-Sandstein des Rotliegenden—eine äolische Sedimentfüllung alter Graben Strukturen. *Zeit. deutsch. geol. Ges.* **133**, 699-725.

Falke, H. (1971) Zur Paläogeographie des kontinentalen Perms in Süddeutschland. *Abh. hess. L-Amt Bodenforsch* **60**, 223-234.

France, D.S. (1975) The geology of the Indefatigable gas field. In: Woodland, A.W. (Ed.) q.v. 233-241.

Galloway, R.W. (1965) Late Quaternary climates in Australia. *J. Geol.* **73**, 603-618.

Glennie, K.W. (1970) *Desert sedimentary environments.* Developments in Sedimentology 14. Elsevier, Amsterdam, 222pp.

Glennie, K.W. (1972) Permian Rotliegendes of North-West Europe interpreted in light of modern desert sedimentation studies. *Bull. Am. Assoc. Petrol. Geol.* **56**, 1048-71.

Glennie, K.W. (1983a) Early Permian (Rotliegendes) palaeowinds of the North Sea. *Sedimentary Geology* **34**, 245-265.

Glennie, K.W. (1983b) Lower Permian Rotliegend desert sedimentation in the North Sea Area. In: *Aeolian Sands.* Developments in Sedimentology. Elsevier, Amsterdam. p. 521-541.

Glennie, K.W., Mudd, G.C. and Nagtegaal, P.J.C. (1978) Depositional environment and diagenesis of Permian Rotliegendes sandstones in Leman Bank and Sole Pit areas of the U.K. southern North Sea. *J. geol. Soc. Lond.* **135**, 25-34.

Glennie, K.W. and Boegner, P. (1981) Sole Pit inversion tectonics. In: Illing, L.V. and Hobson, G.D. (Ed.) q.v. 110-120.

Glennie, K.W. and Buller, A.T. (1983) The Permian Weissliegend of N.W. Europe: the partial deformation of aeolian dune sands caused by the Zechstein transgression. *Sedimentary Geology.* **35**, 43-81.

Gray, I. (1975) Viking gas field. In: Woodland, A.W. (Ed.) q.v. 241-249.

Groen, D.M.W. te and Steenken, W.F. (1968) Exploration and delineation of the Groningen gas field. *Verh. Kon. Ned. Geol. Mijnb.* **25**, 9-20.

Hunter, R.E. (1977) Basic types of stratification in small aeolian dunes. *Sedimentology* **24**, 361-387.

Illing, L.V. and Hobson, G.D. (1981) *The petroleum geology of the Continental Shelf of N.W. Europe.* Hayden, 521 p.

Krinsley, D.H. and Smith, D.B. (1981) A selective SEM study of grains from the Permian Yellow Sands of North-East England. *Proc. Geol. Assoc.* Vol. 92 pt. **3**, 189-196.

Laan, G. van der (1968) Physical properties of the reservoir and volume of gas initially in place. *Verh. Kon. Ned. Geol. Mijnb.* **25**, 25-33.

Laming, D.J.C. (1966) Imbrication, palaeocurrents and other sedimentary features in the Lower New Red Sandstone, Devonshire, England. *J. Sediment. Petrol.* **36** (4), 940-959.

Lith, J.G.J. van (1983) Gas fields of Bergen Concession, The Netherlands. *Geol. Mijnbouw* **62** (1), 63-74.

Lovell, J.P.B. (1983) Permian and Triassic. In: Craig, G.Y. (Ed.) *Geology of Scotland*, 2nd edn. Scottish Academic Press. 325-342.

Lutz, M., Kaasschieter, J.P.H. and Wijhe, D.H. van (1975) Geological factors controlling Rotliegend gas accumulation in the Mid-European Basin. *Proc. 9th World Petroleum Cong.* **2**, 93-97.

Nagtegaal, P.J.C. (1973) Adhesion-ripple and barchan dune sands of the Recent Namib (S.W. Africa) and Permian Rotliegend (N.W. Europe) deserts. *MADOQUA Series 11*, Vol. 2, Nos 63-68, 5-19.

Nagtegaal, P.J.C. (1979) Relationship of facies and reservoir quality in Rotliegendes desert sandstones, Southern North Sea Region. *J. Pet. Geol.* **2**, 145-158.

Nederlandse Aardolie Maatschappij & Rijks Geologische Dienst (1980) *Stratigraphic nomenclature of The Netherlands.* Trans. Royal Dutch Geol. & Mining Soc. Delft. V. 33, 77 p.

Oele, J.A., Hol, A.C.P.J. and Tiemans, J. (1981) Some Rotliegend gas fields of the K and L blocks, Netherlands offshore (1968-1978)—A case history. In: Illing, L.V. and Hobson, G.D. (Eds.) q.v. 289-300.

Peacock, J.D., Berridge, N.G., Harris, A.L. and May, F. (1968) *Geology of the Elgin district.* Memoir Geol. Surv. Scot. H.M.S.O. 165 p.

Pennington, J.J. (1975) The Geology of the Argyll Field. In: Woodland, A.W. (Ed.) q.v. 285-291.

Plein, E. (1978) Rotliegend-Ablagerungen im Norddeutschen Becken. *Zeit. deutsch. geol. Ges.* **129**, 71-97.

Rea, D.K. and Janecek, T.R. (1982) Late Cenozoic changes in atmospheric circulation deduced from North Pacific eolian sediments. *Marine Geology* **49**, 149-167.

Rhys, G.H. (compiler) (1974) *A proposed standard lithostratigraphic nomenclature for the southern North Sea and an outline structural nomenclature for the whole of the (UK) North Sea.* Report No. 74/8, Inst. Geol. Sci. H.M.S.O. 14 p.

Robinson, A.E. (1981) *Facies types and reservoir quality of the Rotliegendes Sandstone, North Sea.* 56th Ann. SPE of AIME Tech. Conf. (San Antonio, Texas). SPE 10303, 10 p.

Roos, B.M. and Smits, B.J. (1983) Rotliegend and Main Buntsandstein gas fields in block K/13. A case history. *Geol. Mijnbouw* **62**, 75-83.

Russell, M.J. (1976) A possible Lower Permian age for the

onset of ocean floor spreading in the northern North Atlantic. *Scott, J. Geol.* **12**, No. 4, 315-323.

Smith, D.B. (1972) The Lower Permian in the British Isles. In: Falke, H. (Ed.), *Rotliegend Essays on European Lower Permian.* Brill, Leiden, 1-33.

Smith, D.B. (1979) Rapid marine transgressions and regressions of the Upper Permian Zechstein Sea. *J. geol. Soc. Lond.* **136**, 155-156.

Smith, D.B., Brunstrom, R.G.W., Manning, P.I., Simpson, S. and Shelton, F.W. (1974) *Permian.* Geol. Soc. Lond. Special Report No. 5, 45 p.

Stäuble, A.J. and Milius, G. (1970) Geology of Groningen gas field. *Am. Assoc. Petrol. Geol. Memoir* **14**, 359-369.

Taylor, J.C.M. (1981) Zechstein facies and petroleum prospects in the central and northern North Sea. In: Illing, L.V. and Hobson, G.D. (Ed.) q.v. 176-185.

Turner, P. (1980) *Continental red beds.* Developments in Sedimentology 29. Elsevier, Amsterdam, 562 p.

Veen, F.R. van (1975) Geology of the Leman Gas field. In: Woodland, A.W. (Ed.) q.v. 223-233.

Voo, R. van der and French, R.B. (1980) Apparent polar wandering for the Atlantic Bordering continents: Late Carboniferous to Eocene. *Earth Sci. Rev.,* **10**, 99-119.

Walker, T.R. (1967) Formation of red beds in modern and ancient deserts. *Geol. Soc. Amer. Bull.* **78** (3), 353-368.

Wijhe, D.H. van, Lutz, M. and Kaasschieter, J.P.H. (1980) The Rotliegend in The Netherlands and its gas accumulations. *Geol. Mijnbouw* **59**, 3-24.

Woodland, A.W. (Ed.) (1975) *Petroleum and the Continental Shelf of North-West Europe.* Vol. 1. Geology. App. Sci. Pub. London. 501 p.

Ziegler, P.A. (1978) North-Western Europe: tectonics and basin development. *Geol. Mijnbouw* **57**, 487-502.

Ziegler, P.A. (1980a) *Geology and hydrocarbon provinces of the North Sea.* Memoir No. 6 Canadian Society of Petroleum Geologists. 653-706.

Ziegler, P.A. (1980b) *North-Western Europe: subsidence patterns of Post-Variscan basins.* Proc. Int. Geol. Congr. Paris, 1980. C3-5.

Ziegler, P.A. (1982) *Geological atlas of Western and Central Europe.* Elsevier, Amsterdam. 130 p.

Chapter 4   Late Permian—Zechstein

J.C.M. TAYLOR

4.1 Introduction

The Zechstein is a complex of evaporite and carbonate rocks of Late Permian (= Thuringian) age, which underlie a substantial area of the North Sea and north-west Europe (Fig. 4.1). Deposition was apparently coeval with the Guadalupian to Ochoan sequences of the Delaware Basin, USA, with which there are interesting similarities.

As one of the world's 'Saline Giants', the Zechstein Basin merits attention for the light it may cast on the origin of other thick evaporite sequences.

So far as the oil industry is concerned, Zechstein rocks are significant on five counts:

4.1.1 Generation of structure

Flow of Zechstein salt is responsible for closures in overlying strata. North Sea fields which owe their existence to this factor include the giant Ekofisk. Subtle structural effects resulting from salt dissolution at depth are also becoming recognised.

4.1.2 Structural information

The top and base of the Zechstein commonly provide important seismic reflectors which give clues to the structure of hydrocarbon-bearing strata above or below.

4.1.3 Cap rocks

The Rotliegend gas fields of the Southern North Sea depend largely on the sealing efficiency of Zechstein salt.

4.1.4 Reservoir rocks

Some Zechstein carbonates have good porosity and permeability, providing commercial reservoirs on the Continent and in the North Sea.

4.1.5 Source rocks

Zechstein carbonates include potential source facies, though adequacy for North Sea conditions has yet to be demonstrated.

In addition, Zechstein evaporites constitute reserves of commercial potash salts, halite, anhydrite and gypsum, in northeast England and on the Continent. The thick halite sections are being considered for leached-cavern storage of natural gas and the disposal of radioactive wastes.

This summary has been compiled largely from the references listed in context. The place of the Zechstein in North Sea petroleum geology was outlined by Kent (1967 a and b) and with special reference to its evaporites by Brunstrom and Walmsley (1969). Correlation and nomenclature have been discussed by Pattison et al. (1973), Smith et al. (1974), Rhys (1974), and Deegan and Scull (1977); the first two references should be consulted for a guide to the extensive earlier literature relating essentially to onshore areas. Revised nomenclature for the English Zechstein strata due to Harwood et al. (1982) is used in this account where appropriate (see Table 4.1).

4.2 Distribution and general character (Fig. 4.1)

The distribution of the Zechstein north of latitude 59°N is poorly known, though from exposures in eastern Greenland (Maync, 1961) and diapirs in the Norwegian Troms Basin (King, 1977) it is believed to extend to about 75°N.

In the south, the Zechstein occupies two E-W trending sub-basins partly separated by the Mid North Sea—Ringkøbing-Fyn High (Ziegler, 1981). Much of the Ringkøbing-Fyn High remained exposed through the Zechstein. Thick salt occupies the basin centres, whereas carbonates and anhydrites are more important round the edges and over the Mid North Sea High. The northern salt basin stretches from the Scottish Firth of Forth across northern Denmark. It is fairly well delineated by seismic, which shows the group to be over 2000 m (6000 ft) thick in the Norwegian-Danish Basin and locally elsewhere (Taylor, 1981; Fig. 2), but relatively little detail is known as much of it is below current drilling depths (Fig. 4.3).

The southern salt basin extends from eastern England through the Netherlands and Germany to Poland and western Russia, and is well-documented as a result of the search for hydrocarbons and other minerals.

A smaller and mainly independent basin—the Bakevellia Basin—occupies parts of the Irish Sea and adjacent Northern Ireland and north-west England (Pattison et al., 1973; Colter and Barr, 1975).

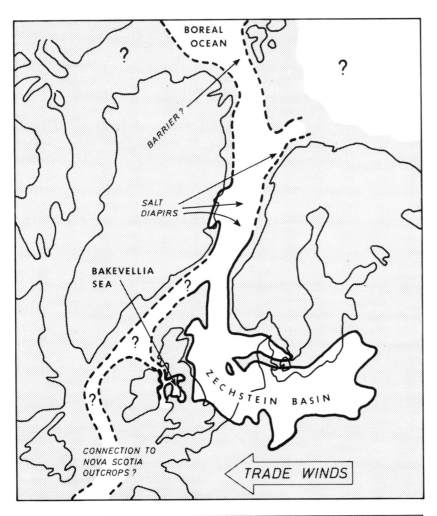

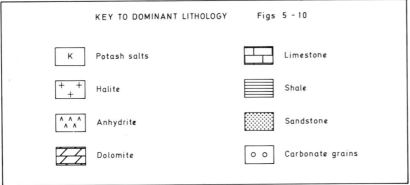

Fig. 4.1. Sketch map of Zechstein basins in relation to pre-drift North Atlantic area.

4.3 Outline of depositional pattern and typical facies

Figure 4.4 is a diagrammatic profile of the Zechstein in the U.K. Southern North Sea showing the principal lithofacies, together with the German nomenclature in common use. The correlation of formation names on-shore in the U.K. is given in Table 4.1. Similar profiles described from the Netherlands (Brueren, 1959; Clark, 1980a; Boogaert and Burgers, 1983), Germany (Richter-Bernburg, 1959; Füchtbauer, 1968), Denmark (Sorgenfrei and Buch, 1964; Clark and Tallbacka, 1980), and Poland (Depowski, 1978; Wagner *et al.*, 1981), testify to the depositional unity of the southern salt basin and lead to a model which is useful for exploration purposes. Examples from as far away

as Poland, where more than 1000 Zechstein wells have been drilled in the past two decades, are therefore relevant to the Southern North Sea. It is believed, though not yet conclusively proved, that this model can also be applied to the northern salt basin (Taylor, 1981). Correlation from the southern to the northern basin across the Mid North Sea High and on to the Moray Firth is illustrated by typical well logs summarised in Figure 4.5.

Towards the close of the Permian, the whole region lay in the Trade Winds belt and the climate was predominantly hot and dry. By the end of Rotliegend time, aridity, combined with lowering and mantling of source areas, appears to have cut sediment supply until deposition lagged behind subsidence, resulting in inland depressions well below ocean level. Zechstein

CYCLES (England)	GROUPS	YORKSHIRE PROVINCE	DURHAM PROVINCE	CYCLES	S. NORTH SEA, GERMANY, NETHERLANDS, S. DENMARK, POLAND
EZ5	ESKDALE GROUP	Saliferous Marl Formation (Permian Upper Marls)	Saliferous Marl Formation; Top Anhydrite Formation; Sleights Siltstone Formation	Z5	Zechsteinletten; Grenzanhydrit
EZ4	STAINTONDALE GROUP	Saliferous Marl Formation (Permian Upper Marls)	Sneaton Halite Formation; Sherburn Anhydrite Formation; Upgang Formation	Z4	Aller Halit; Pegmatitanhydrit
EZ3	TEESSIDE GROUP	Brotherton Formation (Upper Magnesian Limestone)	Carnallitic Marl Formation; Boulby Halite Formation; Billingham Main Anhydrite Formation; Seaham Formation	Z3	Roter Salzton; Leine Halit; Hauptanhydrit; Plattendolomit; Grauer Salzton
EZ2	AISLABY GROUP	Edlington Formation (Permian Middle Marls); Kirkham Abbey Formation	Fordon Evaporites and Seaham Residue; Hartlepool and Roker Formation	Z2	Stassfurt Evaporites; Basalanhydrit; Hauptdolomit; Stinkdolomit, Stinkkalk, Stinkschiefer
EZ1	DON GROUP	Hayton Anhydrite; Sprotbrough Member; Wetherby Member; Cadeby Formation (Lower Magnesian Limestone); Marl Slate	Hartlepool Anhydrite; Ford Formation (Middle Magnesian Limestone); Raisby Formation (Lower Magnesian Limestone); Marl Slate	Z1	Werraanhydrit; Werradolomit & Zechsteinkalk; Kupferschiefer

Table 4.1. Stratigraphic nomenclature and correlation of the Zechstein, after Smith (1980) and Harwood *et al.* (1982). Former names in common use in brackets.

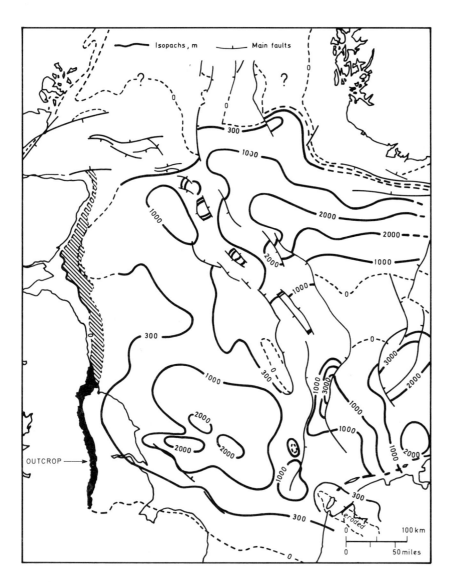

Fig. 4.2. Generalised Zechstein isopachs. Modified from Taylor (1981) with additions. Note uneven contour interval.

sedimentation began when these desert basins were flooded from the Boreal Ocean (Fig. 4.1) perhaps by rifting between Greenland and Spitzbergen together with a rise in sea level.

Five main evaporite cycles can be distinguished—Z1 to Z5—as on the Continent. The 'ideal' cycle reflects the influence of increasing salinity through evaporation following an initial marine incursion, commencing with a thin clastic member, passing upwards in turn through limestone, dolomite, and anhydrite to halite, and finally to highly soluble salts of magnesium and potassium (the 'bitterns'). Whilst this simple scheme is useful as a framework, there are many omissions and reversals of stages in the Zechstein cycles. Some of the exceptions are attributable to the diagenetic origin of the present minerals, but most arise through the subtle interplay of changing climate, runoff, oceanic exchange, water depth, sedimentation rate and subsidence. The overriding cyclic control is thought to have been eustatic variation in ocean level (perhaps largely glacial in origin), amplified by the mediation of a barrier which, whenever it was exposed, allowed evaporative drawdown in the Zechstein basin system behind it (Smith, 1980; Taylor, 1980).

A deep-water, barred-basin model—long inferred

from the sequences in Germany—seems to be required by the geometry of correlatable units in the North Sea, and reflux is needed to explain the imbalance of anhydrite over halite in the early stages. However, the shallow-water/deep-basin and shallow-water/shallow-basin models of evaporite deposition also have application. Intermittent partial or complete desiccation has been deduced for the first cycle, although the bulk of the Werraanhydrit nevertheless appears to have formed when the basin contained a substantial depth of brine (Taylor, 1980); similar processes may have operated in subsequent cycles. Marginal salina and sabkha deposits are recognisable, and progressive shoaling of the whole area is apparent in later cycles as the rate of salt deposition periodically overtook subsidence.

As shown in Figure 4.4, the carbonates overlap their early transgressive shaly beds towards the margin of the basin, where in many areas they pass into continental sandstones. The anhydrite formations do not generally extend so far from the basin centre as the carbonates. Halite formations are less extensive than anhydrite, passing landwards into clays, mudstones and siltstones, while potash zones make their appearance even further from the margin. The resulting

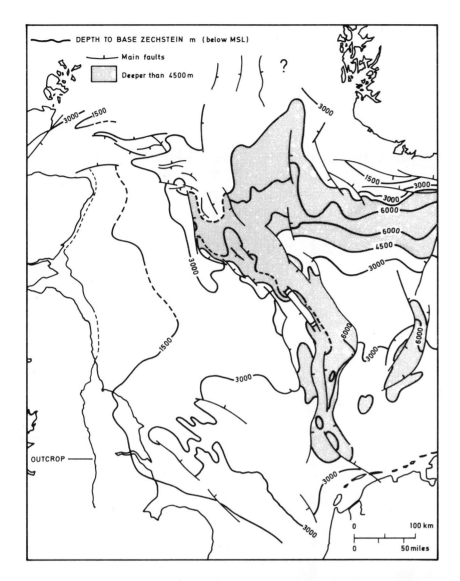

Fig. 4.3. Structure on base of Zechstein. Generalised mainly from Day *et al.* (1981).

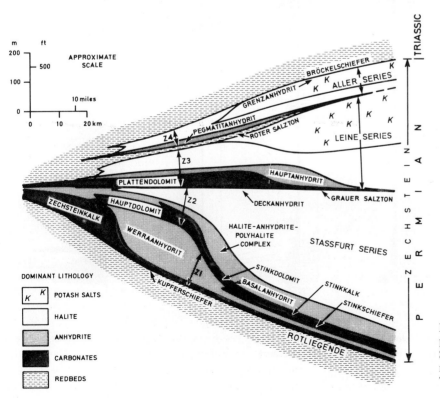

Fig. 4.4. Diagrammatic shelf-to-basin profile of the Zechstein in U.K. Southern North Sea, showing German nomenclature in common use. After Taylor (1981) with additions.

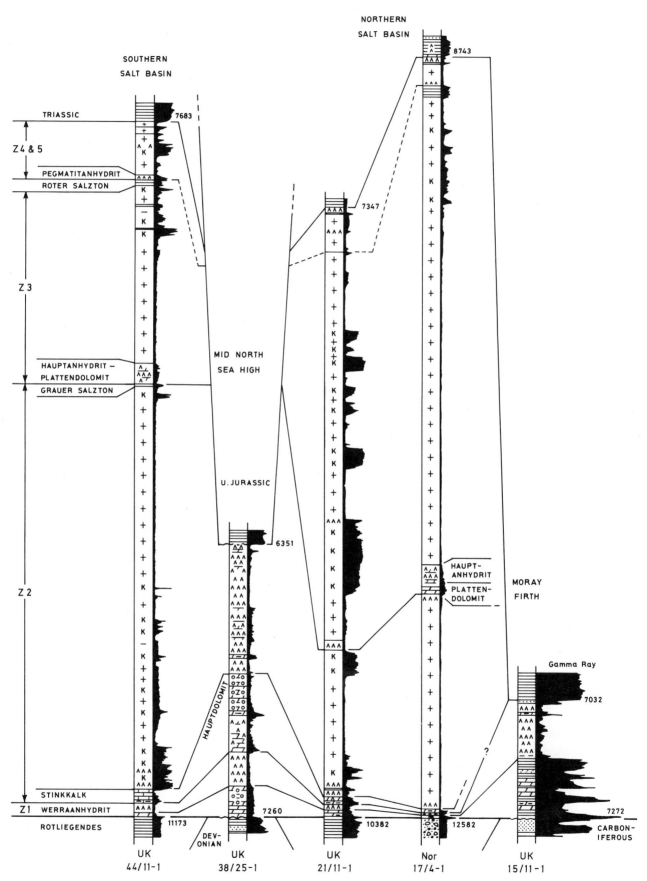

Fig. 4.5. Suggested correlation of typical Zechstein sequences from the Southern North Sea to the Moray Firth. Modified from Taylor (1981). Note: not to uniform scale—short sequences exaggerated for clarity.

'bullseye pattern' (Hsü, 1972) in each cycle is not, however, the simple product of a sea shrinking under the influence of evaporation following a single flooding, for each phase—carbonate, anhydrite, and halite in turn—persisted long enough for many basin-volumes of sea water to yield their precipitate, during which significant basin subsidence took place.

Successive cycles are increasingly 'evaporitic', significant carbonate development being confined to the first three cycles of which only the first possesses a diverse marine fauna. In the U.K. sector, halite is important only in Z2 and later cycles whilst potentially economic potash salts are concentrated in Z3 and Z4.

The marginal carbonates of the first three cycles in the southern salt basin are complex wedges, each thickening for some distance into the basin and then thinning towards the centre (Fig. 4.4). Following each transgressive phase, they formed by the progradation of a variety of near-shore facies across more uniform and finer-grained slope and basin facies.

Carbonate production was most rapid in the well-aerated, warm, well-lit shallow water round the edges of the basin where sedimentation consequently easily outran subsidence, leading to the construction of broad shelves comprising barrier, lagoonal, intertidal and slightly emergent (including sabkha) environments. Reefs were important in Z1. Algae were particularly significant as sediment producers and binders, especially in the later stages of each carbonate phase when salinities rose beyond the tolerance of many other taxa.

These shallow conditions favoured accumulation of grainstones and boundstones with good primary porosity and permeability—properties tending sometimes to be preserved or enhanced by early conversion to microdolomite, and in places augmented by vadose processes.

By contrast, basin carbonates are thin, dark, compact, argillaceous or shaly micrites, mostly deposited well below wave-base and commonly under anoxic conditions, normally offering no reservoir prospects. Their potential as source rocks is limited mainly by their thinness.

Slope carbonates—which represent a considerable proportion of the thickest part of each wedge—are intermediate in texture and composition, becoming progressively paler, less argillaceous and less shaly upwards. They are probably built mainly from the winnowings of the shelf, together with fine skeletal material, and appear to include carbonate turbidites and slumped material. Though commonly dolomitised, the predominant micritic facies tends to have poor inter-crystalline porosity and hence mediocre permeability.

The geometry of the anhydrite phase of each cycle also testifies to centripetal accretion of prograding wedges as the main process of basin-filling, and a similar conclusion has been reached for the mixed halite/anhdrite/polyhalite/kieserite zone in the second cycle (Colter and Reed, 1980). Whether the same principle

applies to later evaporitic stages remains to be determined.

4.4 First Zechstein cycle, Z1 (Figs. 4.6 & 4.7)

The initial flooding of the Zechstein Basin was accompanied by disturbance and minor reworking of uncemented Rotliegend sands (see Glennie, ch. 3; Glennie and Buller, 1983).

4.4.1 Kupferschiefer

The Kupferschiefer (Copper-Shale) which marks the formal base of the Zechstein, is a dark 1 m (3 ft) sapropelic shale distinguished on logs by a strong gamma-ray peak. It formed under anoxic conditions below the influence of waves (though not everywhere under particularly deep water) and drapes minor eminences. It can be recognised across the floor of both southern and northern salt basins, also in the Moray Firth and in some wells in the Viking Graben area, but is absent from parts of the Mid North Sea High and other highs, especially where the Rotliegend is missing. Although organic-rich, it is too thin to be a credible source rock.

4.4.2 Zechsteinkalk and equivalents

The Kupferschiefer rapidly passes up into marine carbonates—the equivalents of the Zechsteinkalk and Werradolomit. The carbonates thicken from a mere 3 m (10 ft) or so of argillaceous and carbonaceous dolomites and limestones on the floor of the southern basin to about 30 m (100 ft) of combined slope and shelf facies 16-32 km (10-20 miles) from the depositional margin. Two transgressive-regressive subdivisions are recognisable. Onshore these are represented by the Raisby and Ford Formations in Durham, and the Wetherby and Sprotbrough Members of the Cadeby Formation in Yorkshire. Both contain an abundant fauna of brachiopods, bivalves, bryozoa, crinoids and foraminifera in the marginal belt, but became impoverished basinwards and upwards.

Reefs

Bryozoan-algal reefs capped by stromatolites are characteristic of the shelf Z1 carbonates in the U.K., Denmark, Germany and Poland, and may therefore be expected in the appropriate setting in areas not yet explored. Small patch reefs are a feature of the lower subdivision at outcrop from Yorkshire to Nottinghamshire (Smith, 1981b), whereas a massive barrier some 100 m (300 ft) high and at least 35 km (20 miles) long was constructed in the upper subdivision of Durham (Smith, 1981a). Sheets of dolomitised and generally porous oolites spread shorewards from the Durham reef, and similar beds surround the patch reefs. Small pinnacle reefs occupying open shelf and fore-reef

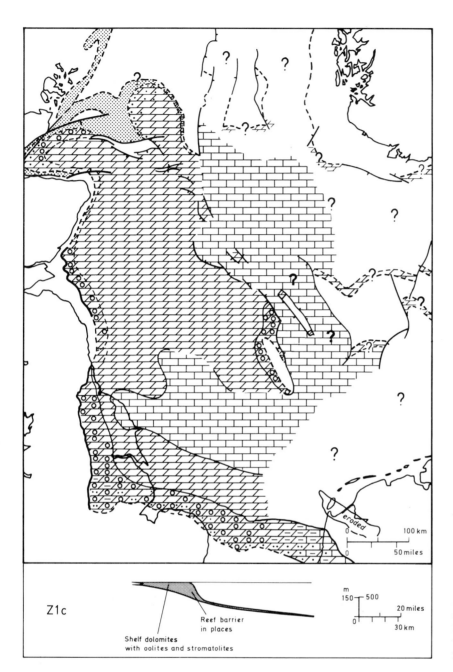

Fig. 4.6. Generalised facies of Z1 carbonates (Zechsteinkalk of North Sea, Ford and Raisby Formations of the Durham province, Cadeby Formation of the Yorkshire province).

positions are also known on the Continent, where they are sited on local topographic highs such as those commonly provided by the stumps of Lower Rotliegend volcanoes (Paul, 1980).

One North Sea area that appears to offer good conditions for reef development is the Mid North Sea High, which provided shoals through much of Z1 and Z2. Jenyon and Taylor (1983) show seismic evidence which may support this view. The possible influence of buried granites on sedimentation over the Mid North Sea High has been discussed by Donato *et al.* (1983).

Reservoir facies

Oolitic and oncolitic shelf dolomites of Z1 form commercial gas reservoirs in Poland (Depowski, 1981). Effective porosities of 6-13% and sometimes up to 30% are recorded, with permeabilities in the range 100-200 mD and rarely 1,000 mD.

Near the depositional margin in Nottinghamshire and off the Norfolk coast, dolomite grainstones pass southwards into terrigenous sands. Further east in Netherlands waters, the Z1 carbonates are seldom more than 45 m (150 ft) thick and are predominantly shaly. A high influx of clastics is noted in this area in later cycles also, and may be related to a major river system draining the London-Brabant massif (Boogaert and Burgers, 1983).

Another area of clastic input is indicated in the northwestern part of the Moray Firth where 90-120 m (300-400 ft) of interbedded limestones, shales, and dolomites, with oolites and terrigenous sands near the top, may correspond with the Z1 carbonates. The lower part of this sequence has been designated the Turbot Bank Formation (Deegan and Scull, 1977). The influence of the initial Zechstein transgression can be recognised in the Hopeman Sandstone near Lossiemouth on the southern shore of the Moray Firth

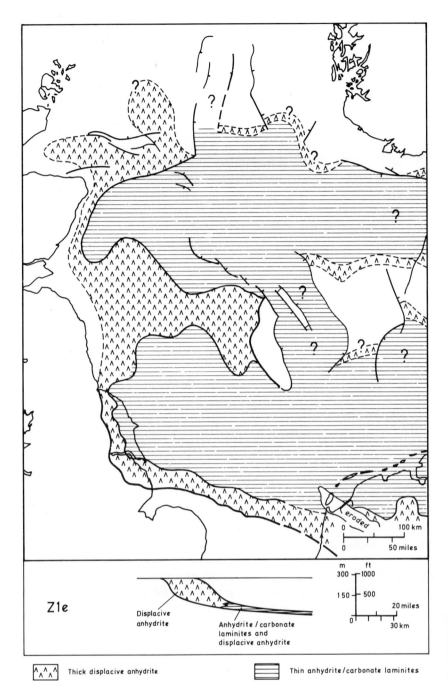

Fig. 4.7. Generalised facies of Z1 evaporites (Werraanhydrit of North Sea, Hartlepool Anhydrite of the Durham province, Hayton Anhydrite of the Yorkshire province).

(Glennie, ch. 3, p. 47 and Fig. 3.7), and carbonates probably appear a short distance offshore to the north.

At the top of the poorly-fossiliferous carbonates across the southern basin floor there is commonly a distinctive thin argillaceous packstone layer containing oncoliths accompanied by foraminifera and crinoids. It occurs from the U.K. (the Trow Point Bed; Smith, in press) to Poland, though probably only on sea-floor eminences, and appears to reflect a temporary fall in water level of 90 m (300 ft) or more. Normal marine conditions were never subsequently re-established on the basin floor during the Zechstein after the period represented by this bed, though restricted marine faunas reappeared on the shelves in the early parts of Z2 and Z3.

4.4.3 Werraanhydrit and equivalents

A sharp break, accompanied by evidence of desiccation, separates the Z1 carbonates from the Z1 evaporites—the Werraanhydrit and equivalents (Table 4.1). As shown in Figures 4.4 and 4.7, the evaporites form a peripheral lens up to 180 m (600 ft) thick round the southern basin, contained mainly within the encircling Z1 carbonates, and consisting of bluish-white displacive anhydrite (with 'nodular', 'chicken-wire' or 'mosaic' structure, and enterolithic bands) in an exiguous brown microdolomite host. This thick development can be traced in wells across the Mid North Sea High to the southern margin of the northern basin. Part, if not all, of the similar thick anhydrite in the

Moray Firth (the Halibut Bank Formation; Deegan and Scull, 1977) is thought by the writer to have been formed at the same time.

A sabkha or shallow-water origin is assumed for the thick marginal Werraanhydrit. Minor bodies of massive halite, probably of salina origin, occur near the top of the Lower Werraanhydrit, and are found locally on the southern shelf off the Norfolk coast, and in the southern Netherlands.

In both the northern and southern basins, the thickness of the peripheral anhydrite lens declines to only about 18 m (60 ft) across the floor, where four or five sub-cycles each commence with a layer of displacive anhydrite and grade up into dark brown to black bituminous, flat, sub-millimetre anhydrite/carbonate/laminites. It has been deduced that these units formed during successive periods of basin recharge following episodes of evaporative drawdown (Taylor, 1980).

The slope facies of the Werraanhydrit has been poorly sampled by wells in the U.K. sector, but on the Continent, a variety of slump, mass flow and turbidite features have been described (Schlager and Bolz, 1977; Meier, 1981), which demonstrate the depth of water in the basin at the time.

A contemporaneous volcanic centre is indicated by over 60 m (200 ft) of tuff in block 29/20.

Reservoir facies

The carbonate layers within the basin laminites are normally tight, but on some highs bordering the Central Graben, Mesozoic uplift, erosion and weathering have dissolved the anhydrite layers, leading to creation of good secondary porosity and permeability in the Auk and Argyll oilfields in blocks 30/16 and 30/24 (Brennand and van Veen, 1975; Pennington, 1975). This development has been designated the Argyll Formation (Deegan and Scull, 1977).

4.5 Second Zechstein cycle, Z2 (Figs. 4.8 & 4.9)

4.5.1 Hauptdolomit and equivalents

On the floor of both the southern and northern basins, the Werraanhydrit passes upwards with rapid transition into dark brown to black, bituminous thinly laminated carbonates. The lower part—the Stinkschiefer—contains many thin shale layers providing a distinctive gamma-ray marker. In the southern basin this unit merges into the overlying Stinkkalk, which is similar but less shaly. Limestone predominates over dolomite in the deeper parts of the basin, typified by large interlocking calcite crystals spreading across several laminae. Despite a promising organic content, early calcitisation together with the thinness of the basin carbonates (only 9-18 m, 30-60 ft) raise doubts about their ability to release large volumes of oil.

Around the edges of the southern basin and across the shelf provided by the Werraanhydrit, the Haupt-

dolomit or Main Dolomite (the Kirkham Abbey Formation of Yorkshire and the Hartlepool and Roker Formation of Durham) consists of 30-90 m (100-300 ft) of shallow-water to intertidal dolomites, typically consisting of very fine ooliths and pelletoids with leached centres. Large pisoliths and algal sheets characterise the barrier facies developed near the break in slope into the basin. A restricted fauna of bivalves, gastropods, foraminifera and ostracods occurs locally, but no true reef-builders are known.

The environments and lithofacies of the Hauptdolomit from wells in the Netherlands are discussed and well-illustrated with cores and photomicrographs by Clark (1980a), and in Denmark by Clark and Tallbacka (1980).

Reservoir facies

The shelf facies of the Hauptdolomit provides commercial oil and gas reservoirs in Poland, East and West Germany, and the Netherlands, principally from oncolithic and oolitic beds in the barrier facies, from local highs on the fore-barrier and in back-barrier lagoons. Good primary porosity is often reduced by cementation, especially by anhydrite or halite, but is locally enhanced by secondary solution. In Poland, porosities of 10-15% and permeabilities of a few to a few hundred milledarcies are considered normal, with porosities sometimes up to 25% accompanied by permeabilities of 1-5 D (Depowski, 1981; Depowski et al., 1981).

In the North Sea, the Hauptdolomit has been penetrated by many wells along the southern margin of the basin off East Anglia. Despite a number of hydrocarbon shows, productivity has commonly been limited by anhydrite and salt-plugging.

The barrier zone has been encountered by a few wells north of the Humber in Eastern England, but so far has generally been missed in wells drilled offshore. However, Antonowicz and Knieszner (1981) have demonstrated that it is possible to map the barrier seismically onshore in Poland. In the Netherlands, porous zones have been identified in the Hauptdolomit, before drilling, with the aid of synthetic seismograms (Maureau and van Wijhe, 1979; van Wijhe, 1981).

The Hauptdolomit is traceable across the Mid North Sea High and presumably occurs round the margin of the northern basin, though it has yet to be described there.

The Z2 slope facies is known mainly from outcrop in Durham and boreholes in north Yorkshire and on the Continent. It consists dominantly of grey-brown dolomitised lime-muds, often pelleted and burrowed, with ostracods, foraminifera, and occasional bivalves, the passage downwards and laterally into the basin facies being marked by calcite concretions (Taylor and Colter, 1975; Smith, 1980). Porosity and permeability are generally poor. Gas production at Lockton in Yorkshire is thought to have depended mainly on fractures, as in some German fields, but gravity-displaced shelf material could offer better reservoir conditions

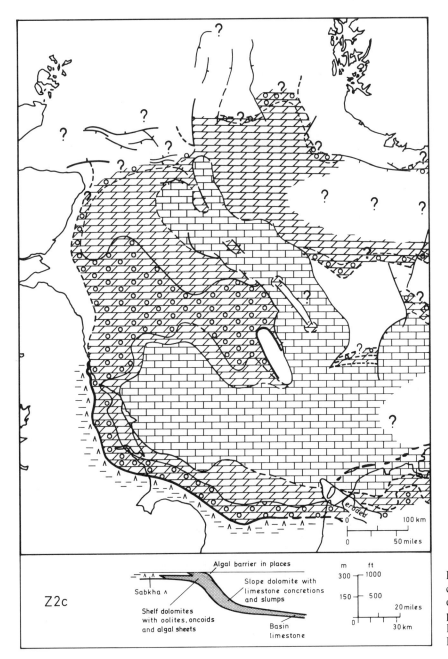

Fig. 4.8. Generalised facies of Z2 carbonates (Hauptdolomit and basinal equivalents in the North Sea, Hartlepool and Roker Formation of the Durham province, Kirkham Abbey Formation of the Yorkshire province).

locally, whilst Clark (1980b) has deduced the creation of secondary porosity by diagenetic processes in carbonates of the lower slope.

4.5.2 Stassfurt evaporites

The bulk of the Z2 evaporites (Stassfurt Salze, Fig. 4.9) consists of halite—the main source of Zechstein salt structures. The original thickness may have been in excess of 1,400 m (4,500 ft) in the middle of the southern U.K. basin (Christian, 1969). At the margins, red-beds pass basinward into anhydrite above the Hauptdolomit, and are joined by fore-setted units of halite, polyhalite, and (beyond the break in slope) kieserite (Colter and Reed, 1980) forming a complex about 90 m (300 ft) thick, which spreads out across the floor beneath the main body of halite (Taylor and Colter, 1975). The

presence of a widespread thin layer of potash minerals at the top of the Z2 evaporites suggests the overtaking of subsidence by salt precipitation, and the construction of a broad shelf close to sea level reaching far towards the basin centre. This levelling was responsible for considerable lateral uniformity during later sedimentation.

According to Day *et al.* (1981) Z2 salt probably filled the Central Graben and also crossed the Mid North Sea High further west (Fig. 4.9). If so, this halite seems to have been too thin to have had pronounced halokinetic effects.

The Z2 salt is assumed to be important in the northern basin, but cannot always be distinguished from evaporites of later cycles there. It is absent from the Moray Firth and has not yet been demonstrated further north into the Viking Graben than 59°30′N.

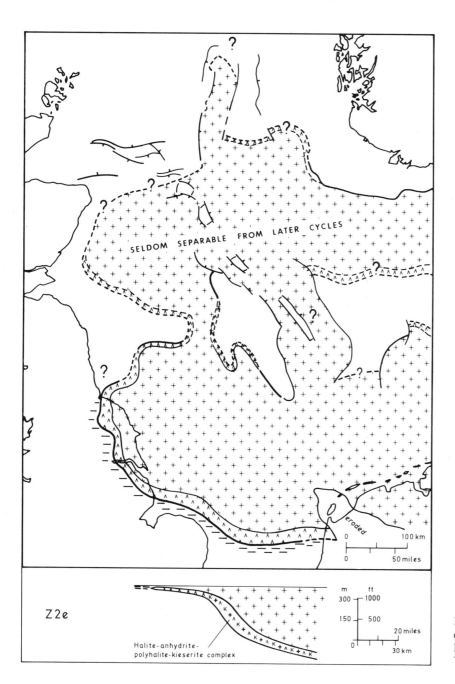

Fig. 4.9. Generalised facies of Z2 evaporites (Stassfurt series of the North Sea; Fordon evaporites of the Yorkshire province).

4.6 Third Zechstein cycle, Z3 (Fig. 4.10)

4.6.1 Plattendolomit and equivalents

The Z3 carbonate (Plattendolomit or Platy Dolomite, equivalent to the Seaham Formation of Durham and the Brotherton Formation—formerly the Upper Magnesian Limestone—in Yorkshire) is best known in the southern basin. It overlaps a 1 m (3 ft) basal shale (Grauer Salzton or Grey Salt Clay) towards the basin margin and thickens inwards to a maximum of 75-90 m (250-300 ft) (Fig. 4.4). The Grauer Salzton provides a sharp gamma-ray peak, though less strongly radioactive than the Kupferschiefer (e.g. well 44/11-1 in Fig. 4.5).

The Plattendolomit consists mainly of grey microcrystalline dolomite with thin shaly layers. Sheets of the tubular calcareous alga *Calcinema permiana* are typical, and microfossils and stunted bivalves are

locally abundant. Calcite concretions, similar to those of the Z2 slope carbonates, occur in the lower part of the formation.

The Plattendolomit appears to have formed in mainly shallow but quiet-water conditions. An algal-rich barrier zone has been recognised (Depowski, 1981) but is probably of lower relief than such features in earlier cycles.

Reservoir facies

Porous grainstones are common at the extreme landward margins of the formation but porosity and permeability elsewhere seldom approach that of the Hauptdolomit. Some gas production has been obtained at Lockton and Eskdale in Yorkshire and various fields on the Continent. It is usually dependent on natural fracturing, but at Alkmaar, northwest of Amsterdam in the Netherlands, production is from leached algal and oolitic facies (Van Lith, 1983).

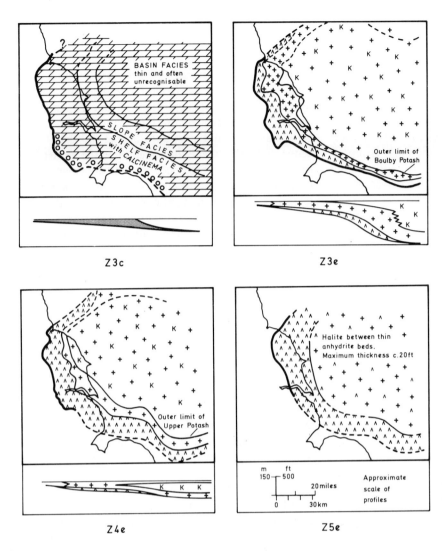

Z3c Z3e

Z4e Z5e

Fig. 4.10. Generalised facies of Z3 carbonates (Plattendolomit of the North Sea, Seaham Formation of the Durham province, Brotherton Formation of the Yorkshire province), Z3, Z4 and Z5 evaporites in UK Southern North Sea (adapted from Taylor and Colter, 1975).

4.6.2 Z3 evaporites

Typical sabkha cycles with algal mats and nodular anhydrite at the top of the Plattendolomit lead upwards into the Hauptanhydrit (Main Anhydrite) which thickens to some 45 m (150 ft) within the Plattendolomit shelf rim. Towards the basin centre the Grauer Salzton, Plattendolomit and Hauptanhydrit are much attenuated and difficult to recognise, particularly when displaced by Z2 salt movements as is commonly the case.

An unfossiliferous but apparently correlative shale-dolomite-anhydrite unit has been found in several wells in the northern basin (Fig. 4.5) but so far it has not been possible to identify the Plattendolomit or Hauptanhydrit over the Mid North Sea High, in the Moray Firth, or the Viking Graben.

The Z3 or Boulby Halite thickens from less than 30 m (100 ft) near its boundary with the encompassing red-beds, to around 120 m (400 ft) in the centre of the southern basin. Near the top of the unit the thin Boulby Potash, consisting mainly of sylvinite, is worked in north Yorkshire. It thickens towards the centre of the basin by development at successively lower levels, forming a variable halite/carnallite/polyhalite/mudstone complex.

The evaporites show evidence of deposition in very shallow to emergent conditions, and are capped by thin, salty, and potassic red-beds (the Carnallitic Marl, Roter Salzton, or Red Salt Clay), the terrigenous components dying out away from the margins. An equivalent unit is apparent in some northern basin wells, where thick zones of mixed potash salts in the underlying halite suggest broad similarity with the third cycle of the southern basin.

Magnesium-bearing salts of this cycle are being exploited in NE Netherlands by solution-mining, using modern drilling and logging techniques developed by the oil industry (Coelewij *et al.*, 1978).

4.7 Fourth and fifth Zechstein cycles, Z4 and Z5 (Fig. 4.10)

A thin tight dolomite or magnesite unit (the Upgang Formation) representing the transgressive base of the fourth cycle in Durham has not yet been recognised offshore, but the overlying Upper Anhydrite or Pegmatit-anhydrit can be followed widely around the edge of the southern basin (Fig. 4.10) and appears to be present as far north as Norwegian waters (well 17/4-1 in Fig. 4.5).

The Upper (Aller) Halite occupies a slightly smaller area, thickening to about 90 m (300 ft) in the middle of the southern basin, where sylvite, carnallite and red mudstone are developed. Equivalents north of the Mid North Sea—Ringkøbing-Fyn High are difficult to correlate with certainty.

The fifth cycle, separated from the fourth by a thin red mudstone, consists in the southern basin of the Grenzanhydrit or Top Anhydrite, which splits basinward to include a thin halite member, the whole totalling only about 6 m (20 ft) in thickness. In the northern basin, what is thought to be the same cycle is represented by up to 60 m (200 ft) of anhydrite, followed in some wells—mainly in Norwegian waters—by shale and dolomite.

Minor sixth and seventh evaporite cycles have been reported in Germany, but are probably of continental rather than marine origin.

4.8 Halokinetics (Figs. 4.11-4.14)

In areas where Zechstein salt has a thickness of at least 130 m (500 ft) and an overburden of 900 m (3,000 ft) or more, salt structures are common, although the amount of diapiric movement is very slight at the basin margins. To judge from the alignment of most diapirs with known regional trends (Fig. 4.11 and Glennie, this volume, Fig. 3.3), diapirism was probably triggered off by fault movement in the underlying crust. Much of the triggering seems to be related to Late Cimmerian earth movements, although locally diapirism can be shown to have started earlier, especially during the Jurassic and even as far back as the end of the Early Triassic (Brunstrom and Walmsley, 1969).

Once triggered into motion, the salt seems to have responded to any imbalance in the overburden by flowing sideways from the area of greater load to one that had a lighter burden. Differential uplift and subsidence ensued, the products of erosion from the uplifted areas being deposited in the adjacent 'rim synclines' or 'peripheral sinks'. Thus the imbalance was accentuated and the diapiric movement of salt enhanced.

The sequence of events was worked out from the well-explored salt structures of North Germany by Trusheim (1960) who noted a progression from pillows through stocks to elongated walls, related to increasing depth to the base of the salt. Stages of development are depicted on Figure 4.12.

It should be noted that complete withdrawal of salt from areas adjacent to large diapirs may provide the opportunity for hydrocarbons to migrate from the Carboniferous to post-Zechstein reservoirs. This appears to be the case in the Triassic gas fields of Quadrant 43.

Rim synclines commonly preserve strata not found round about, and on the northern edge of the southern salt basin lateral flow of salt appears to have preserved a large elongated prism of Mesozoic rocks that are eroded elsewhere (Jenyon, in preparation). It is worth looking at individual cases to see whether potential source rocks might be present in volumes and at depths which could have enabled them to generate significant quantities of hydrocarbons.

Examples of halokinensis are illustrated in Figs. 4.13-4.15. The timing of the movement often differs in adjacent structures, as in Fig. 4.13 where uplift of the overburden has resulted in removal of virtually the total Jurassic sequence from above the left-hand salt pillow, just prior to deposition of the Upper Cretaceous. Development of the right-hand pillow, on the other hand, did not start until during the Late Cretaceous. Salt movement continued after the earlier Tertiary. The apparent absence of salt movement today possibly results from the removal of a thick Pleistocene ice cap. Note that the brittle Plattendolomit does not extend over the greater length of the upper pillow surface. Also note the apparent uplift of the Hauptdolomit beneath the thickest salt section. This pseudostructure ('velocity pull-up') is caused by the wedge of halite, which has a much higher interval velocity than the laterally adjacent sediments.

In Fig. 4.14 differences in the development of two adjacent diapirs are apparent. The stages of their development can be deduced from their flanking rim synclines. There is only minor tensional faulting of the base Tertiary reflector over the left-hand pillow, whereas the crest of the right-hand diapir is more strongly faulted, with considerable relative vertical movement across the fault, accompanied by collapse of the overlying strata probably related to dissolution. By some time in the mid Tertiary, salt movement had proceeded to the point where there was complete withdrawal from the area between the diapirs.

In Fig. 4.15, salt flow has resulted in the creation of a salt wall which extends up almost to the sea floor, where it has an apparent width of some 3 km. The structural relationships outlining the uplift and erosion of the flanking Jurassic strata indicate that the diapir went through a pillow stage during the Early Cretaceous. Salt movement continued well into the Tertiary but, to judge from the absence of salt in the section to the right of the salt wall, it has probably now ceased. To the left of the salt wall, diapirism has resulted in a pre-Cretaceous structural pattern that is almost too complex to decipher with seismic data alone.

Unexpected relationships can arise through potash salts such as sylvite and carnallite flowing more freely than halite. Highly complex examples revealed by mining in the Harz-Thüringer Wald region of Germany are illustrated by Nachsel and Franz (1960).

Conventional seismic processing attenuates signals originating from steeply dipping structures, but May and Covey (1983) describe inverse modelling techniques using very high stacking velocities which give clear definition of diapir flanks, even where slightly overhanging; these techniques were not used in Figures 4.13 to 4.15.

4.9 Structural effects of evaporite dissolution

Dissolution of Zechstein evaporites beneath the Cimmerian unconformity has already been mentioned in

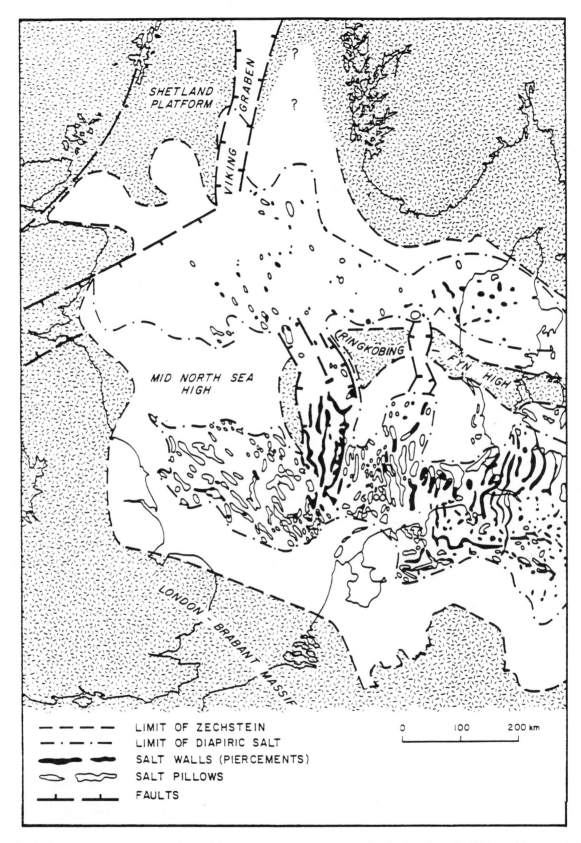

Fig. 4.11. Areas of Zechstein salt diapirism. Modified from Sorgenfrei (1969) after Heybroek *et al.* (1967), and Taylor (1981).

connection with the creation of secondary porosity. Important structural effects can also be produced, especially where dissolution occurred beneath several hundred metres of overburden, as evidence suggests. Little attention has been paid to this aspect since it was noted in the southern North Sea by Lohmann (1972),

but with improving seismic quality it is becoming possible to recognise many subtle indications which distinguish dip-reversal due to dissolution from that due to salt flow (Jenyon, 1984). Substantial drape in overlying beds can result where evaporites are leached around Zechstein carbonate build-ups (Jenyon and

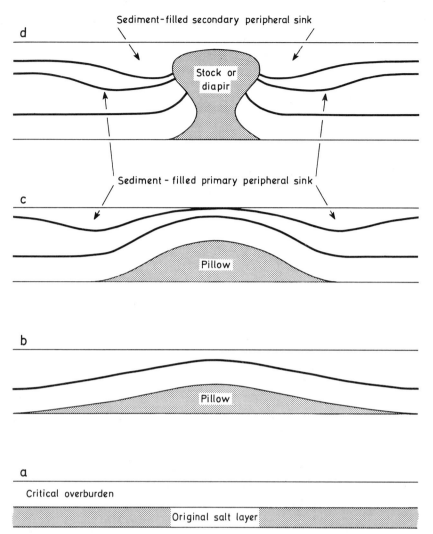

Fig. 4.12. Stages in development of salt pillow and stock, modified from Trusheim (1960).

a. State immediately before movement.

b. With increasing overburden, salt has flowed inwards to form a gentle pillow. Once initiated, the imbalance of sediment load between blanks and crest accentuates the action.

c. As salt withdraws from the flanks a rim syncline or peripheral sink forms. Becoming filled with young sediments it continues to drive the migration of salt inwards.

d. The buoyant salt mass finally breaks through the overlying sediments to form a diapiric plug or stock, or an elongated wall, which may take a variety of cross-sectional forms. The continued withdrawal from the original salt layer forms a secondary peripheral sink inwards from the primary sink.

Taylor, 1983). Dissolution effects are also clearly seen locally in the Central Graben and Norwegian-Danish-Basin.

4.10 Practical problems

4.10.1 Identification of cycles

Because of the cyclic repetition of similar depositional facies in the Zechstein, it can be difficult to identify individual cycles (or even to tell how many are present) where the sequence is incomplete—for instance in marginal areas such as parts of Yorkshire where evaporites do not separate the carbonates, and in central basin regions where shales and carbonates are unrecognisable between salts. The difficulty can be severe when only wireline logs and cuttings are available as is commonly the case (see Drilling Problems below). It is possible that detailed microfossil, palynological, and geochemical work may eventually resolve such problems. Careful tracing of ash bands may prove useful for correlation in Z1.

Cycles can sometimes be distinguished in carbonate sequences by the upward-diminishing radioactivity which reflects their regressive character.

Where present, a diverse marine fauna, especially if including brachiopods, crinoids, and bryozoa, is indicative of the Z1 carbonates, whereas beds consisting largely of *Calcinema* stems are diagnostic of Z3. Polyhalite is the commonest potassium-bearing mineral in Z2, but is often altered to gypsum in cuttings. Carnallite and red mudstones are more abundant in the basin evaporites of Z3 and Z4 than in earlier cycles. Wireline log parameters for identifying the commonest evaporite minerals are given in Table 4.2. Quantitative results can be calculated by linear programming methods (Ford *et al.*, 1974).

4.10.2 Drilling problems

Unless saturated salt—inverse oil emulsion—or oil-based drilling muds are used, hole enlargement occurs in halite sections until the mud has become saturated at formation temperature, and even then potash salts continue to dissolve in any of the water-based muds. Under-compacted clays interbedded with the evaporites also wash out. Large cavities result. The cuttings returned consist only of hard, insoluble lithologies, and may remain unrepresentative and mixed for many hundreds of feet of further penetration. The cavities may later release their cuttings whenever hard formations are encountered; for example, red mudstones are often falsely recorded at the top of the Werra-anhydrit.

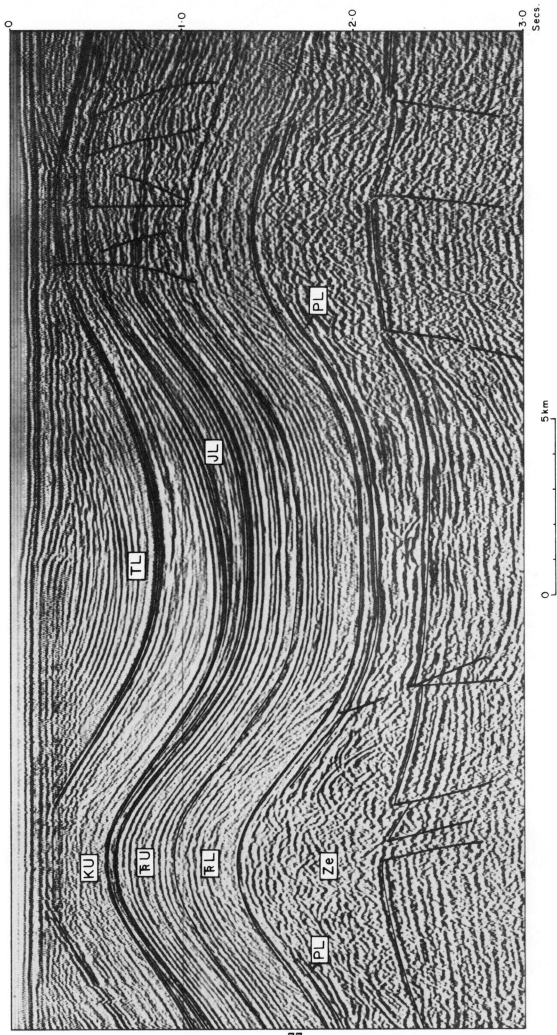

Fig. 4.13. Seismic section, North Sea, showing different times of initiation of salt pillows (ZE). PL = Plattendolomit, TRL = Lower Triassic, TRU = Upper Triassic, JL = Lower Jurassic, KU = Upper Cretaceous, TL = Lower Tertiary, Vertical scale = Two-way time in seconds. See text for further explanation.

77

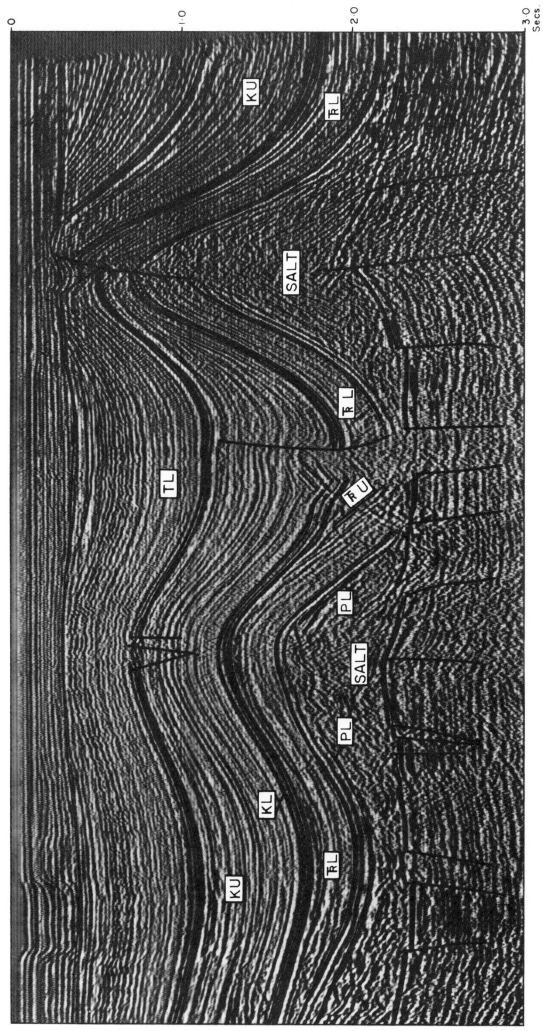

Fig. 4.14. Seismic section, North Sea, showing mid-Tertiary end to diapiric growth caused by complete withdrawal of salt and tensional rupturing of intervening rim syncline. KL = Lower Cretaceous. Other abbreviations and scale as in Fig. 13.

78

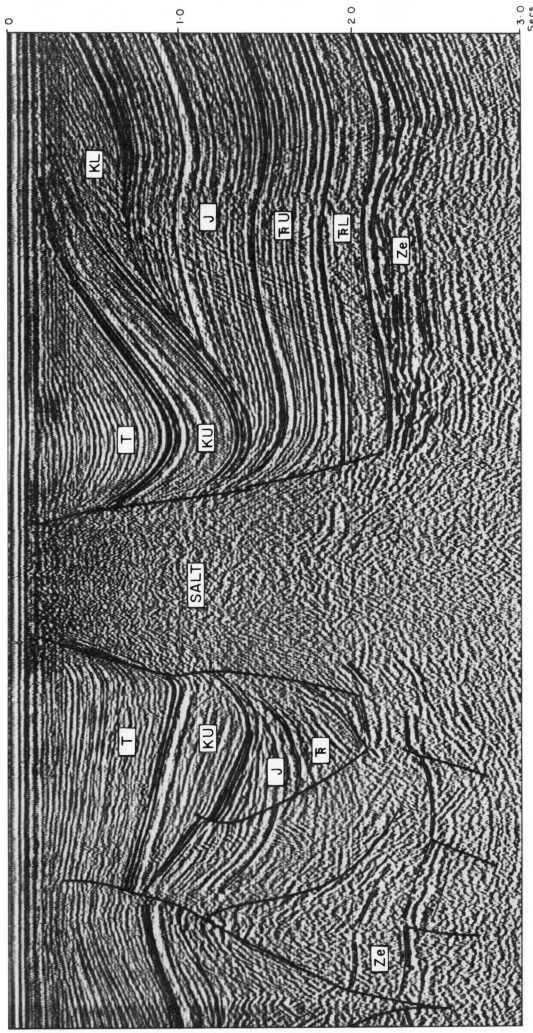

Fig. 4.15. Seismic section, North Sea, showing Cretaceous to Early Tertiary growth of salt wall dated by rim synclines. J = Jurassic, T = Tertiary. Other abbreviations and scales as in Figs. 4.13 and 4.14.

5 km

79

	Apparent Density g/cc	Δt μsec/ft	Apparent Limestone Neutron Porosity %	Gamma Ray API Units
Halite NaCl	2·03	67	0	0
Anhydrite CaSO$_4$	2·98	50	0	0
Gypsum* CaSO$_4$.2H$_2$O	2·35	52·5	49	0
Polyhalite K$_2$SO$_4$.MgSO$_4$.2CaSO$_4$.2H$_2$O	2·79	57·5	15	180
Carnallite KCl.MgCl$_2$.6H$_2$O	1·57	78	65	200
Sylvite KCl	1·86	74	0	500
Kieserite MgSO$_4$.H$_2$O	2·55			0
Kainite 4(KCl.MgSO$_4$).11H$_2$O	2·12		45	225
Langbeinite K$_2$SO$_4$.2MgSO$_4$	2·82	52	0	275

* The stability field of gypsum is such that it is rarely encountered at depths greater than 2000 ft where its place is taken by anhydrite

Table 4.2. Wireline log parameters of common evaporite minerals. From various sources. Blanks indicate no figures available.

Where basinal carbonates have been buried deeply enough for gas generation, local vuggy porosity and fractures can give rise to high pressure/low-volume gas shows.

4.11 Hydrocarbon occurrences

A number of oil and gas discoveries have been made in Zechstein carbonates in and around the North Sea, and there is production from numerous onshore fields in the Netherlands, Germany and Poland. Most finds have been in oncolitic and oolitic rocks of the barrier facies and tend to be concentrated near the basinward margin of the shelf in the southern salt basin; the most significant are shown on Fig. 4.16.

Gas finds predominate, and the Hauptdolomit is the commonest reservoir. Gas has migrated from the Carboniferous to Z1 reservoirs, and to those in Z2 and possibly Z3 where the seat-seals provided by the underlying evaporites have been breached. Oil and condensate discoveries have been confined to Z2 and Z3 carbonates, and are thought from structural and geochemical evidence to have been sourced by the finer-grained basin and slope facies of those cycles. Whilst these sources appear to be adequate for the economics of onshore production, it has yet to be demonstrated that the same would apply in the North Sea.

4.11.1 Poland

Thirteen gas fields and seven small oil fields were discovered in the Zechstein of Poland between 1960 and 1978 as a result of an intensive exploration programme. They are located in the Pomeranian area adjacent to the Baltic, and the Lubuska and Silesian regions on the south side of the basin (Fig. 4.16). Oil production ranges up to 100 tonnes or, exceptionally, 300 tonnes/day from individual wells; gas flows of several thousand m³/day to a few million m³/day have been reported (Depowski *et al.*, 1981). Oils are naphthenic and sulphurous with densities of 0.85-0.87 g/cm³. Gases usually contain H$_2$S and N$_2$.

4.11.2 Germany

There have been several small Zechstein oil finds in Germany. These include discoveries in the Thüringen and Pomeranian areas of East Germany, and others north of the Harz Mountains and in the Schleswig-Holstein region of West Germany. Gas is produced near the Netherlands border and in Lower Saxony where it is generally sour and rich in nitrogen. Reserves in Lower Saxony have been estimated by Hauk *et al.* (1979) to total about 200×10^9 m³ ($7,000 \times 10^9$ ft³); there are some 40 commercial fields, with reserves

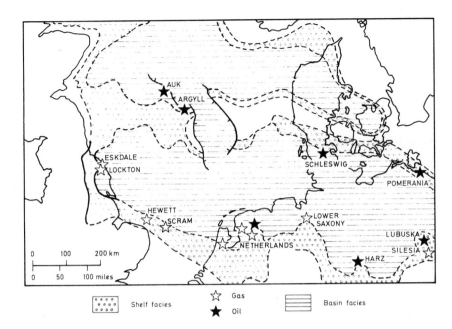

Fig. 4.16. Sketch map showing principal oil and gas occurrences in the western and central parts of the Zechstein basins.

ranging up to about 20×10^9 m^3 (700×10^9 ft^3). The Hauptdolomit is the main reservoir.

4.11.3 Netherlands

Across the border in the eastern Netherlands, gas is produced from the Hauptdolomit in the Drenthe and Twente areas. Facies variations are commonly important in defining the productive areas, (Clark, 1980a; Maureau and van Wijhe, 1979). The absence of Rotliegend reservoirs between the Coal Measures and the Zechstein is presumably a significant factor. The gas contains H$_2$S except in the Coevorden (east) and Schoonebeek fields.

Several gas discoveries have been made in the Z3 carbonates in the Western Netherlands, with production in the Bergen field (Van Lith, 1983) where the Z2 carbonates are in non-reservoir facies. The gas is generally free from H$_2$S.

A significant Zechstein oil discovery was made at Gieterveen-Oost in the northeastern Netherlands in 1982, but no further details are available.

4.11.4 United Kingdom

As smaller fields become economic and seismic exploration more versatile there is little doubt that the Zechstein successes of the Continent will be echoed in the U.K., including the North Sea. So far, results have not been impressive but, as this chapter has shown, broad belts of potential reservoir facies can be mapped. Viable plays require the additional factors of mature source and structural history to be taken into account. Many interesting areas have yet to be tested.

In the North Sea, Auk and Argyll are the only fields so far producing oil from the Zechstein, and are the first Zechstein discoveries known in the northern salt basin. Their structural setting on the edge of the Central Graben, and the relationship of the Zechstein to the probable oil source (the Kimmeridge Clay) are

outlined by Glennie (this volume, ch. 3, p. 57 and Fig. 3.19). The reservoirs are of unusual type. Extensive vugs and fractures sustain production despite low matrix porosity, and appear to have originated mainly as a result of leaching of anhydrite laminae from the tight basinal Werraanhydrit, with collapse, brecciation, and complex diagenesis of associated limestone and dolomite layers. These effects are due to uplift and proximity to the Cimmerian unconformity, which also removed the evaporites of Z2-Z5. Typical well sections are illustrated in Fig. 4.17, and further details are given in Pennington (1975) and Brennand and van Veen (1975). Reserves in this type of reservoir are clearly difficult to determine. Recoverable reserves at Argyll have been quoted as between 20 and 150 million barrels of 36°API oil, 7 wells currently producing 20,300 bbl/day. The Auk field is said to have had reserves of 66 million bbl of 38°API oil. Production at 12,000 bbl/day is believed to be in decline, but the life of the field is said to have been extended to 1985.

In the southern salt basin, sub-commercial gas was found in the Hauptdolomit and Plattendolomit beneath the Hewett Triassic gas field in blocks 48/29 and 48/30, with shows in some nearby blocks. Gas and condensate have since been found in block 53/4, designated the 'Scram' field.

Gas was obtained for a limited period at Lockton and Eskdale in North Yorkshire from tectonically fractured Z2 and Z3 carbonates of relatively tight facies. Gas and condensate have been tested not far offshore in blocks 41/20, 41/24, and 41/25.

4.12 Acknowledgements

I wish to thank colleagues at V.C. Illing & Partners for assistance, particularly Pamela Bass for the drawings and Ruth Marks for the typing. I am grateful to Ken Glennie who provided invaluable advice and help, especially with the section on halokinetics, and to

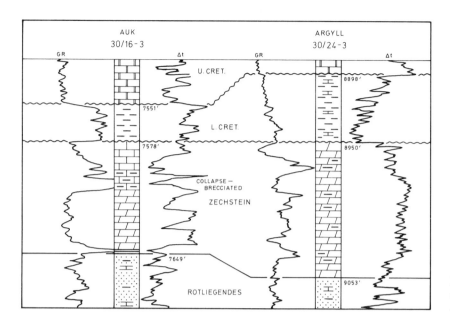

Fig. 4.17. Typical section through Zechstein reservoirs in Auk and Argyll oilfields.

Denys Smith for many past discussions. Most of all I thank my wife Pat, without whose unique assistance no contribution would have been possible.

4.13 References

Antonowicz, L. and Knieszner, L. (1981) Reef zones of the Main Dolomite, set out on the basis of palaeogeomorphologic analysis and the results of modern seismic techniques. *Proc. Int. Symp. Central Europ. Permian* (Jablonna, Poland, 1978) Geol. Inst. Warsaw. pp. 356-368.

Boogaert, H.A. van Adrichem and Burgers W.K.J. (1983) The development of the Zechstein in the Netherlands. *Geol. Mijnbouw* **62**, 83-92.

Brennand, T.P. and van Veen, F.R. (1975) The Auk oilfield. In: Woodland, A.W. (Ed.) q.v. 275-281.

Brueren, J.W.R. (1959) The stratigraphy of the Upper Permian 'Zechstein' formation in the eastern Netherlands. In: *I giacimenti gassiferi dell' Europa Occidentale* 1, Accad. Nazionale dei Lincei, Rome. pp. 243-274.

Brunstrom, R.G.W. and Walmsley, P.J. (1969) Permian evaporites in North Sea basin. *Bull. Am. Assoc. Petrol. Geol.* **53**, 870-883.

Christian, H.E. (1969) Some observations on the initiation of salt structures of the Southern British North Sea. In: Hepple, P. (Ed.) *The exploration for petroleum in Europe and North Africa*. Inst. Petroleum, pp. 231-248.

Clark, D.N. (1980a) The sedimentology of the Zechstein 2 carbonate formation of Eastern Drenthe, The Netherlands. In: Füchtbauer, H. and Peryt, T.M. (Eds.) q.v. 131-165.

Clark, D.N. (1980b) The diagenesis of Zechstein carbonate sediments. In: Füchtbauer, H. and Peryt, T.M. (Eds.) q.v. 167-203.

Clark, D.N. and Tallbacka, L. (1980) The Zechstein deposits of Southern Denmark. In: Füchtbauer, H. and Peryt, T.M. (Eds.) q.v. 205-231.

Coelewij, P.A.J., Haug, G.M.W. and van Kuijk, H. (1978) Magnesium-salt exploration in the northeast Netherlands. *Geol. Mijnbouw* **57**, 487-502.

Colter, V.S. and Barr, K.W. (1975) Recent developments in the geology of the Irish Sea and Cheshire Basins. In: Woodland, A.W. (Ed.) q.v. 61-73.

Colter, V.S. and Reed, G.E. (1980) Zechstein 2 Fordon Evaporites of the Atwick No.1 borehole, surrounding areas of N.E. England and the adjacent Southern North Sea. In: Füchtbauer, H. and Peryt, T.M. (Eds.) q.v. 115-129.

Day, G.A., Cooper, B.A., Anderson, C., Burgers, W.F.J., Rønnevik, H.C. and Schöneich, H. (1981) Regional seismic structure maps of the North Sea. In: Illing, L.V. and Hobson, G.D. (Eds.) q.v. 76-84.

Deegan, E. and Scull, B.J. (1977) A proposed standard stratigraphic nomenclature for the Central and Northern North Sea. *Rept. Inst. Geol. Sci.* 77/25.

Depowski, S. (1978) *Lithofacies-palaeogeographical atlas of the Permian of platform areas of Poland*. Geol. Inst. Warsaw.

Depowski, S. (1981) The geological factors of hydrocarbon accumulations in the Permian in the Polish Lowland. *Proc. Int. Symp. Central Europ. Permian* (Jablonna, Poland, 1978), Geol. Inst. Warsaw, pp. 547-567.

Depowski, S., Peryt, T.M., Piatkowski, S. and Wagner, R. (1981) Palaeogeography versus oil and gas potential of the Zechstein Main Dolomite in the Polish Lowlands. *Proc. Int. Symp. Central Europ. Permian* (Jablonna, Poland, 1978), Geol. Inst. Warsaw, pp. 587-595.

Donato, J.A., Martindale, W. and Tully, M.C. (1983) Buried granites within the Mid North Sea High. *J. geol. Soc. London.* **140**, 825-237.

Ford, M.E., Bains, A.J. and Tarron, R.D. (1974) Log analysis by linear programming—an application to the exploration for salt cavity storage locations. *Trans. 3rd Europ. Formation Eval. Symp.*, paper B. Soc. Prof. Well Log Analysts.

Füchtbauer, H. and Peryt, T.M. (Eds.) (1980) The Zechstein Basin with emphasis on carbonate sequences. *Contr. Sedimentology* 9. Schweizerbart'sche Verlagsbuchhandlung, Stuttgart, 328 pp.

Füchtbauer, H. (1968) Carbonate sedimentation and subsidence in the Zechstein Basin (northern Germany). In: Muller, G. and Friedman, G.M. (Eds.) *Recent developments in carbonate sedimentology in Central Europe*. Springer Verlag, Berlin. pp. 196-204.

Glennie, K.W. and Buller, A.T. (1983) The Permian Weissliegend of N.W. Europe: the partial deformation of aeolian dune sands caused by the Zechstein transgression. *Sedim. Geol.* **35**, 43-81.

Harwood, G.M., Smith, D.B., Pattison, J. and Pettigrew, T. (1982) *Field excursion guide EZ82* (Symposium on the English Zechstein, Leeds University, 1982).

Hauk, V.M., Petersen, H.H.F., Spoerker, H.F. and Moritz, J. (1979) Deep European H_2S is handled with special muds, cement and tubulars. *Oil Gas J.* **77**, 62-69.

Heybroek, P., Haanstra, U. and Erdman, D.A. (1967) Observations on the geology of the North Sea area. *Proc.*

7th World Petrol. Cong. **2**, 905-916.

Hsü, K.J. (1972) Origin of saline giants: a critical review after the discovery of the Mediterranean evaporite. *Earth Sci. Rev.* **8**, 371-396.

Illing, L.V. and Hobson, G.D. (Eds.) (1981) Petroleum geology of the Continental Shelf of North-West Europe. Heyden & Son, London, 521 pp.

Jenyon, M.K. (1984) Seismic response to collapse structures in the Southern North Sea. Marine & Petr. Geol. **1**, 27-36.

Jenyon, M.K. (in preparation) *Basin-edge diapirism and salt movement in the Zechstein of the Southern North Sea.*

Jenyon, M.K. and Taylor, J.C.M. (1983) Hydrocarbon indications associated with North Sea Zechstein shelf features. *Oil Gas J.* **81**, 155-160.

King, R.E. (1977) North Sea joins ranks of world's major oil regions. *World oil* 185, 35-45.

Kent, P.E. (1976a) Outline geology of the Southern North Sea Basin. *Proc. Yorks. Geol. Soc.* **36**, 1-22.

Kent, P.E. (1976b) Progress of exploration in North Sea. *Bull. Amer. Assoc. Petrol. Geol.* **51**, 731-741.

Lohmann, H.H. (1972) Salt dissolution in subsurface of British North Sea as interpreted from seisomograms. *Bull. Amer. Assoc. Petrol. Geol.* **56**, 472-479.

Maureau, G.T.F.R. and Van Wijhe, D.H. (1979) The prediction of porosity in the Permian (Zechstein) carbonate of eastern Netherlands using seismic data. *Geophysics* **44**, 1502-1517.

May, B.T. and Covey, J.D. (1983) Structural inversion of salt dome flanks. *Geophysics* **48**, 1039-1050.

Maync, W. (1961) The Permian of Greenland. In: *Geology of the Arctic* 1, pp. 214-223. Univ. Toronto Press.

Meier, R. (1981) Clastic resedimentation phenomena of the Werra Sulphate (Zechstein 1) at the eastern slope of the Eichsfeld Swell (Middle European Basin)—an information. *Proc. Int. Symp. Centr. Europ. Permian* (Jablonna, Poland, 1978), Geol. Inst. Warsaw. pp. 369-373.

Nachsel, von G. and Franz, E. (1983) Zur Ausbildung der Wippertal-Störungszone im Bereich der Grubenfelder des Kaliwerks "Gluckauf" Sondershausen. (On the construction of the Wippertal fault zone in the region of the Carnallite salt mine "Gluckauf" Sonderhausen). *Z. geol. Wiss. Berlin* 11, 1005-1021.

Pattison, J., Smith, D.B. and Warrington, G. (1973) A review of Late Permian and Early Triassic biostratigraphy in the British Isles. In: Logan, A.V. and Mills, L.V. (Eds.) *The Permian and Triassic Systems and their mutual boundary. Mem. Can. Soc. Petrol. Geol.* 2, pp. 220-260.

Paul, J. (1980) Upper Permian algal stromatolite reefs, Harz Mountains (F.R. Germany). In: Füchtbauer, H. and Peryt, T.M. (Eds.) q.v. 253-268.

Pennington, J.J. (1975) The geology of the Argyll field. In: Woodland, A.W. (Ed.) q.v. pp. 285-291.

Rhys, G.H. (1974) A proposed standard lithostratigraphic nomenclature for the Southern North Sea and an outline structural nomenclature for the whole of the (UK) North Sea. *Rept. Inst. Geol. Sci.* 74/8.

Richter-Bernburg, G. (1959) Zur Palaegeographie der Zechsteins. In: *I giacimenti gassiferi dell Europa Occidentale* 1, Accad. Nazionale dei Lincei, Rome. pp. 88-99.

Schlager, W. and Bolz, H. (1977) Clastic accumulations of sulphate evaporites in deep water. *J. sediment. Petrol.* **42**, 600-609.

Smith, D.B. (1980) The evolution of the English Zechstein basin. In: Füchtbauer, H. and Peryt, T.M. (Eds.) q.v., 7-34.

Smith, D.B. (1981a) The Magnesian Limestone (Upper Permian) reef complex of Northeastern England. *Soc. Econ. Pal. Min.* Spec. Publ. 30, 161-186.

Smith, D.B. (1981b) Bryozoan-algal patch reefs in the Upper Permian Lower Magnesian Limestone of Yorkshire, Northeast England. *Soc. Econ. Pal. Min.* Spec. Publ. 30, 187-202.

Smith, D.B. (in press) The Trow Point Bed—a deposit of Upper Permian marine oncoids, peloids and columnar stromatolites in the Zechstein of North East England. In: Harwood, G.M. and Smith, D.B. (Eds.) *The English Zechstein and related topics.* Spec. Pub. geol. Soc. Lond.

Smith, D.B., Brunstrom, R.G.W., Manning, P.I., Simpson, S. and Shotton, F.W. (1974) Correlation of the Permian rocks of the British Isles. *Spec. Rep. geol. Soc. Lond.* 5.

Sorgenfrei, T. (1969) A review of petroleum development in Scandinavia. In: Hepple, P.W. (Ed.) *The exploration for petroleum in Europe and North Africa.* Inst. Petroleum, London. 191-203.

Sorgenfrei, T. and Buch, A. (1964) Deep tests in Denmark 1935-1959. *Geol. Surv. Denmark* 3rd series, No.36.

Taylor, J.C.M. (1980) Origin of the Werraanhydrit in the Southern North Sea—a reappraisal. In: Füchtbauer, H. and Peryt, T.M. (Eds.) q.v. 91-113.

Taylor, J.C.M. (1981) Zechstein facies and petroleum prospects in the Central and Northern North Sea. In: Illing, L.V. and Hobson, G.D. (Eds.) q.v. pp. 176-185.

Taylor, J.C.M. and Colter, V.S. (1975) Zechstein of the English sector of the Southern North Sea Basin. In: Woodland, A.W. (Ed.) q.v. 249-263.

Trusheim, F. (1960) Mechanism of salt migration in northern Germany. *Bull. Amer. Assoc. Petrol. Geol.* **44**, 1519-1540.

Van Lith, J.G.J. (1983) Gas fields of Bergen concession, The Netherlands. *Geol. Mijnbouw* 62, 63-74.

Van Wijhe, D.H. (1981) The Zechstein 2 carbonate exploration in the eastern Netherlands. *Proc. Int. Symp. Central Europ. Permian* (Jablonna, Poland, 1978). Geol. Inst. Warsaw. pp. 574-586.

Wagner, R., Peryt, T.M. and Piatkowski, T.S. (1981) The evolution of the Zechstein sedimentary basin in Poland. *Proc. Int. Symp. Centr. Europ. Permian* (Jablonna, Poland, 1978). Geol. Inst. Warsaw. pp. 69-83.

Woodland, A.W. (Ed.) (1975) *Petroleum and the Continental Shelf of North-West Europe*, 1, Geology. Applied Science Publishers, London, 501 pp.

Zeigler, P.A. (1981) Evolution of sedimentary basins in North-West Europe. In: Illing, L.V. and Hobson, G.D. (Eds.) q.v. 3-39.

Chapter 5   Triassic

M.J. FISHER

5.1 Introduction

Unlike the underlying Permian or overlying Jurassic sediments, the Triassic succession in the North Sea lacks both major hydrocarbon accumulations and source rocks. Notwithstanding these purely economic considerations, the Triassic encompasses a significant period of earth history. The break-up of Pangea, which began with crustal thinning and rifting along the axis of the incipient Atlantic and the westward extension of Tethys, established a new structural framework in north-west Europe. This resulted in modification of the structural pattern inherited from the Permian by the superimposition of a graben system that controlled deposition throughout the Mesozoic (Fig. 5.1).

In the North Sea, the major new structural element was the Viking-Central Graben. This north-south trending structure, over 1000 km long, transsected the old Northern Permian Basin and breached the Mid North Sea-Ringkøbing-Fyn High. The High was also breached by the rapidly subsiding Horn Graben, which merged with the Central Graben in what had been the Southern Permian Basin. In the east, the fault bounded Polish-Danish Trough was another site of rapid subsidence and accumulation of thick Triassic sediments. Surprisingly, volcanic activity accompanying this phase of rifting has been recorded only where the Central Graben breaches the Mid North Sea High and in south-west Norway (P.A. Ziegler, 1978).

The close of the Permian saw the end of widespread marine sedimentation and a return to dominantly non-marine depositional environments. With the withdrawal of the Zechstein seas from the North Sea basins, the already established continental and paralic environments were extended from the periphery into the centre of the basins. Whenever the basal Triassic overlies uppermost Permian, it does so with apparent conformity but often with abrupt change of facies.

In the Southern North Sea Basin, faulting and uplift at the Permian-Triassic transition rejuvenated the physiographic profile but did not significantly modify the basin geometry. As a consequence, within the overall trend from coarse grained early Triassic sediments to fine grained late Triassic sediments, there is considerable lateral uniformity of facies.

North of the Mid North Sea-Ringkøbing-Fyn High, regional tectonism throughout the Triassic played a much more important role in controlling sedimentation. In some sub-basins, a number of major tectonic

episodes, together with more restricted, local events, resulted in a succession of upward-coarsening cycles with poor lateral facies continuity (Jakobsson et al., 1980). Another tectonic influence that became a significant factor in basin development was halokinesis. The Permian halites were initially mobilised by the Hardegsen tectonism and later by overburden pressure (W.H. Ziegler, 1975; Thomas, 1975).

These tectonic events should also be viewed in the context of global sea level fluctuations, which had a major effect in controlling sedimentation. Vail et al. (1977) and Ormaasen et al. (1980) postulate a gradual rise of global sea level during the Triassic, with regressions in the Lower Triassic, the Middle Triassic and in the early and late Upper Triassic.

In all the North Sea basins, Triassic sediments are dominantly red-beds including representatives of alluvial fan, fluvial, aeolian, sabkha, lacustrine and shallow-marine facies. The relative abundance and the inter-relationship of the various facies is largely dependent on the tectonic setting. In tectonically active fault-bounded basins, alluvial fans will merge with playa lakes or other basin-centre environments. In more stable basins, with lower physiographic profiles, the marginal alluvial fans may be separated from the basin-centre environments by broad flood-plains (cf. Hardie et al., 1978).

The mechanism responsible for the colouration of Triassic red-beds has been the source of much speculation. Turner (1980) has reviewed the current hypotheses and favours a process similar to that described by Walker et al. (1978) in desert sandy alluvium. Briefly, this requires the post-depositional degradation of ferromagnesian minerals supplemented by detrital ferric hydroxides to form the haematite pigment.

Correlation of Triassic red-bed sequences has always presented problems. In north-west Europe the diagnostic ammonite and bivalve assemblages of the Boreal and Tethyan marine provinces are absent. Similarly the vertebrate remains used for correlation in continental facies are rare.

Traditionally, lithostratigraphic correlations have been favoured and although these can be applied in the Southern North Sea Basin where the sediments have remarkable lateral continuity, in the basins north of the Mid North Sea-Ringkøbing-Fyn High, lithostratigraphic correlations are less satisfactory. Here palynological zonations established in the type sections of the Tethyan and Boreal provinces are the most reliable

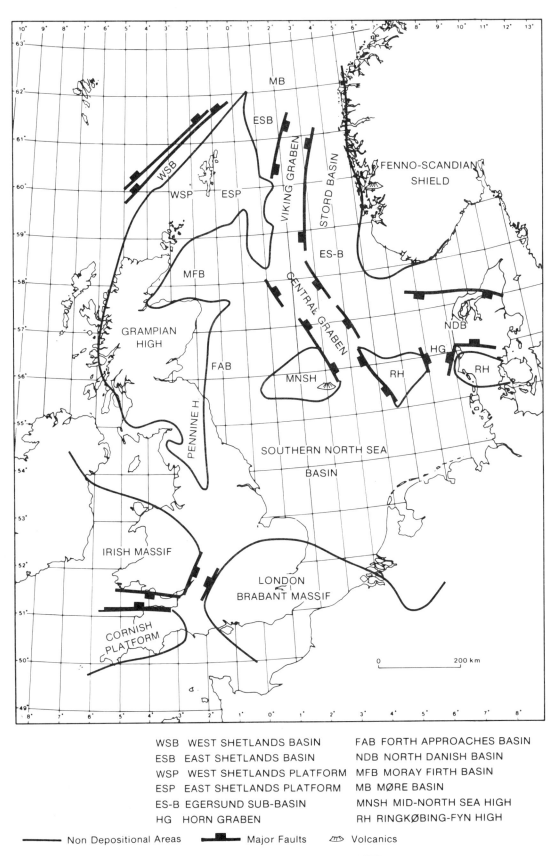

Fig. 5.1. Major Triassic structures. After Brennand (1975) and Ziegler, P.A. (1978).

WSB WEST SHETLANDS BASIN   FAB FORTH APPROACHES BASIN
ESB EAST SHETLANDS BASIN   NDB NORTH DANISH BASIN
WSP WEST SHETLANDS PLATFORM   MFB MORAY FIRTH BASIN
ESP EAST SHETLANDS PLATFORM   MB MØRE BASIN
ES-B EGERSUND SUB-BASIN   MNSH MID-NORTH SEA HIGH
HG   HORN GRABEN   RH RINGKØBING-FYN HIGH

——— Non Depositional Areas   ▄▄ Major Faults   ⟁ Volcanics

means of correlation. One further recent development, which has had encouraging results in correlating monotonous and unfossiliferous mudstone sequences in the Western Approaches, has been the use of diagnostic detrital clay mineral assemblages (Fisher and Jeans,

1982). This technique has now been successfully applied to correlations in the North Sea Basins.

Geophysical characterisation of the Triassic is relatively straight-forward in the Southern North Sea Basin but becomes more conjectural north of the Mid North

Sea-Ringkøbing-Fyn High. Day *et al.* (1981) recognise a strong 'Top Triassic' seismic event which they correlate with the top of the 'Keuper shale' or Rhaetian sandstones, although in parts of the Anglo-Dutch Basin and on the Mid North Sea High thin arenaceous Upper Jurassic sequences cannot be distinguished from the underlying eroded Triassic. Within the Triassic the 'Top Bacton' reflector equates with the base of the Röt Halite and is of variable quality. This reflector is of considerable importance because it can be used, in conjunction with sonic log-derived velocities of the Bunter Shale Formation, to determine regional maximum palaeo-burial depths for the Lower Triassic and Permian reservoirs (van Wijhe *et al.*, 1980).

5.2 The Southern North Sea Basin

The Triassic Southern North Sea Basin occupied roughly the same position as that of the Southern Permian Basin (Fig. 5.1). Residual positive structural features were the London-Brabant Massif to the south, the Mid North Sea-Ringkøbing-Fyn High to the north and the Pennine High to the west. The new tectonic framework was represented by the Central and Horn Grabens, which dissected the Mid North Sea-Ringkøbing-Fyn High. Subsidence continued throughout the Triassic with 1500 m of sediments accumulating in the southern part of the Central Graben and up to 3000 m in the Danish Embayment.

Because of the economic importance of the underlying Permian and to a lesser extent the Triassic itself, this area has provided more information on the Triassic than any of the other North Sea basins. It has also been the subject of two informative studies, by Geiger and Hopping (1968) and Brennand (1975).

The sedimentary succession has been described by Rhys (1974) who proposed the lithostratigraphic subdivision employed in this account (Fig. 5.2). The Triassic succession falls naturally into two major groups: 1. The Bacton Group representing a phase of clastic deposition with red sandstones, shales and mudstones; 2. The Haisborough Group, a largely fine-grained clastic and evaporite sequence with marked cyclicity. The succession is terminated with the Winterton Formation, which reflects the marine transgression that marked the passage from the Triassic to the Jurassic.

5.2.1 The Bacton Group

The Permian-Triassic transition is represented by a distinct facies break where the basal Bunter Shale Formation overlies the basinal carbonates, anhydrites or halites of the Zechstein. At the basin margins, where basal Triassic clastics overlie late Permian clastics, the boundary is less clearly defined. The old Permian basin was, however, markedly featureless and correlation across the basin is facilitated by considerable lateral facies continuity (Fig. 5.3) which is reflected in the uniform petrophysical log characteristics.

The retreat of the Zechstein Sea resulted in the extension of the already established marginal clastic sedimentary environments further into the basin. The basal member of the Bacton Group, the Bröckelschiefer Member, is typically developed as an argillaceous sandstone or siltstone. Local facies variations occur towards the basin margin and, in the south-west of the basin a poorly sorted fine to medium grained quartzose sandstone is recognised as the Hewett Sandstone Member (Cumming and Wyndham, 1975). This and smaller but similar sandy intercalations, were probably derived from local faulting and uplift; in this instance from the London-Brabant Massif to the south.

The overlying Bunter Shale displays considerable consistency in thickness and lithology. It is typically developed as an anhydritic red-brown mudstone with some minor greenish shales. In the upper part of the formation occasional calcareous intercalations with distinctive ferruginous ooliths represent the Rogenstein Member. Towards the basin margin, the Bunter Shale Formation coarsens and thins rapidly and the Rogenstein Member disappears.

This formation reflects the maximum areal extent of an early Triassic playa lake or inland sea which occupied the major part of the basin. The 'lake' margin fluctuated considerably as is evidenced by the often rapid alternations of lacustrine with sheet flood, aeolian or fluvial sediments at the basin periphery. The Bunter shale facies accumulated as lacustrine or flood plain deposits. Onshore in eastern England the contemporaneous sediments were largely deposited by fluvial channels or sheet floods; aeolian or alluvial fan deposits are relatively rare.

With time the marginal clastic sedimentary environments prograded into the centre of the basin so that overall the Bacton Group is represented by a gross upwards-coarsening unit. This progradation was accomplished by a series of progressive encroachments into the basin and, as a consequence, the boundary between the Bunter Shale Formation and the succeeding Bunter Sandstone Formation is markedly diachronous.

The Bunter Sandstone Formation reflects a period of increased tectonic activity. In the western part of the basin the formation is represented by a relatively homogenous upwards coarsening sand complex in which fluvial channels, sheet flood and lacustrine sediments predominate. There is occasional evidence of cyclicity but the individual cycles are poorly defined and may reflect restricted, local tectonic activity rather than widespread, regional disturbances. In the east of the basin up to four depositional cycles, each with a basal sand unit, are widely recognised. These have been correlated with the Volpriehausen, Detfurth, Hardegsen and Solling sequences of Germany (Fig. 5.2). The generating mechanism of these cycles is not entirely clear although it is logical to assume that they resulted from periodic tectonic activity. Although these German sequences are not so clearly defined in U.K. waters, the four classic upward-fining cycles of the Bacton Group are well developed at the northern margin of the basin in the southern Horn Graben (Best *et al.*, 1983).

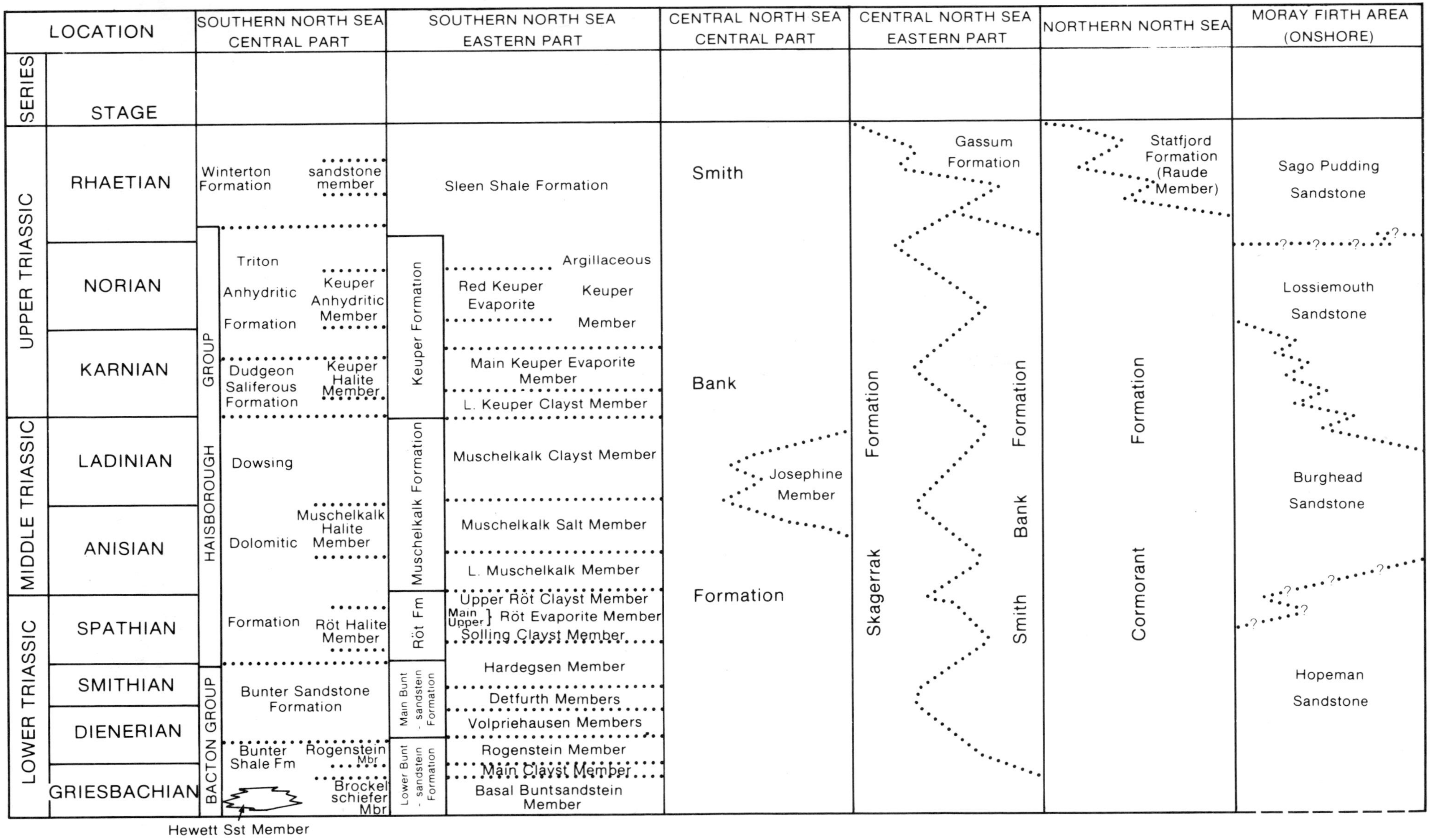

Fig. 5.2. Lithostratigraphic correlation in the North Sea basins. After Rhys (1974), Deegan and Scull (1977), NAM/RGD (1980) and Jakobsson *et al.* (1980).

88

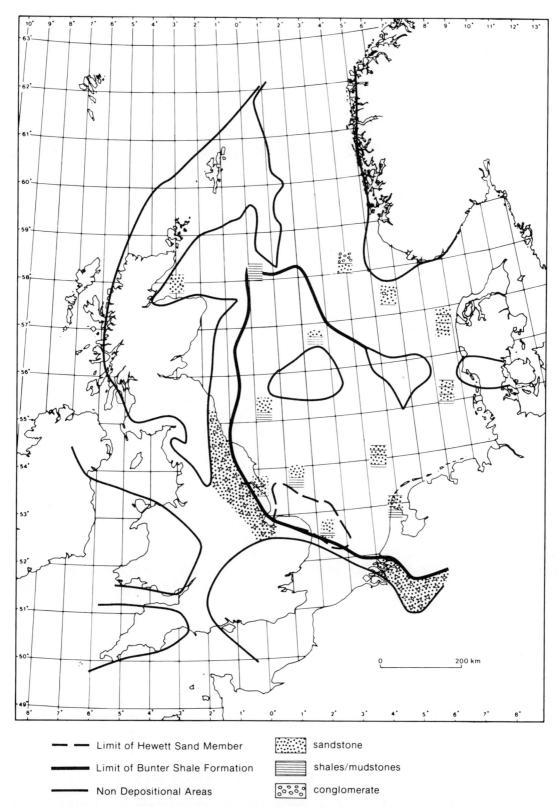

Fig. 5.3. Facies distribution in the Bacton Group and Smith Bank Formation. After Brennand (1975).

--- — Limit of Hewett Sand Member

——— Limit of Bunter Shale Formation

——— Non Depositional Areas

[:::] sandstone

[≡] shales/mudstones

[∘∘∘] conglomerate

The tectonic event of greatest regional significance followed deposition of the unit that correlates with the Hardegsen cycle. Major uplift and subsequent erosion was recorded in the Mid North Sea-Ringkøbing-Fyn High, the Pennine Massif and the London-Brabant Massif. Subsidence was increased in the Central and Horn Grabens. This event may also have initiated the halokinetic events which had a major effect in con-

trolling sedimentation in the North Sea basins (W.H. Ziegler, 1975). The Hardegsen disconformity is of regional significance in continental European sections, although within the sheet sand complex of the western part of the Basin its effect is less discernible. W.H. Ziegler (1975) has suggested that the Hardegsen disconformity could represent the Permian-Triassic boundary. Palynological evidence does not support this view as an assemblage from a presumed Volprie-hausen equivalent is undoubtedly of Triassic, probably Dienerian age (Fisher, 1979).

The effect of the Hardegsen tectonic event had been to generate topographic features, which effectively subdivided the area into minor basins and highs. The period of subsidence that followed saw the erosion of the highs and the accumulation of dominantly fine-grained clastics whose facies and thickness was dependent both on proximity to the sediment source areas and on the local post-Hardegsen relief. It was in this environment that the youngest cycle of the Bacton Group, including the Solling sand equivalent, was deposited.

The Bacton Group undoubtedly accumulated in a relatively arid environment. Clemmensen (1979) has reviewed the available data and concludes that the North Sea basin lay in low northern palaeolatitudes (approximately 20°N) in a central trade-wind zone. Precipitation was probably seasonal and restricted to occasional cloudbursts, which initiated short-lived sheet floods that drained into a large, shallow, saline playa-lake or inland sea. The dominant wind direction, deduced from thin intercalations of aeolian sands with fine-grained lacustrine sediments at Helgoland, was south-easterly. This is partially supported by Mader (1982) who records south-easterly to south-westerly wind directions in the more extensive Middle Bunt-sandstein aeolian sands of the Eifel area.

5.2.2 The Haisborough Group

The Haisborough Group (Fig. 5.2) represents a period of greater tectonic stability. Marine conditions were re-established in the basin and the pronounced lateral facies continuity reflects the overall low relief. Although there is evidence to suggest that erosion of the Mid North Sea-Ringkøbing-Fyn High had proceeded sufficiently to allow a partial sedimentary cover, the High still provided enough relief to form the northern limit of the main basinal deposition. Sedimentation was predominantly in distal floodplain environments alternating with coastal sabkha or shallow-marine environments.

The oldest Member of the Dowsing Dolomitic Formation, the Röt Halite Member, has a thin basal transgressive unit, the Röt Clay which is of considerable lateral extent. The overlying sediments include red and red-brown shales with variably developed halites. One halite, the Main Röt Evaporite, is extensively developed and, in the eastern part of the basin a second, younger Röt halite is also developed. The thickness of the Röt Halite Member is somewhat variable within the basin. These variations result, in part, from thicker salt deposition in the residual depressions generated by the Hardegsen tectonic event. In this context, the origin of the salt is also of interest because it has been postulated that it may be derived from leached and reprecipitated Zechstein halites exposed by the Hardegsen movements (W.H. Ziegler, 1975). Both the Röt trans-

gression and the succeeding transgression represented by the Muschelkalk Halite Member, entered the basin from the east.

The typical shelly limestone facies of the classic Muschelkalk stratotype sections is not represented in the North Sea. In the east of the basin interbedded dolomites and silty shales sandwich a thick halite, but towards the western part of the basin, the dolomites become discontinuous and are eventually replaced by anhydritic mudstones and silty clays. Although the Röt and Muschelkalk halites are of comparable thickness, typically 70 to 150 m, the Röt Halite basin was areally more extensive (Fig. 5.4).

With the retreat of the Muschelkalk sea, clastic sedimentation resumed. Coarse-grained sediments are common only at the basin margins, for example adjacent to the Ringkøbing-Fyn High, and elsewhere uniformly fine-grained sequences of red shales and mudstones predominate. These sediments represent the Dudgeon Saliferous Formation which, in the upper part, contains thick halite beds interbedded with red-brown mudstones. The Keuper Halite Member, the most widespread of these halites, is the thickest halite in the North Sea, exceeding 300 m. It is variable in thickness, possibly due to halokinetic control of subsidence. The lateral impersistence of the thinner halites suggests numerous, scattered saline lakes rather than the more continuous bodies of water envisaged for the Röt and Muschelkalk seas.

The Group terminates with the Triton Anhydrite Formation. This comprises variegated green and red mudstones with common anhydrite and rare dolomite beds. The anhydrite beds become extremely persistent in the upper part of the formation where they constitute the Keuper Anhydrite Member.

Deposition of the Haisborough Group has been considered to have been largely continental. More recent investigations (e.g. Jeans, 1978) suggest that deposition occurred in sub-aqueous environments, commonly in shallow, hypersaline water. These sedimentary studies reveal considerable cyclicity in these sediments in terms of structure, fabric and detailed mineralogy. With additional evidence from the Triassic ichnofacies (Pollard, 1981), it is reasonable to postulate that the Southern North Sea Basin was never totally drained and that the Haisborough Group and possibly the younger part of the Bacton Group accumulated largely under marine or quasi-marine conditions.

5.2.3 The Winterton Formation

The occurrence of diverse marine Rhaetian fossil assemblages, and the disappearance of typically hypersaline clay mineral assemblages in the basal grey shales of the Winterton Formation indicate that more normal marine conditions have resumed. Hallam and El Shaarawy (1982) deduced from the faunal evidence that initially the Rhaetian sea exhibited reduced salinities (25-30‰). It was not until early Liassic times that stenohaline faunas were successfully re-established in

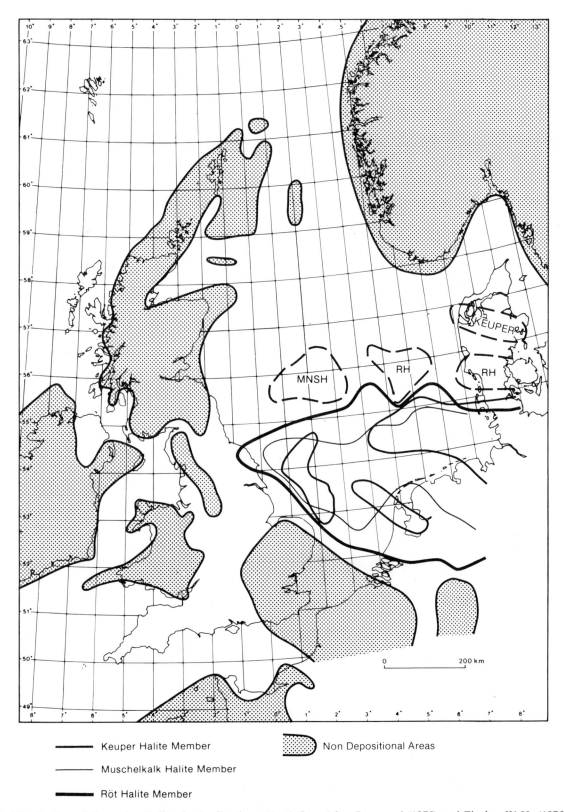

Fig. 5.4. Distribution of the major Halites in the Southern North Sea. After Brennand (1975) and Ziegler, W.H. (1975).

Legend:
— Keuper Halite Member
— Muschelkalk Halite Member
— Röt Halite Member
Non Depositional Areas

the North Sea Basin. The formation represents a significant transgression, however, with sands and shales onlapping in areas of old structural highs. The sandy facies is well developed and is a distinctive correlation marker. It is generally overlain by a dark grey shale, with restricted marine microplankton assemblages, which passes conformably into the overlying Liassic sediments.

5.2.4 Economic geology

The Southern North Sea Basin is an established gas producing area. The major discoveries are in Permian reservoirs, but six, the Hewett, Dotty, K/13, Esmond, Forbes and Gordon fields have Triassic reservoirs. The Hewett Gas Field (Figs. 5.5, 5.6) has been described by Cumming and Wyndham (1975). It is located on a

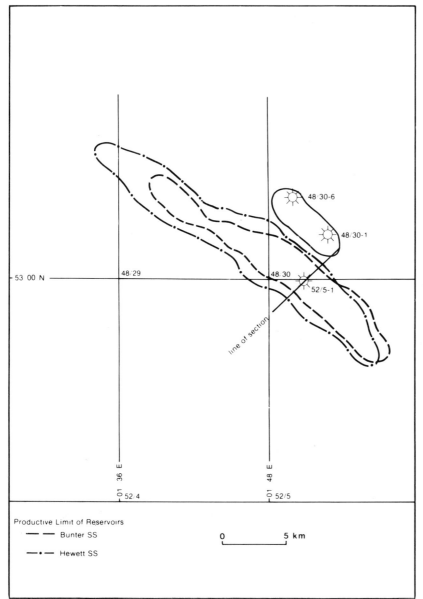

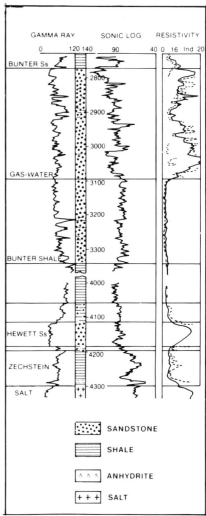

Fig. 5.6. Generalised well log in the Hewett Field. After Cumming and Wyndham (1975).

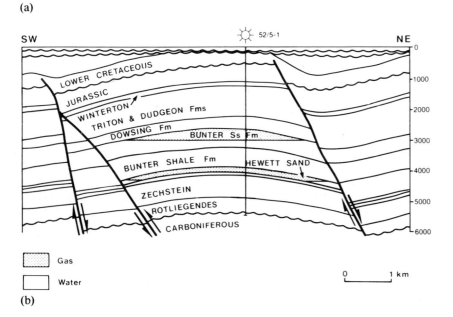

Fig. 5.5. (a) Map of Hewett Field with line of section.
    (b) Cross-section.
After Cumming and Wyndham (1975).

northwest-southeast trending, fault-bounded anticline. There are two reservoirs: the lower reservoir, the Hewett Sand Member has 60 m gross pay with 21.4% porosity and 1310 mD permeability; the upper reservoir, in the Bunter Sandstone Formation, has 98 m gross pay with 25.7% porosity and 474 mD permeability. Recoverable reserves are estimated at 105×10^9 m³ $(3.7 \times 10^{12}$ ft³). The adjacent field, Dotty, also contains gas in the Hewett Sand Member and Bunter Sandstone Formation.

The relative insignificance of the Triassic as an exploration objective in this prolific gas producing area results from the lack of communication between the Westphalian source rocks and the Triassic reservoirs.

Although the Hewett Sand Member is only of local significance, the Bunter Sandstone Formation is an attractive target. It displays good reservoir characteristics (ø 20-25%; k 100-700 mD) over the major part of the basin, and forms large structures, for example, in Quadrants 43 and 44, relatively few of which contain significant volumes of gas. The migration barrier is the Zechstein evaporites, particularly the Z2 halite, which, even when the underlying Rotliegend reservoirs have been severely faulted, have flowed and sealed the fractures. Breaching of the seal was only accomplished by either considerable diapiric withdrawal of salt or, in the Dotty and Hewett Fields on the periphery of the Zechstein evaporite basin, by failure of thin evaporites

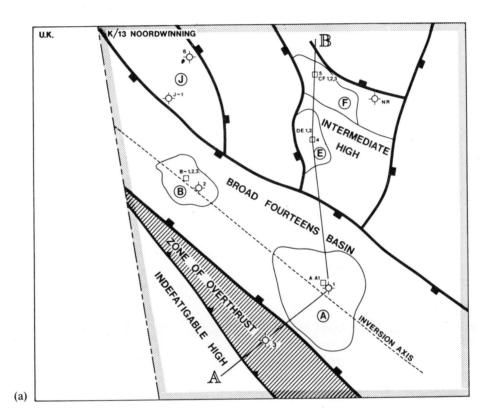

(a)

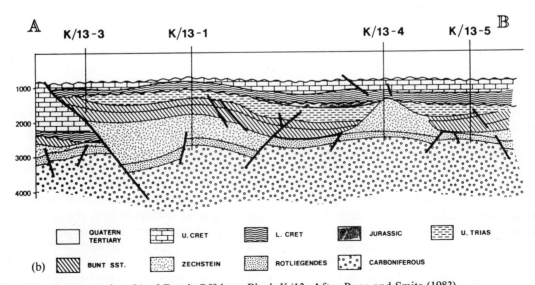

(b)

Fig. 5.7. Map (a) and cross-section (b) of Dutch Offshore Block K/13. After Roos and Smits (1983).

to seal the fault conduits.

The Dutch offshore Block K/13 contains four gas fields: two are reservoired in the Rotliegend and two in the Bunter Sandstone Formation (Fig. 5.7a). Roos and Smits (1983) consider that gas generated from Westphalian source rocks in the late Jurassic to early Cretaceous was originally reservoired in the Rotliegend structures and sealed by Zechstein evaporites. Late Cretaceous inversion movements breached the seal, which had already been thinned by diapiric withdrawal, and gas re-migrated from the Permian reservoirs along large reverse faults into the Bunter Sandstone Formation (Fig. 5.7b).

Salt withdrawal was also a significant factor in the formation of the Esmond (43/8a), Forbes (43/13a) and Gordon (43/15a:43/20a) fields. The Bunter structures are located over salt 'swells' which have generated areas of thinned or absent Zechstein halites. Where the Zechstein seal has been breached by faulting, gas has migrated into the Bunter and has accumulated in the salt controlled Triassic structures. Total combined reserves are estimated to be $15 \times 10^9 \, m^3 \, (0.58 \times 10^{12} \, ft^3)$.

A characteristic feature of some of these Bunter gas accumulations in Quadrants 43 and 44 is that the nitrogen content is generally higher than in equivalent Permian gas accumulations. Nitrogen is a natural product of the Carboniferous source rocks and increases in concentration with increased maturity. The most likely explanantion here, however, is that the higher mobility of nitrogen, compared with methane, results in differential concentration when long migration routes, as in Quadrants 43 and 44, are involved.

In addition to the fields already described, gas has also been tested from the Bunter Sandstone Formation in U.K. Block 44/23 and in the Dutch Block P/6. No other major reservoir units have been recognised offshore although in the Dutch onshore De Wijk Field, Gdula (1983) records gas production from the Volpriehausen Sandstone, Rogenstein and Lower Muschelkalk Members in addition to marginally economic production from the Upper Röt Claystone Member and, in the nearby Wanneperveen Field, from the Main Claystone Member. With the exception of the Volpriehausen Sandstone Member which has good primary reservoir properties, the reservoir potential of the other Members results from leaching of syndepositional anhydrite during phases of Kimmerian erosion.

5.3 The Central North Sea Basins

The Central North Sea Basins lie to the north of the Mid North Sea-Ringkøbing-Fyn High and are located approximately on the site of the Northern Permian Basin (Fig. 5.1). During the Triassic, the Central and Horn Grabens developed and transsected the Mid North Sea-Ringkøbing-Fyn High. Throughout the Triassic, tectonic movements rejuvenated relief and the continuously subsiding basins were filled with cyclical clastics, with evaporites representing only a minor component.

Up to 4000 m of Triassic sediments accumulated in Horn Graben, 2000 m in the Central Graben and 3500 m in the Egersund Sub-basin. Even these figures were exceeded in the North Danish Basin, a grabenal extension of the Polish Trough, where 6000 m of Triassic has been estimated (Kent, 1975). Unlike the Southern North Sea Basin, there are few marker horizons, palyniferous sediments are rare and detailed correlation between the two basins has not been possible. A separate lithostratigraphic breakdown has been proposed by Deegan and Scull (1977) (Fig. 5.2), although Bertelsen (1980) has commented on the inappropriateness of this terminology for the Danish Triassic sequences.

5.3.1 The Smith Bank Formation

This formation represents basinal sediments and is composed predominantly of silty mudstones with rare sandstone stringers and thin anhydrite beds. Towards the north-east margins of the basin, conglomeratic beds are developed, whilst approaching the Ringkøbing-Fyn High the sandstone stringers become thicker and more continuous. At the base of the formation a persistent, transgressive, sandy unit is commonly present and there is apparent conformity with the underlying Zechstein.

In the area of the Josephine structure (30/13) a brown, fine to medium-grained sandstone interbedded with red and green shales occurs in the lower part of the formation. This has been termed the Josephine Member.

In the Forth Approaches Basin the Smith Bank Formation appears to be the only Triassic facies represented. Thicknesses are variable and uplift and erosion during the Lower and Middle Jurassic are considered to have removed a considerable volume of Triassic sediments from this area. Palynology suggests an age range of Griesbachian to possibly Karnian for these sediments, an indication of the stability and relative distance of this basin from areas of high relief.

Detailed information on Triassic sedimentation in the northern part of the Central Graben is derived exclusively from the Dansk Nordsø Q-1 well, where over 300 m of coarse clastics have been referred to the Smith Bank Formation. In the southern part of the graben, however, the pattern of sedimentation is similar to that of the Southern North Sea Basin. Surprisingly, in view of the proximity of what are presumed to have been emergent areas, coarse clastic deposits are relatively rare. The successions described by Jacobsen (1982) and Michelsen and Andersen (1983) suggest that significant deposition only occurred when lacustrine or marine sedimentation extended beyond the normal northern limits of the Southern North Sea Basin. The Bacton Group equivalent is predominantly composed of reddish-brown, occasionally silty, anhydritic and calcareous claystone with minor siltstone and sandstones. The equivalent of the Dowsing Dolomite Formation includes representatives of the Röt and Muschelkalk halites but otherwise comprises variegated dolomitic claystones with anhydrite interbeds.

The Dudgeon Saliferous Formation equivalent also includes a halite, corresponding to the Keuper Halite Member, which grades laterally into grey marlstone, dolomites and anhydrites. The remainder of the Formation is represented by dominantly red-brown or green-grey calcareous claystones. The equivalent of the Triton Anhydrite Formation is a variegated red-brown and green-grey claystone with an anhydritic middle unit that is correlative with the Keuper Anhydrite Member. The Triassic succession terminates with a dark grey claystone of Rhaetian age. Overall, the sequence may be up to 2000 m thick.

After an initial phase of rifting and rapid subsidence in the Permian, the Central Graben was a relatively stable tectonic element throughout the Triassic. Towards the end of Triassic sedimentation, however, halokinesis of Zechstein evaporites commenced and thereafter differential subsidence increased and became more widespread. The restriction of coarse clastic Triassic sediments to the northern part of the Graben suggests more pronounced relief than in the south. There, the dominance of fine-grained clastics and evaporites, and the evidence of progressive onlap onto the southern flanks of the Mid North Sea-Ringkøbing-Fyn High, indicates progressively reduced relief during the Triassic.

Three well sequences in the centre and at the northern and southern ends of the Horn Graben suggest that conditions of deposition were similar to that of the Central Graben, the Bunter sequence, especially, being coarser in the northern well (Olsen, 1983, Fig. 2). This major period of sedimentation and associated taphrogenesis initiated halokinesis in the underlying Zechstein evaporites at the bounding faults and in the graben centre. Growth of these salt structures continued during deposition of the Muschelkalk, culminating in diapiric development in the late Triassic.

The variations in depositional history of these two major grabens reflect fundamental differences in their structural evolution. In the Horn Graben a major phase of rifting and subsidence occurred during the early Triassic, when up to 4200 m accumulated in the centre of the Graben. In contrast, the Central Graben was relatively quiescent throughout the Triassic.

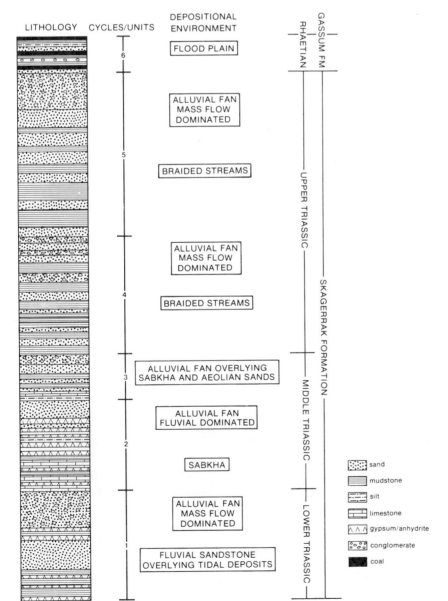

Fig. 5.8. Typical lithological succession of the Skagerrak Formation in the Egersund sub-basin. After Jakobsson *et al.* (1980).

5.3.2 The Skagerrak Formation

Deegan and Scull (1977) imply that the Skagerrak Formation is, in part, younger than and, in part, of equivalent age to the upper section of the Smith Bank Formation (Fig. 5.2). Regional evidence suggests that the Skagerrak Formation represents the lateral facies equivalent of the Smith Bank Formation. The interbedded conglomerates, sandstones, silts and shales accumulated primarily at the basin margins and adjacent to fault scarps in a laterally extensive alluvial fan system. Jakobsson *et al.* (1980) interpret the sedimentary succession in the Egersund Sub-basin as comprising six tectonically-induced, coarsening upwards cycles. The depositional environments described are similar to those outlined by Glennie (1972) to account for the Rotliegendes sedimentary facies in north-west Europe. The cyclical sequence (Fig. 5.8) commences with regional basinal subsidence and a marine transgression. Alluvium which accumulated during the Zechstein was re-deposited in central lakes or marginal marine environments as uniform, brick-red shales. Continued subsidence and erosion of the newly generated fault blocks stimulated rapid progradation of alluvial fans with the deposition of poorly sorted conglomerates. The second cycle commences with a northerly extension of the Muschelkalk transgression through the North Danish Basin. Slow progradation of marginal alluvial fans is reflected in the transition from the interbedded evaporite, sabkha and fluvial sediments into a sandy conglomerate. The third cycle was dominated by aeolian sands and sabkha deposits, which were succeeded by a new fan outgrowth, represented by coarse-grained massflow conglomerates. The fourth and fifth cycles each represent braided stream deposits, terminated by prograding alluvial fan deposits. The sixth and final cycle, which should properly be assigned to the Gassum Formation, reflects the Rhaetian marine transgression. This entered the Central North Sea Basins via the North Danish Basin and resulted in extensive deposition of fluvial sands and silts in the Egersund Sub-basin.

It is difficult to apply the lithostratigraphic nomenclature of Deegan and Scull (1977) to the sediments in the Moray Firth Basin. The onshore succession of the aeolian Hopeman and Lossiemouth Sandstones separated by the fluvial sandstones of the Burghead Beds, extends offshore with the Burghead Beds becoming more argillaceous eastwards. Onshore the succession is terminated by the Sago Pudding Sandstone which is represented offshore by a sequence of calcareous sandstones, sandy limestones and mudstones of probable lacustrine origin.

5.3.3 Economic geology

There are no major hydrocarbon reservoirs in the Triassic of the Central North Sea Basins. Fine-grained clastic sediments predominate and sands are only developed near basin margins or adjacent to isolated 'highs' within the basins.

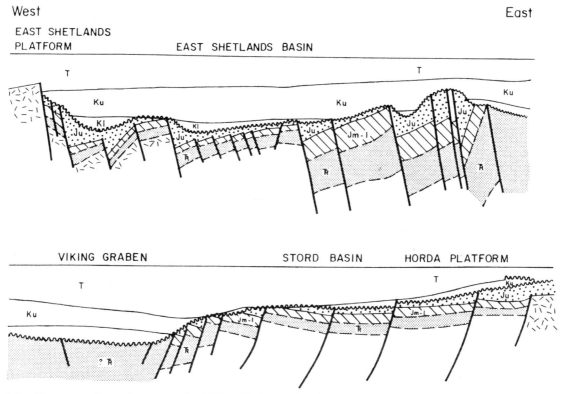

Fig. 5.9. Diagrammatic section across the Viking Graben and adjacent structures to show relationship of Upper Jurassic source rocks and Triassic Reservoirs.

The Josephine structure in Block 30/13 represents just such an intrabasinal high. It contains oil in the Josephine Member, but this has not yet been confirmed as a commercial discovery. The source of the oil is the Kimmeridge Clay Formation and the structural setting, Triassic sands in fault communication with downthrown Jurassic source rocks, is repeated in commercial fields in the Northern North Sea Basin (Fig. 5.9).

5.4 The Northern North Sea Basins

The structure of the Northern North Sea Basins is dominated by north-south faulting which formed deep and well defined grabens (Fig. 5.1). P.A. Ziegler (1975; 1978) suggested that faulting in the Viking Graben commenced in the Permian and that the western fault moved throughout the Triassic. Hay (1978) however suggests that only downwarping occurred over a comparable period. The variations in thickness of the Triassic sediments in the Graben indicate differential subsidence and the facies relationships are compatible with fault-controlled sedimentation.

There is no published evidence that complete sequences of thick Triassic sediments have been penetrated and the relationship of the Triassic with the underlying sediments and the effect of the Hardegsen movements on basin formation are therefore imperfectly understood. In the Beryl Embayment, Ormaasen *et al.* (1980) indicate that in 9/17-1 deposition of continental arenaceous red-beds with interbedded shales was continuous from the late Permain into the early Triassic. It can be assumed, however, that in the basins bordering the Viking Graben, the East Shetland Basin to the north-west and the Stord or Horda Basin to the east (Fig. 5.1), deposition would have been dominated by alluvial fans at the fault-bounded margins, with finer-grained fluvial or lacustrine sediments in the 'lows'. Pending penetration of more complete Triassic sections and the release of these data by industry, Deegan and Scull (1977) have recognised one Triassic formation in the Northern North Sea Basins, the Cormorant Formation (Fig. 5.2).

5.4.1 The Cormorant Formation

The Cormorant Formation is typically composed of pinkish or white, fine or medium-grained argillaceous sandstone with some red-brown siltstones and shales. Thicker sequences with coarser-grained sandstone and conglomerates are developed towards the East Shetlands Boundary Fault. These are overlain by more typical Cormorant Formation facies. Because sedimentation was controlled by relatively local tectonic activity, variations in thickness from one fault block to another are considerable and may be of the order of 2000 m. Both lithostratigraphic and biostratigraphic correlations are difficult in these tectonically controlled sediments. Palynological evidence indicates a late Norian to early Rhaetian age for a relatively widespread argillaceous horizon near the top of the forma-

tion, and a Rhaetian age for the higher beds (Brennand, 1975).

It is difficult to generalise about the depositional environments of the various facies in the Cormorant Formation because of the relatively complex tectonic setting which controlled patterns of sedimentation in the Viking Graben. Some of the more widespread and laterally persistent cyclical sandstones and shale facies, however, appear to have been deposited under shallow-water conditions. The absence of marine organisms suggests a fluvio-lacustrine environment. The sandstones are usually poorly sorted with an argillaceous matrix and often show transitional boundaries with the interbedded shales. This poor sorting, together with the occurrence of oxidised plant debris in the shales and in the argillaceous matrix is consistent with deposition in a low-energy environment. There is no evidence of winnowing or prolonged reworking as would be found, for example with flash flood or alluvial fan processes. It is therefore reasonable to assume that access to the fault-controlled basins from the higher energy basin margin and fault-terrace areas was relatively restricted. A sedimentary model outlined by Clemmensen *et al.* (1980) is compatible with these observations. They suggest marginal alluvial fans, feeding through braided streams and stabilised distributary channels into a central, northwards draining, elongate basin or coalescing series of basins, which would have included lacustrine and sabkha environments.

Towards the end of the Triassic, a transgression from the Boreal Sea to the north resulted in the establishment of marine environments in the central part of the basins at the northern-most end of the Viking Graben and of fluviodeltaic environments along the margins (Clemmensen *et al.*, 1980; Jakobsson *et al.*, 1980). In the central northern part of the Viking Graben there is, therefore, a conformable passage from the Cormorant Formation to the overlying Statfjord Formation.

5.4.2 The Statfjord Formation

The Statfjord Formation was orginally recognised by Bowen (1975) although here it is used as defined by Deegan and Scull (1977). The transition from the Cormorant Formation to the Statfjord Formation is indicated in the type section by an upwards coarsening succession of variegated grey, green and red shales interbedded with thin siltstones, sandstones and dolomites. This unit, which is 60 m thick, constitutes the Raude Member (Fig. 5.2). The marine transgression was a relatively slow, progressive event. The continental pattern of deposition established in the southern and more elevated parts of the basin persisted whilst the Statfjord Formation slowly onlapped from the north with pronounced diachroneity. The most recently published study of the type Statfjord Formation (Chauvin and Valachi, 1980) describes the depositional environment as a floodplain with some meandering streams. This is closely comparable with the interpretation

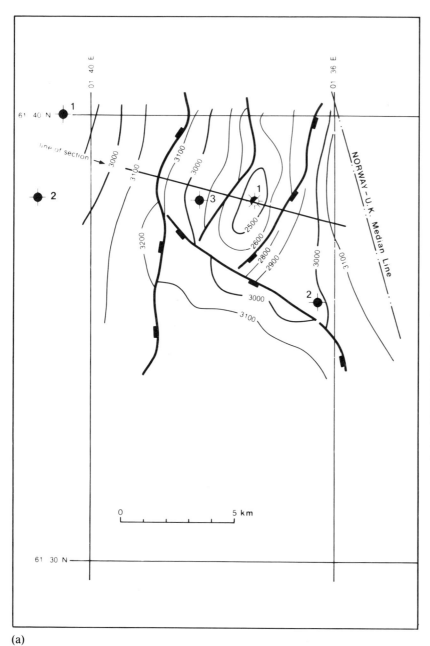

(a)

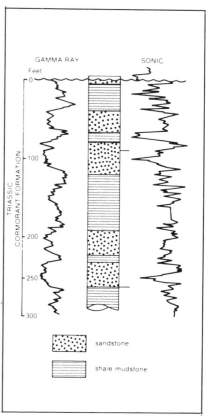

Fig. 5.11. Generalised well log of 211/13-1. After Brooks (1977).

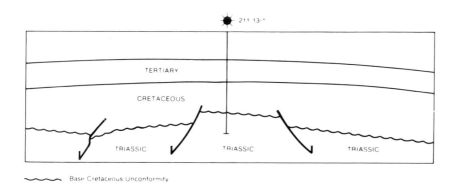

(b)

Fig. 5.10. (a) Map of 211/13-1 discovery with line of section.
(b) Cross-section.
After Brooks (1977).

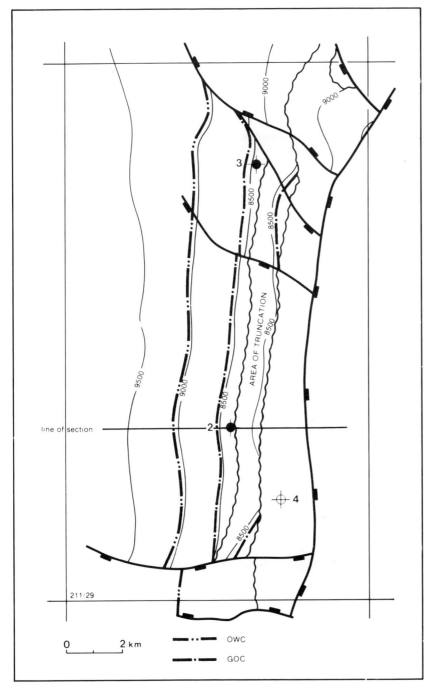

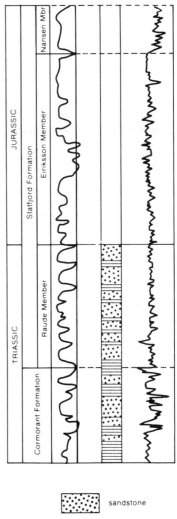

Fig. 5.13. Generalised well log of 211/29-3.

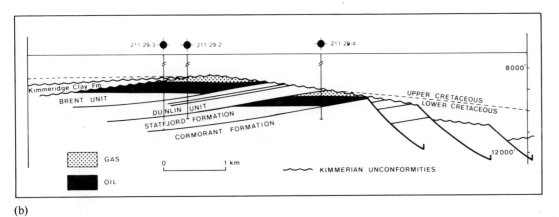

Fig. 5.12. (a) Map of Brent Field with line of section.
(b) Cross-section.
After Bowen (1975).

of the contemporaneous Gassum Formation in the Egersund Sub-basin (Jakobsson *et al.*, 1980).

The age of the Raude Member is generally accepted as Rhaetian, probably mostly late Rhaetian, but the paucity of diagnostic palynofloras and the problems inherent in defining the Triassic-Jurassic boundary on palynological criteria alone (Fisher and Dunay, 1981) do not allow an accurate dating of the upper boundary of the member.

5.4.3 Economic geology

The Northern North Sea Basins have proved to be a highly successful offshore exploration area. Gas condensate has been recorded in the Cormorant Formation of 211/13-1 (Brooks, 1977). The well, located on a northeast-southwest trending horst, has 52 m of gross pay with 21 m of sand (Figs. 5.10, 5.11).

The Beryl Field also contains some oil in sands of the Cormorant Formation although the major reservoirs are of Jurassic age (Marcum *et al.*, 1978).

Two large fields, Statfjord and Brent located on tilted fault blocks have significant oil reserves in the lower part of the Statfjord Formation (Figs. 5.12, 5.13). Both fields have major hydrocarbon reserves in the Jurassic Brent Unit and the Statfjord Formation, the two reservoirs being separated by the impermeable Dunlin Unit. In the Brent Field, the Statfjord Formation has a gas cap of 140 m and an oil column of 130 m. Reservoir characteristics are good, with porosity of 10 to 26% and permeability of up to 5500 mD (Bowen, 1975).

The source of the oil reservoired in the Triassic is the Kimmeridge Clay Formation. Oil generation, which is still in progress, probably commenced in the late Upper Cretaceous to Late Tertiary in the East Shetland Basin and Viking Graben (Goff, 1981). The characteristic tilted fault blocks, with Triassic sediments faulted above Jurassic source rocks, therefore pre-date the main phase of oil generation, and the faults themselves provide the migration pathways (Fig. 5.9).

5.5 Adjacent areas

North of 62° our knowledge of the Triassic decreases rapidly except in the areas of East Greenland, Svalbard and the Barents Sea. From these areas we know that marine sedimentation was more or less continuous in the Boreal Sea and the rich source rocks in the Middle Triassic Botneheia Formation of Svalbard (Mørk and Worsley, 1979) must be considered in any assessment of regional prospectivity in northern waters.

To the west of the Shetland Platform, Triassic sedimentation was active in the narrow, fault-bounded, southwest-northeast trending West Shetland Basin (Fig. 5.1).

Ridd (1981) has postulated over 1000 m of Permo-Triassic sediments adjacent to the Shetland Spine Fault System. The dominant lithologies are thin red-brown siltstones and shales interbedded with white and grey, commonly calcareous sandstones. Coarser clastics, probably sandstone and conglomerates, are encountered near the basin margins. If this is analogous with the Central and Northern North Sea Basins, deposition probably took place in a rapidly subsiding asymmetric graben. Coarse clastic material accumulated in marginal alluvial fans, and finer clastic material, fed via fluvial channels across extensive flood plains, possibly with aeolian sands, accumulated in basinal lakes.

5.6 Summary and conclusions

Throughout the Triassic, sedimentation was predominantly of clastic red-beds accumulating in continental basins. South of the Mid North Sea-Ringkøbing-Fyn High the sedimentary succession displays considerable lateral uniformity and includes thick halites. North of the High the sedimentary basins, which were fault-bounded, lack persistent halites and display little lithostratigraphic continuity.

Rarely in the North Sea has the Triassic realised its potential as a major exploration objective. In the Southern North Sea Basin the Zechstein evaporites have effectively sealed off the Bunter Sandstone reservoirs from the Westphalian source rocks. In the Central North Sea Basins significant sand accumulations have not been located where migration pathways from the Jurassic source rocks have been established. In the Northern North Sea Basins the structural relationship between source rocks and reservoirs and the timing of trap formation relative to oil generation are excellent. Unfortunately the sands in the Cormorant Formation are commonly thin, impersistent and with poor reservoir characteristics. Only in the Statfjord Formation, where sand quality is good, have appreciable hydrocarbon reserves been tested.

At the end of the Triassic, peneplantation was almost complete. The basins contained thick sediments which onlapped the eroded 'highs'. The early Jurassic transgression rapidly invaded the vast continental floodplains and tidal flats of the North Sea basins and re-established epicontinental marine conditions in northwest Europe.

5.7 Acknowledgements

I would like to thank my colleagues in Britoil for their advice, and especially Dr J.A. Miles for critically reviewing the manuscript.

This contribution is published with the permission of the directors of the company.

5.8 References

Bertelsen, F. (1980) Lithostratigraphy and depositional history of the Danish Triassic. *Danm. geol. Unders., Ser.* B4, 1-59.

Best, G., Kockel, F. and Schöneich, H. (1983) Geological history of the Southern Horn Graben. *Geol. Mijnbouw* **62**, 25-33.

Bowen, J.M. (1975) The Brent Oilfield. In: Woodland, A.W. (Ed.) q.v. 353-360.

Brennand, T.P. (1975) The Triassic of the North Sea. In: Woodland, A.W. (Ed.) q.v. 295-311.

Brooks, J.R.V. (1977) Exploration status of the Mesozoic of the U.K. northern North Sea. *N.P.F. Mesozoic Northern North Sea Symposium, Oslo.* MNNSS/2, 1-28.

Chauvin, A.L. and Valachi, L.Z. (1980) Sedimentology of the Brent and Statfjord Field. *N.P.F. The sedimentation of the North Sea Reservoir Rocks.* Geilo XVI, 1-17.

Clemmensen, L. (1979) Triassic lacustrine red-beds and palaeoclimate : The 'Buntsandstein' of Helgoland and the Malmros Klint Member of East Greenland. *Geol. Rundsch.* **68**, 748-774.

Clemmensen, L., Jacobsen, V. and Steel, R. (1980) Some aspects of Triassic sedimentation and basin development, East Greenland, North Sea. *N.P.F. The sedimentation of the North Sea Reservoir Rocks.* Geilo XVII, 1-21.

Cumming, A.D. and Wyndham, C.L. (1975) The geology and development of the Hewett Gas Field. In: Woodland, A.W. (Ed.) q.v. 313-325.

Day, G.A., Cooper, B.A., Andersen, C., Burgers, W.F.J., Rønnevik, H.C. and Schöneich, H. (1981) Regional seismic structure maps of the North Sea. In: Illing, L.V. and Hobson, G.D. (Eds.) q.v. 76-84.

Deegan, C.E. and Scull, B.J. (1977) *A standard lithostratigraphic nomenclature for the Central and Northern North Sea.* Report No. 77/25 Inst. Geol. Sci., H.M.S.O. 36 p.

Fisher, M.J. (1979) The Triassic palynofloral succession in the Canadian Arctic Archipelago. *AASP Contrib. Ser.* 5B, 83-100.

Fisher, M.J. and Dunay, R.E. (1981) Palynology and the Triassic-Jurassic boundary. *Rev. Palaeobot. Palynol.* **34**, 129-135.

Fisher, M.J. and Jeans, C.V. (1982) Clay mineral stratigraphy in the Permo-Triassic red-bed sequences of BNOC 72/10-1A, Western Approaches, and the south Devon coast. *Clay Minerals* **17**, 79-89.

Gdula, J.E. (1983) Reservoir geology, structural framework and petrophysical aspects of the De Wijk gas field. *Geol. Mijnbouw* **62**, 191-202.

Geiger, M.E. and Hopping, C.A. (1968) Triassic stratigraphy of the Southern North Sea Basin. *Phil. Trans. Roy. Soc. Lond.* Ser. B. **254**, 1-36.

Glennie, K.W. (1972) Permian Rotliegendes of North-West Europe interpreted in light of modern desert sedimentation studies. *Bull. Am. Assoc. Petrol. Geol.* **56**, 1048-1071.

Goff, J.C. (1981) Timing of hydrocarbon generation and mechanism of hydrocarbon migration in the E. Shetland Basin and Viking Graben of the Northern North Sea. Abstr. *10th Intl. Meeting Organic Geochem., Bergen.* 3-4.

Hallam, A. and El Shaarawy, Z. (1982) Salinity reduction of the end-Triassic sea from the Alpine region into northwestern Europe. *Lethaia* **15**, 169-178.

Hardie, L.A., Smoot, J.P. and Eugster, H.P. (1978) Saline lakes and their deposits: a sedimentological approach. *Spec. Publ. Int. Ass. Sediment.* **2**, 7-41.

Hay, J.T.C. (1978) Structural development in the northern North Sea. *J. Petroleum Geol.* **1**(1), 65-77.

Illing, L.V. and Hobson, G.D. (Eds.) (1981) *Petroleum geology of the continental shelf of N.W. Europe.* The Institute of Petroleum, London. 521 p.

Jacobsen, F. (1982) Triassic. In: Michelsen, O. (Ed.): Geology of the Danish Central Graben. *Danm. Geol. Unders.,* Ser. B8, 32-37.

Jakobsson, K.H., Hamar, G.P., Ormaasen, O.E. and Skarpnes, O. (1980) Triassic facies in the North Sea north of the Central Highs. *N.P.F. The sedimentation of the North Sea Reservoir Rocks.* Geilo XVIII, 1-10.

Jeans, C.V. (1978) The origin of the Triassic clay assemblage of Europe with special reference to the Keuper Marl and Rhaetic of parts of England. *Phil. Trans. Roy. Soc. Lond.* **289** (1365), 549-639.

Kent, P.E. (1975) Review of North Sea Basin development. *J. Geol. Soc. Lond.* **131**, 435-468.

Mader, D. (1982) Aeolian Sands in continental red-beds of the Middle Buntsandstein (Lower Triassic) at the western margin of the German Basin. *Sediment. Geol.* **31**, 191-230.

Marcum, B.L., Al-Hussainy, R., Adams, G.E., Croft, M. and Block, M.L. (1978) Development of the Beryl 'A' Field. EUR-97. *E.O.P.C.E.* 1978, 319-321.

Michelsen, O. and Andersen, C. (1983) Mesozoic structural and sedimentary development of the Danish Central Graben. *Geol. Mijnbouw* **62**, 93-102.

Mørk, A and Worsley, D. (1979) The Triassic and Lower Jurassic succession of Svalbard: a review. *N.P.F. Norwegian Sea Symposium, Tromsø.* NSS/29, 1-22.

Nederlandse Aardolie Maatschappij & Rijks Geologische Dienst (1980) *Stratigraphic nomenclature of the Netherlands.* Trans. Royal Dutch Geol. and Mining Soc. Delft. 77 p.

Olsen, J.C. (1983) The structural outline of the Horn Graben. *Geol. Mijnbouw* **62**, 47-50.

Ormaasen, O.E., Hamar, G.P., Jakobsson, K.H. and Skarpnes, O. (1980) Permo-Triassic correlations in the North Sea area north of the Central Highs. *N.P.F. The sedimentation of the North Sea Reservoir Rocks.* Geilo XIX, 1-15.

Pollard, J.E. (1981) A comparison between the Triassic trace fossils of Cheshire and South Germany. *Palaeontology* **24**, 555-588.

Rhys, G.H. (1974) *A proposed standard lithostratigraphic nomenclature for the Southern North Sea and an outline structural nomenclature for the whole of the (UK) North Sea.* Report No. 74/8. Inst. Geol. Sci. H.M.S.O.

Ridd, M.F. (1981) Petroleum geology west of the Shetlands. In: Illing, L.V. and Hobson, G.D. (Eds.) q.v. 414-425.

Roos, B.M. and Smits, B.J. (1983) Rotliegend and main Buntsandstein gas fields in Block K/13—a case history. *Geol. Mijnbouw* **62**, 75-82.

Thomas, J.B. (1975) *The geology of the southern North Sea.* OE-75, **213**. 1-12. Offshore Europe '75.

Turner, P. (1980) Continental red beds. *Developments in Sedimentology,* **29**, 562 pp.

Vail, P.R., Mitchum, R.M. Jr. and Todd, R.G. (1977) Eustatic model for the North Sea during the Mesozoic. *N.P.F. Mesozoic Northern North Sea Symposium, Oslo.* MNNSS/12, 1-35.

van Wijhe, D.H., Lutz, M. and Kaasscheiter, J.P.H. (1980) The Rotliegend in The Netherlands and its gas accumulations. *Geol. Mijnbouw* **59**, 3-24.

Walker, T.R., Waugh, B. and Crone, A.J. (1978) Diagenesis in first-cycle desert alluvium of Cenozoic age, southwestern United States and northwestern Mexico. *Geol. Soc. Am. Bull.* **89**, 19-32.

Woodland, A.W. (Ed.) (1975) *Petroleum and the continental shelf of northwest Europe, I. Geology.* Applied Science Publishers, London. 501 pp.

Ziegler, P.A. (1975) Geologic evolution of North Sea and its tectonic framework. *Bull. Am. Assoc. Petrol. Geol.* **59**, 1073-1097.

Ziegler, P.A. (1978) North-western Europe: Tectonics and basin development. *Geol. Mijnbouw* **57**, 509-626.

Ziegler, W.H. (1975) Outline of the geological history of the North Sea. In: Woodland, A.W. (Ed.) q.v. 165-187.

Chapter 6 Jurassic

STEWART BROWN

6.1 Introduction

From an economic viewpoint the Jurassic is the most important single stratigraphic system in the North Sea basins. Reservoir sands occur at a number of stratigraphic horizons and represent depositional environments ranging from fluvial to submarine fan. Oil-bearing sands have been encountered in diverse local structural positions, including the upthrown edge of tilted fault-blocks, the downthrown side of normal faults and arched over salt-induced 'highs'. In addition, the Upper Jurassic organic-rich, marine shales of the Kimmeridge Clay Formation and its lateral equivalents provide the principal source rocks in the central and northern North Sea. Furthermore, tectonic activity during the Jurassic played a significant role in the construction of traps in many fields, especially through fault-block rotation. Argillaceous strata of Jurassic age commonly contribute to the sealing of oil-bearing structures.

The Jurassic has proved to be most productive within the main North Sea graben system, consisting of the Viking Graben, Central Graben and Moray Firth Basin (Fig. 6.1) . The critical factor in the location of this fairway is the depth of burial and favourable maturation state of the Upper Jurassic source rocks. Potential reservoirs and traps have been located outside these basins but limited availability of a mature source rock has restricted exploration success to date.

6.1.1 Stratigraphic terminology

A number of lithostratigraphic schemes have been published for the North Sea Jurassic, with each restricted in most cases to parts of basins delimited by international boundaries (Fig. 1.1). Summaries of the various schemes are presented here (Fig. 6.2 and 6.3); for fuller accounts the following should be consulted, viz. Rhys (1974) for the U.K. southern North Sea, Deegan and Scull (1977) for the U.K. central and northern North Sea, Michelsen (1978) for the eastern Norwegian-Danish Basin, Nederlandse Aardolie Maatschappij B.V. and Rijks Geologische Dienst (N.A.M. and R.G.D., 1980) for the Dutch southern North Sea and Michelsen (1982) for the Danish Central Graben.

Jurassic lithostratigraphy in the Norwegian sector is currently under review (Norwegian Nomenclature working party, in press) and a number of alterations and additions to the formerly accepted scheme of Deegan and Scull (1977) are anticipated. Olsen and Strass (1982) and Hamar et al. (1983) provide some details of the new Norwegian proposals.

Differences in the usage of chronostratigraphic terminology occur in both published and confidential accounts of the North Sea Jurassic. In this account, following Cope et al. (1980b), the Aalenian is taken as the basal stage of the Middle Jurassic and the Callovian is taken as its topmost stage.

A more complex and confusing problem concerns stage nomenclature in the uppermost Jurassic. Provincialism in ammonite faunas has led biostratigraphers to erect different schemes in different parts of Europe (Riley, 1977), each of which have on occasion been applied to the North Sea succession (see Table 6.1, also Ofstad, 1983). Following Riley (1977) and Rawson et al. (1978), a Boreal scheme, with the Volgian the uppermost Jurassic stage, is used in this account. The Ryazanian then becomes the corresponding basal Cretaceous stage.

6.2 Regional setting

Jurassic strata in the North Sea occur for the most part in fault-bounded basins (see Fig. 10.3, sections 1, 2 and 3, and regional cross-sections of Ziegler, 1982, Figs. 9.1, 9.2 and 9.3) related to the development of a complex graben system which was initiated in the Permian (see Glennie, this volume, 2.4.2). The Jurassic was a period of active faulting. Differential fault-controlled subsidence, contemporaneous with sedimentation, had a marked influence on stratigraphic thicknesses and facies, especially during the late Jurassic.

Of faults moving during the Jurassic, many can be assigned to older structural trends. Evidence of reactivated Precambrian, Caledonian and Variscan lineaments has been reported from different parts of the North Sea (see Rønnevik et al., 1975; Glennie and Boegner, 1981; Johnson and Dingwall, 1981; Threlfall, 1981). Displacements have mostly been along normal faults, many of which are considered to have listric geometry (Halstead, 1975; Gibbs, 1983b; Jackson and McKenzie, 1983). No large scale strike-slip or reverse movements occurred during the Jurassic but there are indications of minor lateral displacements within the overall tensional regime (Hay, 1978; McQuillin et al., 1982; Ziegler, 1982b; Skjerven et al., 1983).

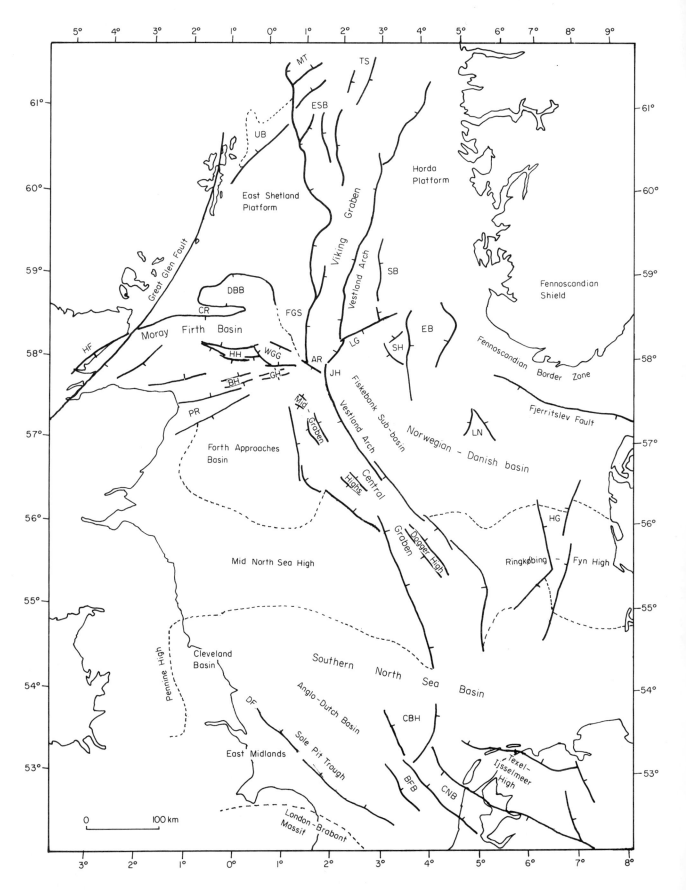

Fig. 6.1. Structural elements of the North Sea basins.
Abbreviations:

AR	Andrew Ridge	ESB	East Shetland Basin	LN	Lista Nose
BFB	Broad Fourteens Basin	FGS	Fladen Ground Spur	MT	Magnus Trough
BH	Buchan Horst	GH	Glenn Horst	PR	Peterhead Ridge
CBH	Cleaver Bank High	HF	Helmsdale Fault	SB	Stord Basin
CNB	Central Netherlands Basin	HG	Horn Graben	SH	Sele High
CR	Caithness Ridge	HH	Halibut Horst	TS	Tampen Spur
DBB	Dutch Bank Basin	JH	Jaeren High	UB	Unst Basin
DF	Dowsing Fault	LG	Ling Graben	WGG	Witch Ground Graben
EB	Egersund Basin				

'TETHYAN' SCHEME	ONSHORE ENGLAND SCHEME	PREFERRED 'BOREAL' SCHEME FOR NORTH SEA	DINOCYST ZONATION — ZONES	SUB-ZONES	AMMONITE ZONES
Berriasian	Ryazanian	Ryazanian (U)	*Dingodinium spinosum* V	ⅤA	*albidum*
				ⅤB	*stenomphalus*
				ⅤB	*icenii*
		(L)		ⅤC	*kochi*
				ⅤC	*runctoni*
?			*'Imbatodinium' villosum* VI (U)	ⅥA	*lamplughi*
	Portlandian	Volgian		ⅥB	*preplicomphalus*
				ⅥB	*primitivus*
?				ⅥC	*oppressus*
Tithonian			*Muderongia* sp. A VII	ⅦA	*anguiformis*
		(M)		ⅦB	*kerberus*
				ⅦB	*okusensis*
				ⅦB	*glaucolithus*
				ⅦB	*albani*
				ⅦC	*fittoni*
				ⅦC	*rotunda*
			Pareodinia mutabilis VIII	ⅧA	*pallasioides*
				ⅧB	*pectinatus*
	Kimmeridgian (U)		*Gonyaulacysta longicornis* IX	ⅨA	*hudlestoni*
	(L)			ⅨA	*wheatleyensis*
				ⅨA	*scitulus*
				ⅨA	*elegans*
?				ⅨB	*autissiodorensis*
					eudoxus
Kimmeridgian		Kimmeridgian	*Gonyaulacysta cladophora* X		*mutabilis*
					cymodoce
			Scriniodinium crystallinum (part) XI	ⅪA	*baylei*

Table 6.1. Approximate equivalence of stages around the Jurassic-Cretaceous boundary and an approximate correlation between the preferred North Sea stage nomenclature and dinoflagellate and ammonite zonal schemes. (Sources: Casey, 1973; Riley, 1977; Rawson *et al.*, 1978; Cope *et al.*, 1980b; Rawson and Riley, 1982; Ofstad, 1983.)

Early interpretations of the origin of the main graben system suggested a plume-generated uplift centred on the Middle Jurassic volcanic pile which lies at the intersection of the Moray Firth Basin, the Central Graben and the Viking Graben (Fig. 6.1). The basins in turn were regarded as arms of a failed triple junction (Whiteman *et al.*, 1975). More recent interpretations, emphasising the presence of anomalously thin continental crust under much of the main grabens (Christie and Sclater, 1980; Donato and Tully, 1981), have relied on an hypothesis of regional lithospheric stretching (after McKenzie, 1978), with the main phase of extension occurring in the mid-Jurassic (Wood and Barton, 1983). This model envisages stretching and fault-controlled subsidence throughout the Jurassic and into the mid-Cretaceous when, as extension ceased, thermally-induced subsidence became the dominant influence on basin formation ('sag' phase). Stratigraphic evidence points to broad regional uplift of the central North Sea during the mid-Jurassic (see section 6.4).

Over a large part of the central and northern North Sea, Jurassic strata suffered little structural change during this 'sag' phase. Therefore, structures associated with Jurassic reservoirs were generally in existence, in much their present form, long before the peak of hydrocarbon generation in the latest Cretaceous to early Tertiary. Post-Jurassic influences on present disposition are most marked in the southern North Sea where late Cretaceous basin inversion (Hancock, this volume 7.1.1) has resulted in significant uplift and erosion of Jurassic strata (see Fig. 2.13, 3.16 and Section 6.3).

In addition to tectonism (taphrogenesis), the second important influence on Jurassic sedimentation and stratigraphy was sea-level change. The Jurassic period was characterised by an overall, if pulsed rise in sea-level until the Kimmeridgian (Hallam 1978, 1981) followed by a net fall in the Volgian and into the Ryazanian. Rawson and Riley (1982) indicate a change from transgressive to regressive conditions in the North Sea at the close of the early Volgian. Vail and Todd (1981) and Rawson and Riley (1982) have emphasised the importance of sea-level changes in the formation of stratigraphic discontinuities within the North Sea Jurassic. However, the interplay between rates of basin subsidence and rates of sea-level change often remain difficult to disentangle.

6.2.1 Outline of stratigraphic evolution

The Lower Jurassic, consisting predominantly of marine, broadly transgressive, argillaceous deposits, has the most restricted distribution of the three Jurassic series in the North Sea basins. Strata of this age are absent or at best patchily distributed over the eastern Moray Firth Basin, much of the Central Graben, the south Viking Graben and the north-western part of the Norwegian-Danish Basin. The present distribution is related to the regional upwarping of the central North Sea and probable widespread erosion of Lower Jurassic

strata during the mid-Jurassic (see Section 6.4.2). Where best developed, away from this uplifted area, the Lower Jurassic section reaches c.900 m in the Sole Pit Trough (Kent, 1980) and a similar thickness in the eastern Norwegian-Danish Basin (Michelsen, 1978).

The boundary with the underlying Triassic is usually a conformable one, with the first indications of depositional conditions characteristic of the early Jurassic frequently seen in Rhaetian (uppermost Triassic) strata. The boundary with the overlying Middle Jurassic is variable, commonly an unconformity but of different magnitude in different parts of the North Sea.

The dominantly arenaceous Middle Jurassic strata, deposited in non-marine to paralic environments, are more widespread but remain rather poorly documented in much of the Central Graben. Sands were shed from the updomed central North Sea to accumulate in peripheral basins where they form important hydrocarbon reservoirs. Middle Jurassic sandstone sequences reach c.300 m in the north Viking Graben and at least 150 m in the western Moray Firth Basin. The pile of Middle Jurassic basaltic lavas at the intersection of the three main grabens exceeds 750 m in thickness (Howitt *et al.*, 1975).

Marine conditions persisted throughout the mid-Jurassic in the extreme north and south of the North Sea but their more widespread re-introduction, following the regressive, paralic facies of the Middle Jurassic, occurred at different times in different areas. Marine shales of Bathonian age rest on deltaic sandstones in the north Viking Graben for example, whereas marine conditions were not fully established in the Central Graben until the Oxfordian.

Upper Jurassic strata are marine throughout except for part of the Dutch southern North Sea (see Section 6.3.2). Although consisting of dark argillaceous deposits, including organic-rich shales, over most of the North Sea, the Upper Jurassic also includes important reservoir sands. These consist of transgressive, shallow-marine sediments, notably in the eastern Moray Firth, and fault-controlled sand and conglomerate deposits adjacent to basin margins or significant intra-basin 'highs' in the Central and Viking Grabens. The organic shale facies of the Upper Jurassic can be traced into the overlying Ryazanian, in places with no apparent stratigraphic break (see Section 6.9).

6.3 Southern North Sea

6.3.1 Structural elements of the Jurassic basins

The distribution of Jurassic strata is delimited, in much the same way as the Triassic, by the Mid North Sea—Ringkøbing-Fyn High (hereafter the Central Highs) to the north, to the west by the Pennine High and to the south by the London-Brabant Massif. A first-order sub-division of the area recognises (1) the Anglo-Dutch Basin, elements of which extend onshore into the U.K. and the Netherlands, and (2) the southern end of the North Sea Central Graben, projecting beyond the Central Highs into the Dutch offshore sector (terminology of Rhys, 1974). There was an important continuation of the Jurassic basin northwards through the Central Graben, but the Horn Graben, dissecting the Central Highs further east and a major site of Triassic sedimentation, may have been relatively inactive. Only thin patches of Upper Jurassic claystones have been encountered (Best *et al.*, 1983; Olsen, 1983).

The Anglo-Dutch Basin consists of a number of smaller, narrow NW-SE trending basins and ridges. Differential subsidence and, locally, uplift of these elements were important during the Jurassic, but the magnitude and timing of events seems to have varied across the area. Two additional tectonic factors are crucially important to a reconstruction of Jurassic history, namely the effect of halokinesis, with mobilisation of Permian salt from the Triassic onwards, and extensive post-Jurassic erosion, most notably that associated with late Cretaceous basin inversion. Erosion has effectively isolated the Jurassic strata now preserved in the U.K. part of the Anglo-Dutch Basin from the development in the Dutch sector.

6.3.2 Stratigraphy and facies

Lower-Middle Jurassic of the Anglo-Dutch Basin

Marine conditions characteristic of the early Jurassic were initiated in the Rhaetian. Rhaetian strata locally rest unconformably on the underlying continental Triassic (N.A.M. and R.G.D., 1980; Bodenhausen and Ott, 1981) but Ziegler (1982b), from a broader perspective, considers that a base Rhaetian ('Early Cimmerian') unconformity only occurs along the edges of stable 'highs'. Sedimentation across the Triassic-Jurassic boundary was continuous.

A predominantly argillaceous Lower Jurassic sequence, including organic-rich shales (Posidonia Shale Member of Toarcian age, in Dutch sector), occurs widely. There is evidence of regression at the top of the Lower Jurassic section in the Dutch area (N.A.M. and R.G.D., 1980) but the boundary with the overlying Middle Jurassic shallow-marine shales and sandstones is a conformable one. Bodenhausen and Ott (1981) point to evidence of regressive sands within the Middle Jurassic section of the Rijswijk Province, onshore Netherlands, which locally contain oil (Werkendam Sandstone Member). This mixed sandstone-shale sequence is Bajocian in age and is overlain conformably by a sequence of shallow-marine limestones and marls, dated as Bathonian to Oxfordian (see Fig. 6.2).

In contrast, the Middle Jurassic in the U.K. sector, consisting of marine shales with minor sandstones for the most part, rests unconformably on the Lower Jurassic. This discontinuity is associated with the regional uplift of the central North Sea. Laterally equivalent strata in the Cleveland Basin of Yorkshire consist of more arenaceous, broadly deltaic deposits (Hemingway, 1974) which have been compared to the Brent Group of the northern North Sea (Hancock and Fisher, 1981).

From a study of the Broad Fourteens Basin area, Oele *et al.* (1981) argue for little lateral variation in thickness and facies of Lower-Middle Jurassic strata, indicating relatively uniform subsidence. In sharp contrast, Kent (1980) demonstrates marked differences in early to mid-Jurassic subsidence between the Sole Pit Trough, the Cleveland Basin and the adjacent East Midlands Shelf.

Upper Jurassic of the Anglo-Dutch Basin

In the Dutch sector, sedimentation of shallow-marine marls and limestones was continuous from the mid-Jurassic to the late Oxfordian, when a phase of block-faulting and general uplift occurred. Although earlier, minor fault activity can be recognised, it was this late Oxfordian episode which gave greater individual identity to the NW-SE trending structures, including, most prominently, the uplifted Texel-Ijsselmeer High (Heybroek 1974; van Staalduinen *et al.*, 1979). This episode also markedly affected Jurassic stratigraphy, resulting in an unconformity between Kimmeridgian and older strata.

It should be noted that van Staalduinen *et al.* (1979) refer these late Oxfordian movements to the 'Late Kimmerian' (sic) tectonic phase whereas N.A.M. and R.G.D. (1980) include the same in a 'Mid-Kimmerian' phase. This illustrates the confusion which may arise from using terms which, as Ziegler (1982b) points out, do not correspond to discrete tectonic events. This terminology for tectonic phases, derived from Stille (1924), is in common use in North Sea literature and is used even in cases where sea-level changes rather than tectonic events may be the dominant control (see critique by Kent, 1975, and Badley in discussion of Ziegler, 1975).

Sedimentation in the Dutch sector resumed in the Kimmeridgian, in the narrow NW-SE basins least affected by uplift. Paralic sandstones (Delfland Group) form the remainder of the Upper Jurassic section and extend into the Ryazanian. Roos and Smit (1983) record uplift and erosion of the margins of the Broad Fourteens Basin at the end of the Jurassic, whereas N.A.M. and R.G.D. (1980) place the peak of uplift and erosion in the late Ryazanian-Early Valanginian. Ziegler (1980b) also records intra-Ryazanian erosion but attributes this to emergence of certain areas during regression.

Marine argillaceous sediments, including the Kimmeridge Clay Formation, dominate the Upper Jurassic of the U.K. sector. The sequence is conformable, lacking the sub-Kimmeridgian unconformity described above. Kent (1980) however, demonstrates a change in subsidence pattern in the late Jurassic, recording thicker sequences over the East Midland Shelf relative to the adjacent troughs, and argues for a degree of basin inversion. This change seems to have been initiated in the late Oxfordian. In some offshore sequences

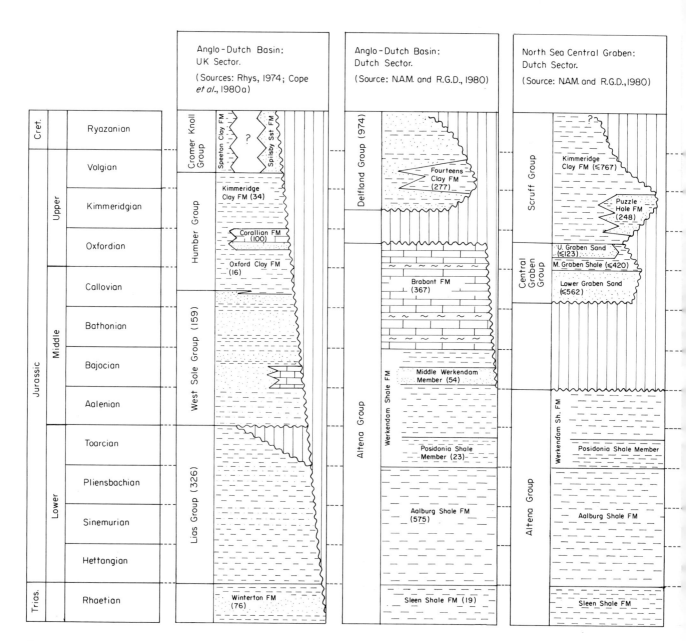

Fig. 6.2. Generalised outline of Jurassic lithostratigraphy in the Anglo-Dutch Basin, Central Graben and Norwegian-Danish Basin (For Key see Fig. 6.3).

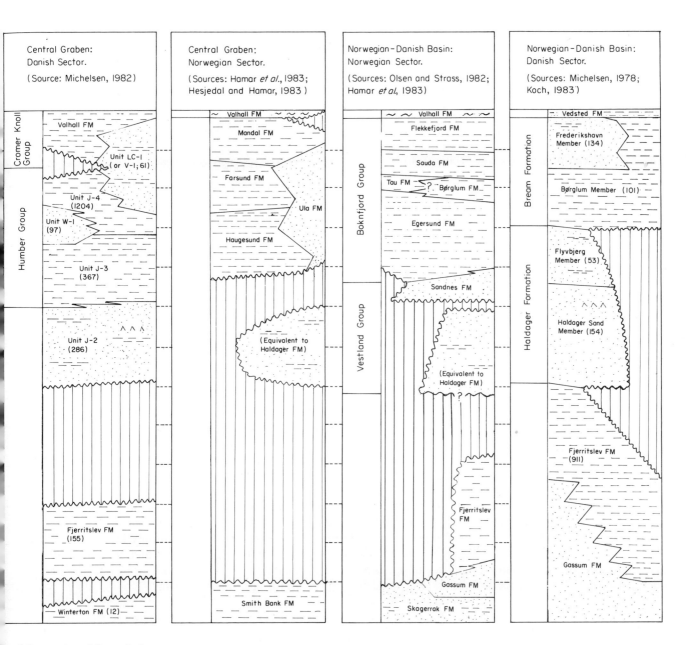

Fig. 6.2. *continued from facing page*

the topmost Jurassic is represented by a sandstone unit that persists into the Lower Cretaceous (Cope *et al.*, 1980a). More often however, less complete sequences occur capped by an unconformity which records erosion in the latest Jurassic-earliest Cretaceous or, often more significantly, following late Cretaceous inversion.

Jurassic of the southern Central Graben

The Jurassic succession in the southern extension of the North Sea Central Graben into the Dutch area differs from that in the Anglo-Dutch Basin. Paralic sandstones and shales of Callovian-Oxfordian age rest directly on Lower Jurassic marine shales. The remainder of the succession consists largely of the marine, argillaceous Kimmeridge Clay Formation but with a tongue of Kimmeridgian sands developed in the south. The sands pass to shales northwards and reflect the northward limit of the paralic conditions more evident in the Dutch part of the Anglo-Dutch Basin during the late Jurassic. The sands are therefore in part laterally equivalent to the Delfland Group. The Kimmeridge Clay Formation is overlain unconformably by Lower Cretaceous strata (see N.A.M. and R.G.D., 1980).

6.3.3 Economic geology

The Jurassic has not been a major exploration target in the southern North Sea to date, its principal importance lying in the source potential of its shales. The Posidonia Shale is the main source of oil found in Lower Cretaceous reservoirs in the Netherlands (Bodenhausen and Ott, 1981). This unit is also found in the Dutch offshore and laterally equivalent organic-rich shales are present in the U.K. area (e.g. the Jet Rock of the Cleveland Basin). (Barnard and Cooper (1983) have shown that Lower and Middle Jurassic shales are locally mature for oil generation in parts of the Cleveland Basin.).

Shales of the Upper Jurassic Kimmeridge Clay Formation also have source potential, but are likely to be immature except perhaps in the southern Central Graben. Bodenhausen and Ott (1981) suggest that non-marine shales in the Upper Jurassic Delfland Group may be the source of high wax crudes in Dutch onshore fields.

The Posidonia Shale is thought by Bodenhausen and Ott (1981) to have reached maturity in the northern part of the Rijswijk Province during the Cretaceous, prior to basin inversion. Any oil reaching pre-inversion traps however, probably had a low preservation potential. Post-inversion burial may locally have been sufficient to bring the shales elsewhere in the Dutch area into the oil-generating 'kitchen'.

Sandstones of the Upper Jurassic Delfland Group and Middle Jurassic Werkendam Formation are oil-bearing in the Rijswijk Province, onshore Netherlands. Ziegler (1980) also reports onshore Jurassic oil reservoirs in the Gifhorn Trough and Lower Saxony Basin of Germany. The best potential offshore in terms of

lithological character also lies in the Delfland Group. The Middle Jurassic deltaic sands of the onshore Cleveland Basin extend only a short way offshore into the U.K. sector (Hancock and Fisher, 1981).

6.4 Central Graben

6.4.1 Structural elements

The Central Graben, a broad trough bounded by normal, rotational and often en echelon faults, has a N-S trend where it breaches the Central Highs but swings to the NW-SE when followed northwards. The Graben loses its identity south of the Central Highs, towards the Dutch coast (see Section 6.3), and terminates in the north where a thick pile of Middle Jurassic volcanic rocks lies at its intersection with the Viking Graben and eastern Moray Firth Basin.

To the north of the Central Highs, the Central Graben is bounded to the east by the Vestland Arch, a complex of asymmetric, rotated fault-blocks which dip off eastward into the Norwegian-Danish Basin. On the western flank of the Graben there is a broad shelf coincident with the late Palaeozoic Forth Approaches Basin. Jurassic strata are thin or absent over these marginal structures.

There is a series of prominent but narrow mid-basin 'highs' within the Graben which parallel its axis. These have various origins including upfaulted basement blocks and late Jurassic basin inversion features (Skjerven *et al.*, 1983). Salt movement has occurred widely in the Central Graben, influencing local Jurassic subsidence both through uplift and salt withdrawal, and contributing to the formation of prospective structures within the Jurassic section.

The late Jurassic in the Central Graben was characterised by major fault-controlled subsidence. Hansen and Mikkelsen (1983) demonstrate that rates of subsidence in the Danish sector at this time were only exceeded at one other period in basin history, during the Neogene.

6.4.2 Stratigraphy and facies

Lower-Middle Jurassic

On presently available data, Lower Jurassic deposits have a very restricted distribution in this basin. They are best documented in the Danish sector, adjacent to the Ringkøbing-Fyn High, where a sequence of marine shales is recorded resting on Rhaetian claystones (Koch *et al.*, in Michelsen, 1982; see also Fig. 6.2).

The lack of Lower Jurassic strata results from up-doming and erosion of the Central North Sea initiated, according to Eynon (1981), during the late Toarcian. Ziegler (1982b) postulates that the dome encompassed an area stretching from the Ringkøbing-Fyn High to the Pennine High, a distance of c.600 km, and from the northern end of the present Central Graben southwards for c.450 km, with a maximum structural relief of 2,000-3,000 m. Leeder (1983) argues for a less dramatic uplift, of the order of c.60 m (see also Kent in discussion of Ziegler 1982a). The Sinemurian palaeogeographic map presented by Skarpnes *et al.* (1980) appears to suggest the alternative hypothesis that the central North Sea uplift began earlier, influenced early Jurassic sedimentation and caused local emergence and the development of a shoreline facies. The possibility of more Lower Jurassic strata being found in the Central Graben, preserved in fault-bounded 'lows', cannot be discounted (cf. Fig. 2.1).

The Bajocian-Bathonian stages are represented by a heterolithic paralic sequence, with coals, in the Danish sector of the Central Graben, resting unconformably on Pliensbachian shales (Koch *et al.*, in Michelsen, 1982). Hamar *et al.* (1983) record sandstones of similar age from the Norwegian sector, resting directly on Triassic. In the U.K. Central Graben in addition to the Middle Jurassic volcanics in the north (see Section 6.6.2), there is some evidence elsewhere of non-marine to paralic sand-dominated deposits of rather uncertain age. The reported occurrences of Bajocian-Bathonian strata within the Central Graben appear at variance with the corresponding palaeogeographic map of Ziegler (1982b) which suggests no significant deposition within the area of the dome. A minor basin within the central North Sea dome is shown in an earlier version of this map (Ziegler, 1981). It seems likely that fault-controlled subsidence was sufficient, at least locally, to receive Middle Jurassic deposits.

Open marine conditions returned in the Callovian, recorded by shales in the Danish and Norwegian areas. In the U.K. sector the first marine shales are probably of Oxfordian age. Marine transgressive sands may be present in places but non-marine, coal-bearing strata remain characteristic of the Callovian, where it can be differentiated, in the U.K. area.

Upper Jurassic

Marine conditions persisted from the Oxfordian to the end of the Jurassic. Argillaceous deposits dominate the Upper Jurassic succession, including the organic-rich hydrocarbon source rocks of the Kimmeridge Clay Formation (terminology of Deegan and Scull, 1977). These shales, commonly highly radioactive ('hot'), form the upper Oxfordian to Volgian stages and indeed frequently persist, with little or no stratigraphic break, into the Ryazanian. In the Norwegian sector, equivalent shales are assigned to the Farsund Formation (Kimmeridgian-Volgian) and the Mandal Formation (Volgian-Ryazanian), the upper unit distinguished on the basis of its higher gamma ray log response and lower density (Hamar *et al.*, 1983).

Contemporaneous with shale deposition, sands accumulated along both margins of the Central Graben and also adjacent to some intra-basin 'highs' (see for example Koch *et al.*, in Michelsen, 1982). Sands of the Fulmar Formation on the western margin and the Ula Formation (Hamar *et al.*, 1983) on the eastern side are

important reservoir units. The Ula sands have been interpreted as shallow-marine, tidal deposits by Bailey *et al.* (1981) but gravity-flow sediments may also be expected in other places. The deposition of these sand units was closely controlled by synchronous movement on adjacent faults.

Upper Jurassic strata are thinly and discontinuously developed over the 'highs' bounding the Central Graben. Over the shelf to the west a basal transgressive sand is commonly overlain by shales of the Kimmeridge Clay Formation. The Vestland Arch to the east probably acted as at least a partial barrier between the Central Graben and Norwegian-Danish Basin during much of the late Jurassic. However, during the Volgian-Ryazanian, despite a tendency towards regression regionally (Rawson and Riley, 1982), a thin veneer of marine shales was deposited over the 'highs'. Subsidence at this time seems to have outpaced sea-level fall.

6.4.3 Economic geology

The upper Oxfordian-Ryazanian organic-rich shales of the Kimmeridge Clay Formation and its lateral equivalents play the principal role in hydrocarbon generation in the Central Graben, forming a mature oil source over much of the area. In a more detailed discussion of source characteristics in the Danish Central Graben, Lindgreen *et al.* (in Michelsen, 1982) conclude that significant variations exist in the amount of organic matter and its degree of maturity. The depth of burial of Upper Jurassic shales across the North Sea can be seen in the 'Base Cretaceous' depth map of Day *et al.* (1981; see also Fig. 9.18) which shows that Upper Jurassic shales are buried deeper than 3,000 m within the Central Graben but less on the flanking 'highs'. For oil generation in this area a minimum depth of burial of c.3,000 m is required. Therefore Upper Jurassic shales on 'highs' bordering the Graben are immature.

Oil-bearing Upper Jurassic sands are developed along both graben margins, for example, in the Fulmar (U.K. Block 30/16), Clyde (U.K. Block 30/17) and Ula (Norwegian Block 7/12) Fields. The extent and interconnection of sand bodies along the graben edge faults, whether isolated lobes or more laterally continuous fringing developments, and the nature of sand distribution away from the graben margins remains to be fully determined, making the Central Graben one of the main areas of current exploration interest.

Of traps which may be available at Jurassic level in the Central Graben, the following can be cited, viz. (1) sub-conformity traps at crests of tilted fault-blocks, (2) anticlinal closures over deeper fault- or salt-induced 'highs' and (3) stratigraphic trapping at lateral facies changes from Upper Jurassic proximal sands to distal shales (see also Koch, in Michelsen, 1982). Bailey *et al.* (1981) describe the Ula Field structure as an elongate dome at top Jurassic level, formed over a Zechstein salt swell. Upper Jurassic shales provide the seal. Gibbs

(1983a) describes the Clyde Field as a structural 'high' developed over the toe of an underlying, shallow dipping, listric fault which soles out in the deeper Zechstein salt. It remains to be seen whether this interpretation, which contrasts with the conventional view of tilted fault-block terraces in the area, can be sustained and provide a valid model for use in exploration elsewhere in the North Sea grabens.

6.5 Norwegian-Danish Basin

6.5.1 Structural elements

The Norwegian-Danish Basin lies to the east of the Central and Viking Grabens. Separated from the Grabens by the fault-blocks of the Vestland Arch, it is bounded to the south by the Ringkøbing-Fyn High, to the east and north-east by the Fennoscandian Shield and to the north by the Horda Platform. The latter element is a shallow basement feature, most prominent during the Triassic, which subsided locally in the late Jurassic to form the N-S trending Stord Basin as a northward extension of the Norwegian-Danish Basin Rønnevik *et al.*, 1975).

The Norwegian-Danish Basin can be divided into north-west and south-east parts, the Norwegian and Danish Sub-basins respectively, at the Lista Nose, a SSE-plunging structural 'high'. Michelsen (1979) emphasises the importance of this feature during the Jurassic in separating areas of contrasting tectonic and depositional history.

Differential fault-controlled subsidence especially during the late Jurassic, commonly occurred along pre-established fault lines and caused further differentiation into smaller sub-basins and intervening 'highs' whose definition and nomenclature are outlined by Rønnevik *et al.* (1975) and Hamar *et al.* (1980). Notable among these elements are the Fiskebank Sub-basin, which occupies much of the Norwegian sector east of the Vestland Arch, and the Egersund Sub-basin, a N-S trending basin lying further to the north-east (see Fig. 6.1). The Egersund Sub-basin, which had its main development during the late Jurassic, was also the site of an early Jurassic dyke intrusion and of early to mid-Jurassic volcanism (Furnes *et al.*, 1982).

The Danish Sub-basin is crossed by the WNW-ESE Fjerritslev Fault which influences shifts in the locus of maximum subsidence during the Jurassic (Michelsen, 1978). Two further factors should be noted, firstly the widespread influence of Zechstein salt movement in the Norwegian-Danish Basin on local subsidence and stratigraphy, and secondly the reference to as yet little documented occurrences of strike-slip fault displacements by Skjerven *et al.* (1983).

6.5.2 Stratigraphy and facies

Lower-Middle Jurassic

The thickest and stratigraphically most complete Jurassic succession in the Norwegian-Danish Basin

occurs in the Danish Sub-basin, where c.1,200 m of Jurassic strata are developed in places (Michelsen, 1978). Lower Jurassic strata, though probably once widely distributed, are now only found as local erosional remnants in the Norwegian sector. Middle Jurassic strata mostly rest unconformably on Lower Jurassic or older beds in the Norwegian area.

The Triassic-Jurassic boundary where preserved is conformable. It lies in a transition from fluvio-deltaic to shallow-marine deposits within the Gassum Formation (Rhaetian-Pliensbachian). The boundary between the Gassum sands and the overlying marine shales of the Fjerritslev Formation (Hettangian-Aalenian) is diachronous, younging to the north-east and reflecting the transgressive nature of the sequence (Michelsen, 1978; see also Bertelsen, 1978; Skarpnes *et al.*, 1980).

Lower Jurassic strata in the Norwegian area tend to be preserved only in downfaulted 'lows'. As well as marine shale, Olsen and Strass (1982) report volcanic material of Lower-Middle Jurassic age from the Egersund Sub-basin, locally reaching 800 m in thickness.

The Middle Jurassic sands of the Danish Sub-basin, assigned to the Haldager Formation (Bajocian-Oxfordian) by Michelsen (1978), have been interpreted by Koch (1983) as a braided alluvial plain succession (Haldager Sand Member, Bajocian-Callovian) overlain by meandering floodplain deposits (Flyvberg Member, Callovian-Oxfordian). Michelsen (1978) interpreted the upper Member as an interdistributary bay or tidal flat deposit. In the north-east, towards the Fennoscandian Border Zone, both units pass laterally to an undifferentiated coal-measures succession. Koch (1983) describes floodplain deposition persisting until the late Oxfordian when marine conditions were re-established

In the Norwegian Sub-basin, Olsen and Strass (1982) and Hamar *et al.* (1983) record an upward passage from a non-marine, broadly deltaic sequence of mostly Bajocian-Bathonian age to shallow-marine Callovian sandstones. The marine deposits rest unconformably on older strata. Marine conditions persisted with a siltstone and shale sequence spanning the Callovian-Oxfordian boundary. The lowermost unit, laterally equivalent to the Haldager Formation of the Danish sector, has been interpreted by Dypvik and Vollset (1980) as the product of small prograding delta lobes building out from intra-basinal 'highs' which acted as sediment source areas.

The lithostratigraphy of the Jurassic in the Norwegian-Danish Basin is summarised in Fig. 6.2, but the Norwegian scheme should, in advance of the findings of the Norwegian Nomenclature working party, be regarded as provisional. It is an attempt to marry recently published schemes, which claim to anticipate the working party's proposals (Olsen and Strass, 1982; Hamar *et al.*, 1983).

Upper Jurassic

Following the deposition of the Flyvberg Member, marine conditions prevailed throughout the Danish

Sub-Basin during the remainder of the Jurassic. Michelsen (1978) records a shale sequence (Børglum Member, Kimmeridgian-Volgian) overlain by nearer-shore siltstones and sandstones (Frederikshavn Member, Volgian-Ryazanian). The coarser sediment was derived from the Fennoscandian Border Zone and passes laterally, towards the south-west, to age-equivalent shales (Koch, 1983).

Hamar *et al.* (1983) show that the Upper Jurassic in the Norwegian area is dominantly marine, often anaerobic and sometimes radioactive, shales. Sands may occur in a narrow zone fringing the basin to the east and north-east. Sands are also reported by Olsen and Strass (1982) along the southern margin of the Horda Platform, in the Ling Graben. Deposition of marine shales continued into the Ryazanian without any major stratigraphic break. Radioactive shales within the Norwegian-Danish Basin are best developed in the Norwegian sector where they are assigned by Olsen and Strass (1982) to the Tau Formation (Kimmeridgian-lower Volgian).

Most structural 'highs' within the Norwegian-Danish Basin were submerged during the late Jurassic. Marine communication with the Central and Viking Grabens was established at least locally.

Olsen and Strass (1982) assign Middle Jurassic sands in the Norwegian Sub-basin to the Vestland Group and Upper Jurassic-lowermost Cretaceous shales to the Boknfjord Group. Sub-units within the latter are picked largely on their degree of radioactivity, as shown by gamma ray log responses, and the amount of silt present in these predominantly argillaceous strata. Analysis of the nature of the organic matter within the Boknfjord Group shales also reveals differences, both in total organic carbon content and in the nature of the kerogen. Variations in the amount of terrestrially derived kerogen aid palaeogeographic reconstruction but also have a crucial bearing on the likely hydrocarbon product following kerogen maturation.

6.5.3 Economic geology

The best reservoir sands within the Norwegian-Danish Basin belong to the Middle Jurassic (Haldager Formation and its equivalents). The Gassum Formation may also have favourable characteristics in places. Oil has been found in Middle Jurassic sands within the small Bream and Brisling Fields (both in Norwegian Block 7/12).

The Bream Field is located over a salt swell within the Egersund Basin, whereas the Brisling Field lies to the west, on the flank of a bounding fault block (the Sele High). The source of the oil is thought to be Upper Jurassic shales which, although immature in most well sections in the area, are more deeply buried in the southern, axial part of the Egersund Sub-basin.

Analyses of source rock potential in the Norwegian Sub-basin given by Olsen and Strass (1982) indicate that the Kimmeridgian-lower Volgian 'hot' shales are rich source rocks for oil (see also Al-Kasim *et al.* 1975). Other Upper Jurassic shales, rich in terrestrial organic

matter, may contribute waxy oils and gas. In general, however, the hydrocarbon potential of the Norwegian-Danish Basin as a whole is relatively low on presently available evidence, due largely to insufficient burial of local source rocks over much of the area, and perhaps also to a deterioration in the source-rock characteristics of Upper Jurassic sediments south-eastwards into the Danish Sub-basin (Ziegler, 1980).

6.6 Moray Firth Basin

6.6.1 Structural elements

The Moray Firth area stretches from the intersection of the Central and Viking Graben in the east, westwards to the Scottish coast where Jurassic strata crop out onshore. The northern boundary of the Jurassic basin is formed in part by the Caithness Ridge, an E-W trending fault-bounded 'high', but further east a large embayment in the East Shetland Platform, called the Dutch Bank Basin, extends the basin area northwards (Fig. 6.1). Both Ridge and Platform consist of Devonian rocks. The southern margin of the basin is formed by the Peterhead Ridge, a NE-SW trending basement 'high' and, further east, by the Buchan and Glenn Horsts, similarly-trending Devonian blocks. In the east, the Jurassic basin is narrowed by the southward projection of another Devonian 'high', the Fladen Ground Spur. Within the Moray Firth area the most prominent structural element is the E-W Halibut Horst, a persistent up-faulted block of mostly Devonian rocks with a patchy Upper Cretaceous chalk cover.

There is a fundamental distinction to be made between the eastern Moray Firth Basin, in particular the NW-SE trending Witch Ground Graben, which is underlain by anomalously thin continental crust (Christie and Sclater, 1980) and the western or Inner basin where a more normal thickness occurs. Interpreting the origin of the Inner Moray Firth Basin, McQuillin *et al.* (1982) envisage 8 km of dextral transcurrent movement along the NE-SW Great Glen Fault, mostly during the Jurassic. This displacement, accomplished by the adjustment of rigid, regional crustal blocks in the central and northern North Sea to varying extension in the Viking and Central Grabens, led in turn to limited extension and basin subsidence in the western Moray Firth.

Fault trends within the Moray Firth area also show differences; a NE-SW alignment is dominant in the extreme west, changing through E-W to NW-SE in the east. Under the hypothesis of McQuillin *et al.* (1982), the Witch Ground Graben can be considered as a northward continuation of the Central Graben (see also Johnson and Dingwall 1981; Threlfall 1981).

Within the Inner Moray Firth Basin, Lower-Middle Jurassic strata are broken up by a large number of relatively small normal faults, many of which terminate upwards in the Upper Jurassic section. Later, larger scale movement occurred along fewer faults, contemporaneous with late Jurassic sedimentation.

These faults define major half-grabens in the Inner basin (Chesher and Bacon, 1975; Chesher and Lawson, 1983). From the distribution of Middle Jurassic volcanic strata in the eastern area, Woodhall and Knox (1979) demonstrate the importance of fault-controlled subsidence during the Bajocian-Bathonian in the development of the Witch Ground Graben. Major half-grabenal features are much less evident in the east.

6.6.2 Stratigraphy and facies

Onshore succession

Onshore outcrop provides a useful window on the subsurface development of Jurassic strata in the Inner Moray Firth Basin. The Lower-Middle Jurassic can be interpreted broadly as a transgressive-regressive-transgressive sequence on the evidence of sedimentary facies. Shallow-marine, stable-shelf sedimentation continued into the Oxfordian, but the re-activation of the nearby NE-SW Helmsdale Fault, led to the deposition of marine gravity-flow sandstones and 'boulder beds' during the Kimmeridgian and Volgian.

The Lower Jurassic Dunrobin Bay Formation (Hettangian-Pliensbachian), resting locally on a thin (?) Rhaetian conglomerate, consists of shales with minor sandstones and records an upward transition from non-marine to open-marine conditions. The Toarcian-Bajocian is unrepresented in onshore exposures; the next unit exposed, the Brora Coal Formation, is dated as Bathonian. The latter is a sequence of interbedded sandstones and shales, capped by a laterally persistent coal (the Brora Coal), and probably represents an estuarine or coastal plain environment with infrequent marine incursions. The Callovian Brora Argillaceous Formation, resting on a minor erosion surface, consists of glauconitic, fossiliferous shales and subordinate sandstones and records a return to fully marine conditions. Similar conditions persisted into the Oxfordian with the deposition of the Brora Arenaceous Formation, a unit of shallow-marine bar sands. This is succeeded locally by the middle Oxfordian Ardassie Limestone, more strictly calcareous sandstone notably rich in spicules from the sponge Rhaxella.

The youngest Jurassic cropping out onshore are Kimmeridgian-middle Volgian strata (Lam and Porter, 1977). They consist mostly of the Helmsdale Boulder Beds (Linsley, 1972) in which coarse breccias, conglomerates and sandstones, deposited by diverse gravity flow processes, are interbedded with dark marine shales.

The brief summary of this much studied succession can be supplemented by reference to Lee (1925), Berridge and Ivimey-Cook (1967), Neves and Selley (1975) and Sykes (1975).

Lower-Middle Jurassic offshore

The Inner Moray Firth Basin has the more complete Jurassic succession, as Lower Jurassic strata appear to be absent further east. This can once more be attributed to upwarping of the central North Sea.

The succession described by Linsley *et al.* (1980) from the Beatrice Field (U.K. Block 11/30) compares closely with the onshore Lower-Middle Jurassic. A Rhaetian-Lower Jurassic transgressive sequence, consisting of shales with minor sandstones, rests on a distinctive, thin cherty limestone of Triassic age. Interpreted as a fossil soil horizon, this widely distributed unit is probably responsible for the strong 'base Jurassic' seismic reflection over much of the Inner basin. A coarsening-upward, sandy sequence formed by progradation in a broadly deltaic setting, represents the Toarcian-Bajocian and is succeeded by a heterolithic Bathonian unit, capped by a coal (Brora Coal). These alluvial plain sediments are overlain in turn by marine Callovian strata but of more arenaceous character than the age-equivalent deposits onshore. The shallow-marine sandstones of the Callovian provide the main reservoir in the Beatrice Field. There is a rapid change to marine shales in the uppermost Callovian of Block 11/30, with shales then characterising the remainder of the Jurassic succession. The top of the shallow-marine sandstone facies, represented by deposits as young as Oxfordian onshore and by Callovian in the Beatrice Field, is therefore strongly diachronous in the Inner Moray Firth Basin (see Fig. 6.3).

The Lower-Middle Jurassic sequence in the Beatrice Field is c.300 m thick and may be thicker off structure. In general, however, this part of the Jurassic sequence in the Inner Moray Firth is characterised by lateral uniformity in thickness and also in gross lithological character.

Middle Jurassic strata in the eastern Moray Firth rest unconformably on Triassic continental sediments, and consist of basaltic lavas and tuffs (Fall *et al.*, 1982), the products of subaerial eruptions. These volcanic rocks have been dated as Bajocian-Bathonian (Howitt *et al.*, 1975; Deegan and Scull 1977) but the age of the oldest flows remains somewhat uncertain. The volcanic pile is thickest and dominantly of lava flows between the Forties Field (U.K. Block 21/10) and the Piper Field (U.K. Block 15/17) and passes to the north and west into a sequence of interbedded volcaniclastic sediment, tuffs, some coal seams and infrequent lava flows (Woodhall and Knox 1979). Volcanic activity appears to have ceased by the Callovian which is represented by a thin, patchy development of sandstones, shales and coals.

Upper Jurassic offshore

Upper Jurassic strata in the Beatrice Field consist predominantly of shales, with only minor beds of sandstone interpreted by Linsley *et al.* (1980) as distal submarine-fan deposits. The coarse clastic facies found onshore is absent. Upper Oxfordian-Volgian shales (equivalent to the Kimmeridge Clay Formation) are enriched in organic matter relative to the underlying Callovian-Oxfordian shales. However, in contrast to

Upper Jurassic organic-rich shales elsewhere in the central and northern North Sea, much of the Beatrice section lacks high radioactivity, (i.e. has gamma-ray log responses of less than 100 API units). Thin 'hot' intervals do occur and can be useful in local correlations.

Vail and Todd (1981) recognise a number of minor stratigraphic discontinuities within the Upper Jurassic of the Inner Moray Firth from reflection terminations on seismic sections. In addition, they equate a strong intra-Ryazanian seismic event, marked by overlying onlap, with the 'Late Cimmerian' unconformity. Mapping of certain seismic discontinuities on a regional scale allows definition of wedge-shaped packages of Upper Jurassic sediment thickening towards major NE-SW trending faults. Comparison of the Beatrice Field succession with the Upper Jurassic onshore reveals significant lateral changes in sediment facies, in large part related to the supply of coarse clastics from an active fault scarp to a marine basin. Comparable processes associated with major intra-basinal faults may have produced other heterolithic wedges of Upper Jurassic sediment elsewhere in the basin.

In the eastern Moray Firth, the Upper Jurassic consists largely of two widely distributed units, namely the shallow-marine sandstones of the Piper Formation (Williams *et al.* 1975; Maher 1980) and the overlying organic-rich marine shales of the Kimmeridge Clay Formation. The shales are partly replaced adjacent to certain intra-basin 'highs' by sands, mostly of Volgian age, deposited by gravity flow processes.

6.6.3 Economic geology

Proven reservoir sands occur at a number of horizons in the Moray Firth Basin. The pay zone in the Beatrice Field spans the Sinemurian to Callovian stages, whereas Oxfordian-Kimmeridgian sands are productive in the Piper Field (see Fig. 6.4 and Table 6.2) and Tartan Field (U.K. Block 15/16). Volgian sands also have reservoir qualities in the eastern area. Of prospects for further discoveries, the Upper Jurassic deposits in half-grabenal settings within the Inner Basin may yield significant sand deposits adjacent to major faults which pass laterally, updip across tilted fault-blocks, to shales.

The Tartan and Piper oil accumulations are associated with tilted, fault-bounded blocks (Fig. 6.5). Linsley *et al.* (1980) describes the Beatrice structure as a faulted anticline draped over a deeper fault-block. Upper Jurassic shales seal the reservoirs in all cases except in the Piper Field where the reservoir was finally sealed by the deposition of Campanian marls.

The Kimmeridge Clay Formation is the generally accepted source of oil in the eastern Moray Firth. Fisher and Miles (1983) present data on the likely hydrocarbon product from Upper Jurassic shales in the area based on kerogen composition and maturation. They report that samples from most well sections proved to be at an early mature state for oil generation but predict optimum conditions off-structure, e.g. near the axis of the Witch Ground Graben. It is noted that Fisher and Miles (1983) also predict condensate generation locally, providing a source for condensate in Lower Cretaceous reservoirs south of the Fladen Ground Spur (Brooks, 1977).

The source of oil in the Beatrice Field is more controversial. The organic-rich Upper Jurassic shales in the Inner Moray Firth tend to lie at depths of 1,000-3,000 m in contrast to the eastern basin where the shales commonly lie below 3,000 m. McQuillin *et al.* (1982), however, argue for c.1,000 m of uplift of the Inner Moray Firth Basin, with accompanying erosion, during the Tertiary. Pearson and Watkins (1983), from a study of maturity levels of the organic-rich shales, suggest possible uplift of the basin by 500-700 m locally. The Upper Jurassic shales in general, however, are seen to be less mature than equivalent strata in the Witch Ground Graben. Pearson and Watkins (1983) indicate that if oil generation has been achieved in the Inner Basin from Upper Jurassic shales the volumes involved will be small. Furthermore, it is suggested that the Upper Jurassic shales are an unlikely source of a high wax crude such as found in the Beatrice Field (in discussion of Pearson and Watkins, 1983). Possible source rocks may also be found in the shales of the Lower-Middle Jurassic, intimately associated with the Beatrice reservoir sands (Barnard and Cooper, 1981). The uplift of the basin during the Tertiary may also affect any oil once held in pre-Tertiary structures, either through leakage updip to the west or by the influx of freshwater with resultant biodegradation of the hydrocarbons in place.

6.7 Viking Graben

6.7.1 Structural elements

The Viking Graben is a N-S trending linear trough straddling the boundary between the Norwegian and U.K. sectors of the northern North Sea. The East Shetland Platform, with Tertiary strata resting directly on Devonian red beds for the most part, lies to the west of the Graben. To the east is the Vestland Arch, a narrow fault-bounded ridge, with a thin Jurassic cover, which separates the Viking Graben from the Stord Basin (Fig. 6.1).

The axis of the Viking Graben shifts northwards to a NNE-SSW trend diverging from the edge of the East Shetland Platform with the development of an intermediate area of Mesozoic sedimentation termed the East Shetland Basin (Fig. 10.3, Section 1 cross-section from Ziegler, 1982, Fig. 9.1). The East Shetland Basin, the site of the Brent oil province, is characterised at Jurassic level by a complex of tilted fault-bounded blocks whose differential subsidence played a crucial role in trap construction. Both NE-SW and more nearly N-S faults cut Jurassic strata, the former paralleling an old, Caledonide tectonic trend. The NE-SW structures tend to occur in the north-western part of the East Shetland Basin where they define the 'deep' Magnus

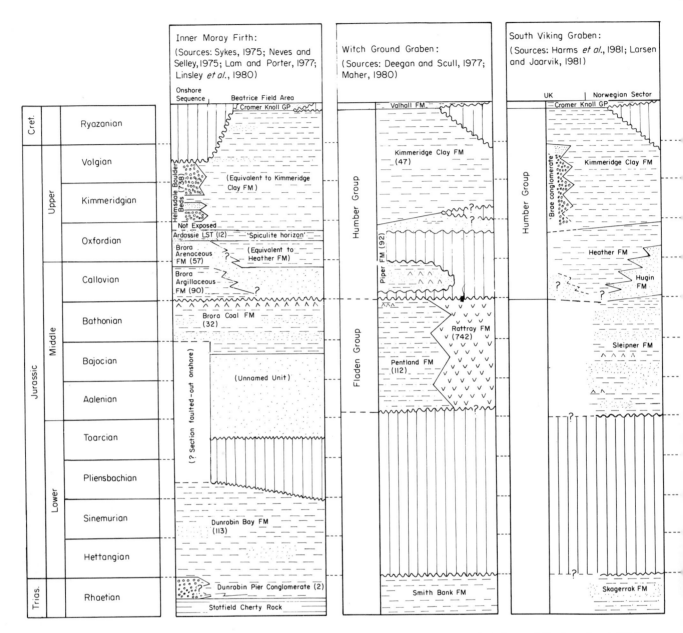

Fig. 6.3. Generalised outline of Jurassic lithostratigraphy in the Inner Moray Firth Basin, Witch Ground Graben, South Viking Graben and East Shetland Basin.

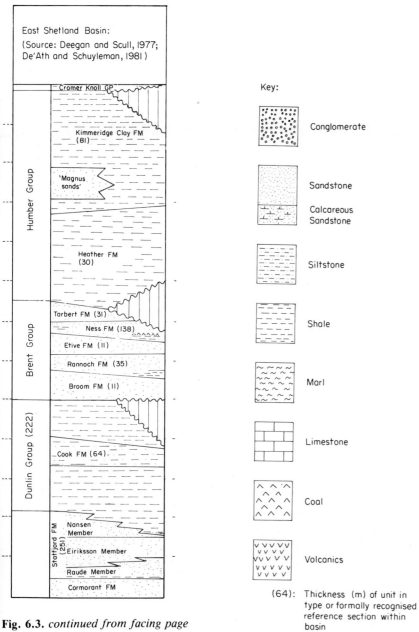

East Shetland Basin:
(Source: Deegan and Scull, 1977;
De'Ath and Schuyleman, 1981)

Cromer Knoll GP

Kimmeridge Clay FM (81)

'Magnus sands'

Heather FM (30)

Humber Group

Tarbert FM (31)

Ness FM (138)

Etive FM (11)

Rannoch FM (35)

Broom FM (11)

Brent Group

Cook FM (64)

Dunlin Group (222)

Nansen Member

Eiriksson Member

Raude Member

Statfjord FM (251)

Cormorant FM

Key:

Conglomerate

Sandstone

Calcareous Sandstone

Siltstone

Shale

Marl

Limestone

Coal

Volcanics

(64):   Thickness (m) of unit in type or formally recognised reference section within basin

Fig. 6.3. *continued from facing page*

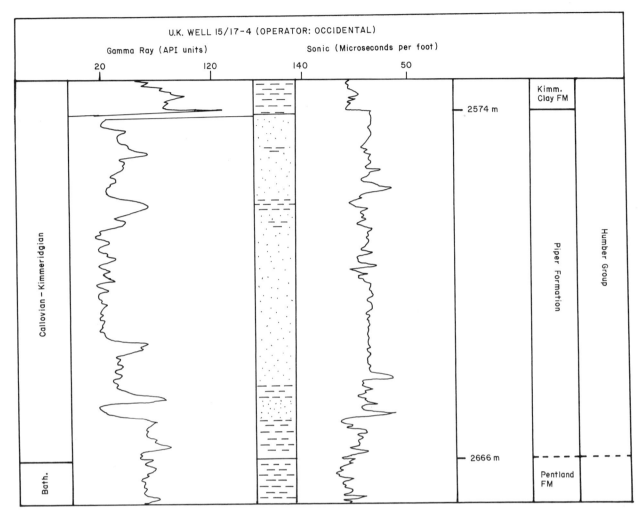

Fig. 6.4. Typical well log through the Piper Formation, eastern Moray Firth Basin (for Key see Fig. 6.3).

Table 6.2. Some reservoir parameters from North Sea Jurassic fields. (Sources: 1. Gray and Barnes, 1981; 2. Nadir and Hay, 1978; 3. De'Ath and Schuyleman, 1981; 4. Brekke *et al.*, 1981; 5. Larsen and Jaarvik, 1981; 6. Maher, 1980; 7. Bailey *et al.*, 1981.)

Field	Location	Depth to reservoir	Average porosity	Average permeability	Nature of hydrocarbons
Heather[1]	East Shetland Basin	c. 2900 m	12–19%	'Good'	35° API oil
Thistle[2]	East Shetland Basin	c. 2650 m	23–28%	500–650 mD	37° API oil
Magnus[3]	East Shetland Basin	c. 2900 m	19–25%	100–500 mD	39° API oil
Troll[4]	North–East Viking Graben	c. 1400 m	25–40%	c. 110 mD	Gas
Sleipner[5]	South Viking Graben	c. 3400 m	18–21%	——	Gas–condensate
Piper[6]	Eastern Moray Firth Basin	c. 2280 m	19–30%	c. 4000 mD	37° API oil
Ula[7]	Eastern Central Graben	c. 3400 m	18–25%	650–850 mD	38° API oil

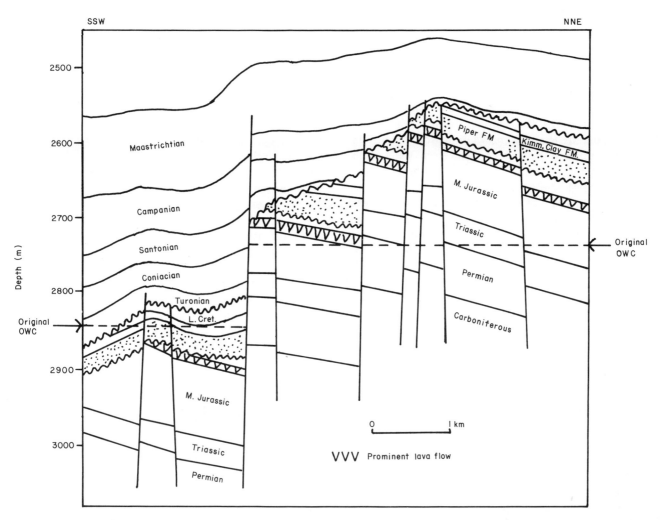

Fig. 6.5. Structural cross-section through Piper Field (U.K. Block 15/17), Outer Moray Firth Basin. (After Maher, 1980)

Trough (Ridd 1981) and can be seen for example in the Heather Field (U.K. Block 2/5; Gray and Barnes, 1981). A strike-slip component of movement on NE-SW faults during during the Mesozoic was postulated by Hay (1978) who suggested that the resulting rotation of individual fault-blocks would explain their observed variation in tilt direction. The area of the Basin to the south is structurally simpler, being dominated by a N-S or NNE-SSW fault trend, with most faults throwing down to the east and blocks dipping to the west. Two prominent and economically important sub-parallel structural alignments occur; the more westerly one being the location of the Ninian Field (U.K. Blocks 3/3 and 3/8; Albright *et al.*, 1980), Hutton Field (U.K. Block, 211/28), Dunlin Field (U.K. Blocks 211/23 and 211/24; van Rijswijk *et al.*, 1980) and the Murchison Field (U.K. Block 211/19 and Norwegian Block 33/9; Davies and Watts, 1977; Simpson and Whitley, 1981), that to the east having the Brent Field (U.K. Block 211/29; Bowen 1975) and the Statfjord Field (U.K. Block 211/24 and Norwegian Blocks 33/9 and 33/12; Kirk, 1980).

The Viking Graben further south has an asymmetric broadly half-grabenal form in east-west section (Fig. 10.3, Section 2 cross-section from Ziegler, 1982, (Fig. 9.2), with the major basin-margin faults located on the western side. The N-S trending western margin has been interpreted by Johnson and Dingwall (1981) as being composed of a series of en echelon NE-SW faults, re-activated Caledonide structures, each forming the edge of minor half-grabens which plunge towards the Graben axis. The hypothesis remains controversial (compare map in Harms *et al.*, 1981) especially because of difficulties in seismic interpretation close to the fault zone, but is an interesting contribution for example to any attempt to understand the localisation of Upper Jurassic sand fans along the margin of the Viking Graben.

The movement of Zechstein salt influences structures at Jurassic level in the south Viking Graben and is largely responsible for the construction of the traps in the Sleipner Field complex (Norwegian Block 15/6; Larsen and Jaarvik, 1981).

Finally, Hamar *et al.* (1980) recognise the NW-SE fault-bounded Andrew Ridge, a feature running between the Vestland Arch (Jaeren High) and the southern tip of the Fladen Ground Spur, as forming the southern limit of the Viking Graben during the Jurassic. The area to the south belongs to the Central Graben-Witch Ground Graben basin trend.

6.7.2 Stratigraphy and facies

Lower-Middle Jurassic

Lower Jurassic strata, consisting predominently of marine shales above an important basal sand, are widely distributed throughout much of the Viking Graben but are less well known in the south. This may in part reflect a paucity of data from wells aimed mostly at Upper Jurassic targets but there is some evidence of Middle Jurassic resting on Triassic (Larsen and Jaarvik, 1981). The unconformity can once again be related to mid-Jurassic upwarping of the central North Sea. Further north, in the East Shetland Basin, a minor discontinuity between Lower and Middle Jurassic strata has been detected in places (Hallet, 1981).

The oldest Jurassic deposits in the East Shetland Basin are sands of the Statfjord Formation (Rhaetian-Sinemurian; see Fig. 6.6). A similar unit, the Gassum Formation, spans the Triassic-Jurassic boundary in the Norwegian-Danish Basin. The Statfjord Formation in its type section in the Statfjord Field records a transition from non-marine to marginal marine conditions, with a sequence of floodplain, sinuous stream, braided stream and, at top, coastal plain sediments recognised by Kirk (1980) and Chauvin and Valachi (1981). Roe and Steele (1981) favour a fan delta model for the equivalent sands found to the north, in the area of the Tampen Spur. The Statfjord Formation is most fully developed towards the axial parts of the north Viking Graben and onlaps westwards towards the edge of the East Shetland Platform. Successively higher units within the Formation rest unconformably on Triassic (Deegan and Scull, 1977). Basal Jurassic sands, equivalent to the Statfjord Formation, can be traced southwards into parts of U.K. Quadrant 9 where they contribute to the reservoir in the Beryl Field (U.K. Block 9/13; Lower Beryl Sands of Marcum *et al.*, 1978).

The Statfjord Formation is overlain by marine shales which occur throughout most of the remaining Lower Jurassic sequence. This boundary, which is diachronous, ranges in age from Hettangian to Sinemurian and youngs towards the west. Thin sands (Cook Formation, Pliensbachian-Toarcian) do occur within the Lower Jurassic shales and form one of the reservoir units in the Gullfaks Field (Norwegian Block 34/10; Hazeu, 1981).

Middle Jurassic strata consist of a sand-dominated, broadly deltaic sequence overlain by marine shales which continue into the Upper Jurassic. The boundary between the two is locally an unconformity, especially near the crests of tilted fault-blocks in the East Shetland Basin where the upper parts of the mostly Aalenian-Bajocian, Brent Group may be removed by erosion. On a regional scale the boundary is diachronous with the establishment of open marine conditions over the whole north Viking Graben during the Bathonian (Calloman, 1975, 1979), whereas in the south the first marine shales (Heather Formation) appear in the Cal-lovian or, as in parts of the Sleipner Field, in the Oxfordian (Larsen and Jaarvik, 1981).

In the extreme south, Middle Jurassic strata contain traces of volcaniclastic material derived from the volcanic centre situated nearby at the intersection of the main North Sea grabens. Additional evidence of mid-Jurassic volcanism elsewhere in the Viking Graben comes from reports of ash bands in the Brent Group within the Statfjord and Murchison Fields (Malm *et al.*, 1979; Morton and Humphreys, 1983).

The Brent Group is probably the single most productive reservoir unit in the North Sea and as such merits a full descriptive account. Deegan and Scull (1977) defined five widely distributed sub-units, namely the Broom, Rannoch, Etive, Ness and Tarbert Formations (see Figs. 6.7 and 6.8; compare with the less formal, more descriptive terminology of Bowen, 1975). The Broom Formation, at the base, is a thin, poorly sorted and arkosic sandstone of limited reservoir importance, interpreted variously as a fluvial deposit (Kirk, 1980), a beach deposit (Eynon, 1981) and, more broadly and less controversially, as a sub-littoral sand sheet (Budding and Inglin, 1981). The overlying Rannoch Formation sandstones are rich in mica (Hodson, 1975) and are generally interpreted as shallow marine, delta-front or shoreface deposits. The micas are concentrated along low-angle bedding surfaces and inhibit vertical fluid flow through reservoir sections. The massive, well sorted sands of the succeeding Etive Formation are assigned to sub-environments of a barrier-bar complex by Budding and Inglin (1981; see Fig. 6.9). This unit has horizons of very high permeability possibly related to the position of tidal-inlet channel sands, which can lead to anomalously rapid flow rates locally within the reservoir during oil production. Both the Rannoch and Etive units form extensive, readily correlatable sand sheets.

The Ness Formation, deposited in a back-barrier lagoonal and alluvial plain environment (see Parry *et al.*, 1981), is lithologically more varied, consisting of interbedded sandstones, shales and coals. Sand units are frequently discontinuous, even within fields, but where stacked fluvial channel sands occur the Ness Formation makes an important contribution to effective reservoir volume. A prominent shale horizon, the Mid-Ness Shale, is recognised in the northern part of the East Shetland Basin. Deposited during a period of widespread lagoonal conditions, possibly reflecting a transgressive episode, it is regarded by Budding and Inglin (1981) as an isochronous unit and of value as a stratigraphic marker horizon.

The Tarbert Formation, the uppermost unit of the Brent Group, is dominantly arenaceous, sometimes micaceous, and has minor shale intercalations. It probably represents the first signs of the transgression which finally drowned the Brent delta (Bowen, 1975; Hallet, 1981). It is commonly thin or completely absent through erosion in well sections from crestal positions on tilted blocks.

From a regional perspective, there are divergent

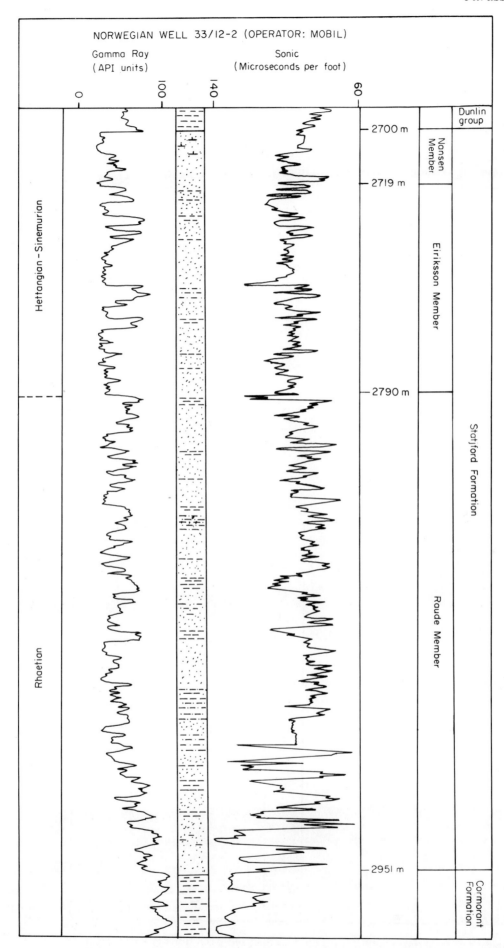

Fig. 6.6. Typical well log through the Statfjord Formation, East Shetland Basin (for Key see Fig. 6.3).

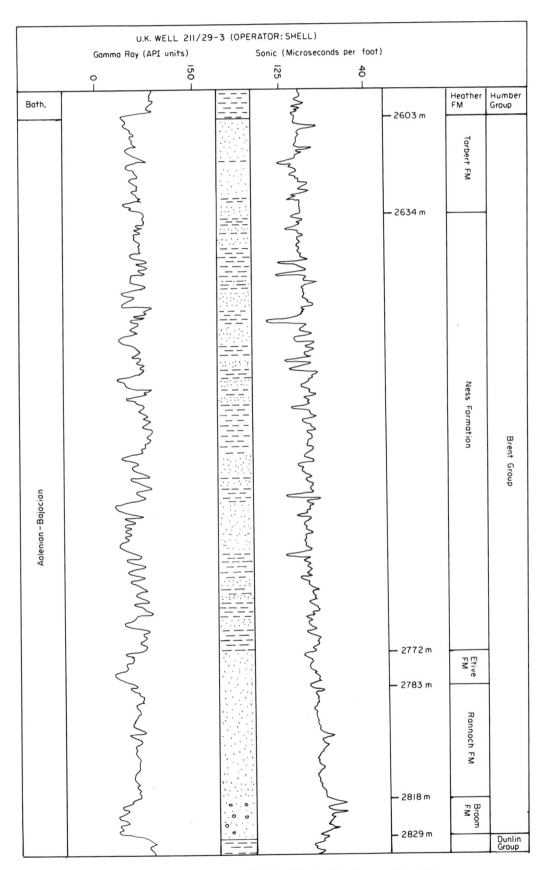

Fig. 6.7. Typical well log through the Brent Group, East Shetland Basin (for Key see Fig. 6.3).

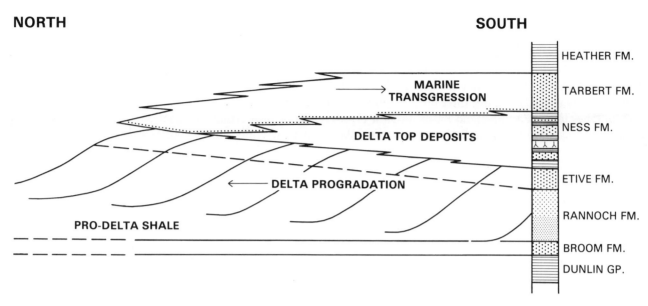

Fig. 6.8. Facies relations in the Brent Delta.

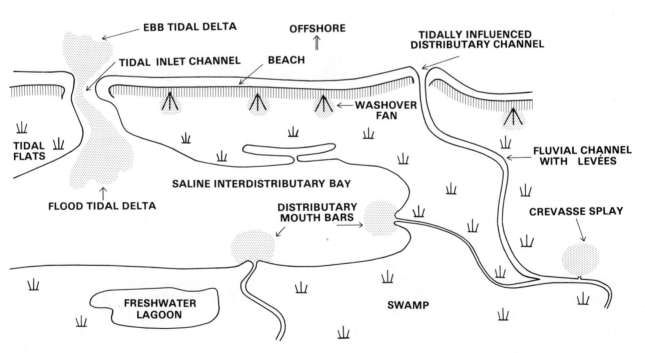

Fig. 6.9. Depositional environments of the Etive and Ness Formations. (After Budding and Inglin, 1981).

views on Brent Group depositional models and palaeo-geography. Eynon (1981) envisages the Middle Jurassic deltaic deposits prograding northwards and passing to marine shales at the extreme north end of the Viking Graben. In his account, the five-fold lithostratigraphic sub-division of the Brent Group is abandoned in favour of a scheme which recognises three distinct prograding delta lobes in the East Shetland Basin. Proctor (1980) also recognises a number of distinct prograding lobes but in a fan delta system building-out into a marine basin lying to the east. Budding and Inglin (1981), utilising the five units of Deegan and Scull (1977), present a more widely accepted view of a northwardly prograding, wave-influenced delta, also passing laterally to a fully marine environment situated to the north.

Two further aspects of Brent Group geology are relevant. Firstly, the thickness of individual formations, and indeed of individual beds, may vary across the area of a field as a result of faulting contemporaneous with sedimentation, as recorded by Hallet (1981) from the Thistle Field (U.K. Blocks 211/18 and 211/19).

Secondly, the diagenetic changes undergone by Brent Group sands have been much studied (see Blanche and Whitaker, 1978; Hancock and Taylor, 1978; Sommer, 1978; Hallet, 1981; Morton and Humphreys, 1983). To summarise, the production of secondary porosity appears to have played only a minor role in influencing reservoir character, being largely restricted to the dissolution of detrital feldspars and heavy mineral grains. Diagenetic damage to reservoir sands occurred in three main and successive phases: devlopment of syntaxial quartz overgrowths, growth of kaolinite and, lastly and of most harm to reservoir permeability, growth of fibrous illite. Although inhibited by the presence of hydrocarbons in the host rock, illite growth and its timing is a most important factor since it seems to have occurred penecontemporaneously with hydrocarbon migration into the reservoirs.

Outside the East Shetland Basin, Middle Jurassic deltaic sediments have been encountered to the east of the Viking Graben axis, for example in the Oseberg Field situated on the Vestland Arch (Norwegian Block 30/6; Larsen *et al.*, 1981). A largely arenaceous sequence dated as uppermost Toarcian to Bajocian has been tentatively assigned to the Etive and Ness Formations. Brekke *et al.* (1981) describe deltaic Aalenian-Bajocian sands in the Troll Field (Norwegian Block 31/2) overlain by shallow marine Bathonian-Callovian sands. The Middle Jurassic marine deposits form part of a thick sequence of hydrocarbon-bearing marine sands which extends into the Volgian. Further south in the Beryl Field, Marcum *et al.* (1978) recognise two units within a non-marine to paralic sand-dominated sequence, termed the Middle and Upper Beryl Sands. Broadly similar in terms of depositional environment to the Ness and Tarbert Formations respectively, the Middle Beryl Sand is equivalent in age to most of the Brent Group in the East Shetland Basin whereas the

Upper Sand is of Bathonian age and equivalent to marine shales further north. Similar deposits can be traced into the south Viking Graben where, in the Sleipner Field, Larsen and Jaarvik (1981) record a heterolithic coal-bearing sequence of (?)Aalenian-Bathonian age (the Sleipner Formation) overlain by clean, delta front to shoreface sands of mostly Callovian age (Hugin Formation). The Callovian sands pass laterally and vertically to marine shales of the Heather Formation.

Upper Jurassic

The Upper Jurassic of the Viking Graben, including the East Shetland Basin, consists mostly of marine shales but with important coarser clastic sequences developed locally, especially along the western margin where fault-controlled fan deposits occur. The argillaceous sequence includes the organic-rich, highly radioactive Kimmeridge Clay Formation (upper Oxfordian-Ryazanian) which rests on shales of the Heather Formation (Bathonian-Oxfordian). The boundary is often sharp, marked on geophysical well logs by an upward increase in gamma ray response and by an upward decrease in both seismic velocity and density. There may be a minor stratigraphic discontinuity at this sharp boundary.

Within the East Shetland Basin, Kimmeridgian sands interpreted as submarine fan deposits derived from the west, form the reservoir in the Magnus Field (U.K. Blocks 211/7 and 211/12; De'Ath and Schuyleman, (1981). Submarine fan sediments of a coarser, more proximal nature occur along the western graben-margin fault in the south Viking Graben, where they form the reservoir in the Toni, Thelma and Tiffany structures (U.K. Block 16/17). This sediment was derived from the adjacent Fladen Ground Spur.

Also in the south Viking Graben, the reservoir in the Brae Field (U.K. Block 16/7) consists of sandstone and conglomerate, dated as Kimmeridgian-Volgian. The coarse clastic deposits are overlain by a veneer of Kimmeridge Clay Formation and pass laterally, distally, to sandy siltstone-shale intercalations ('striped' facies) and finally to beds of Kimmeridge Clay Formation shale towards the axis of the Graben. Two depositional models for the reservoir section in the South Brae structure have been proposed. Harms *et al.* (1981) favour a fan-delta model in which the coarse sediment is deposited subaerially on the fan surface by fluvially-dominated processes ('topset unit' of Harms and McMichael, 1983) and the fringing interbedded 'fines' are marine ('foreset to bottomset unit'). In contrast, Stow *et al.* (1981) propose a submarine fan model analogous to that derived by Surlyk (1978) for fault-controlled, shallow, small-basin fans of similar age in East Greenland. One point of distinction between the two models emphasised by Harms and McMichael (1983) is the nature of the lateral facies change from coarse clastics (reservoir facies) to fines; the fan-delta model has an abrupt change at the shoreline of the delta whereas the gravity flow processes characteristic of a submarine fan environment would be capable of transporting coarse debris to form lobes within the distal shales and therefore produce less abrupt facies changes and a less well defined limit to the reservoir.

Upper Jurassic Oxfordian sands contribute to the pay zone in the Beryl Field, (Bruce Formation; Marcum *et al.*, 1978). The gas reservoir in the giant Troll Field situated on the north-west margin of the Horda Platform, consists of Bathonian-Volgian sands deposited in a shallow marine environment (Brekke *et al.*, 1981).

6.7.3 Economic geology

The Viking Graben and especially the East Shetland Basin is probably the most thoroughly explored part of the north-west European continental shelf and is currently the most productive in hydrocarbons. There are areas of the basin yet to be fully explored however, the Norwegian sector of the north Viking Graben being the most obvious example.

Reservoir rocks occur in all three Jurassic sub-systems, the Statfjord and Cook Formation sands of the Lower Jurassic, the Brent Group and its lateral equivalents of the Middle Jurassic and the Magnus, Bruce and Brae Formations plus the Troll Field sands of the Upper Jurassic (see Table 6.2). Middle Jurassic sands are most widely distributed. Upper Jurassic sands tend to be more localised and their regional distribution probably less well known at present.

The Kimmeridge Clay Formation, the generally accepted principal source rock, lies at depths greater than 3000 m over most of the area (Day *et al.* 1981). From the present maturity gradient, Goff (1983) calculates that the oil generating window extends from 2,500 m-4,500 m with peak generation at c.3,250 m. Therefore the Upper Jurassic shales are mature over much of the basin area. Peak generation occurred during the Palaeocene in the Viking Graben and during

the late Eocene in the deeper parts of the East Shetland Basin (Goff, 1983). However, over the crests of certain fault-blocks within the East Shetland Basin the Upper Jurassic shales are in an early-mature or immature state. In the axial parts of the north Viking Graben the Upper Jurassic shales may have been the source of gas in the Troll Field. Fisher and Miles (1983) show that both oil and condensate may come from Upper Jurassic shales in the U.K. south Viking Graben.

Of additional hydrocarbon source potential, Heritier *et al.* 1981) proposed a Lower Jurassic source for oil and gas in the Frigg Field (U.K. Block 10/1, Norwegian Block 25/1). Goff (1983) suggests that dry gas is likely from coals within the Brent Group, and that the Heather Formation shales, where thickly developed in axial parts of the Viking Graben, may also be capable of yielding significant amounts of gas. Larsen and Jaarvik (1981) postulate that thick coal beds in the Sleipner Formation sourced the gas in the Sleipner Field and that the Heather Formation shales sourced the condensate.

The structures associated with Statfjord Formation and Brent Group oil reservoirs in the East Shetland Basin consist of sub-unconformity traps at the up-tilted edge of the fault blocks, sealed by Upper Jurassic or Lower Cretaceous shales (Fig. 6.10). Goff (1983) favours oil migration from overpressured Kimmeridge Clay Formation, situated in half-grabenal troughs along high lateral fluid pressure gradients to Brent Group reservoirs which are at most only slightly over-pressured (see Lindberg *et al.*, 1980 and Chiarelli and Duffaud, 1980 on overpressures in the northern North Sea). This model envisages primary migration from the source rock across the stratigraphically lower Heather Formation shales to the reservoir situated updip on the inclined flank of a tilted-block. Migration out of the troughs and across faults or along a path related to the unconformity surface, to the up-tilted edges of reservoir units on immediately adjacent 'highs' may also be a feasible route in some cases.

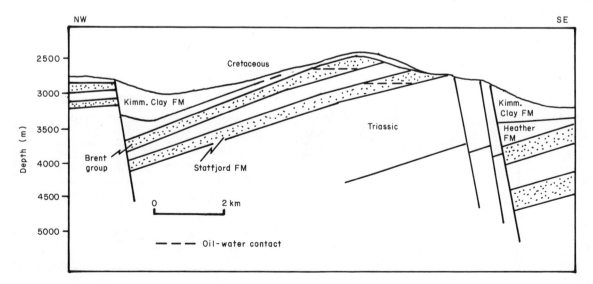

Fig. 6.10. Simplified structural cross-section through the Statfjord Field (U.K. Block 211/24 and Norwegian Blocks 33/9 and 33/12), East Shetland Basin. (Modified from Kirk, 1980)

In the Magnus Field the structure has a similar, tilted fault-block geometry. The Upper Jurassic reservoir however, was finally sealed by Upper Cretaceous marls (De'Ath and Schuyleman, 1981). In the south Viking Graben, the Brae Field also has an Upper Jurassic reservoir, developed on the downthrown side of the graben margin fault zone (Fig. 6.11). The reservoir, with an oil column c.500 m thick, is capped by Upper Jurassic shales but, in addition, is sealed to the west by the fault zone which brings the Jurassic into juxtaposition with Devonian strata (Harms *et al.*, 1981). Finally, Middle Jurassic reservoirs in the Sleipner Field (Fig. 6.12) occur in anticlinal structures over salt swells, with the reservoir capped by Upper Jurassic shales (Larsen and Jaarvik, 1981).

6.8 Depositional environment of the Kimmeridge Clay Formation

The Kimmeridge Clay Formation within the North Sea grabens belongs to the upper Oxfordian-Ryazanian and consists of black shales, silty shales, brown oil shales and some thin limestones (Barnard and Cooper, 1981) deposited in a low energy, marine environment. The shales are organic-rich, with total organic content averaging between five and ten per cent, and are often highly radioactive. In addition to the variations in the amount of organic material present, there are also variations, especially laterally, in its composition. The proportion of terrestrially-derived kerogen to marine algal material reflects palaeogeographic location within the basin, the former decreasing in abundance towards graben axes. This distribution can influence the nature of the hydrocarbon product yielded upon maturation. The high radioactivity of the shales is largely due to their uranium content, the uranium being absorbed onto organic matter on stagnant sea-floors in a process described by Bjorlykke *et al.* (1975).

The depositional environment of the Kimmeridge Clay Formation has been discussed by a number of authors, most basing their interpretations on the study of equivalent strata cropping out in southern England (Gallois, 1976; Tyson *et al.*, 1979; Irwin, 1979; Barnard and Cooper, 1981). A full account of this work is beyond the scope of the chapter, but two divergent views are summarised. Gallois (1976) envisages a high productivity of algae (algal blooms) resulting in the temporary deoxygenation of marine waters and poisoning of the existing benthonic fauna. This, it is argued, temporarily produces the anaerobic bottom conditions necessary for the preservation of algal and other organic matter and the formation of organic-rich sediments.

In contrast, Tyson *et al.*, (1979) hold that algal blooms are a result of widespread anaerobic conditions and not their cause. The preservation of organic ma-

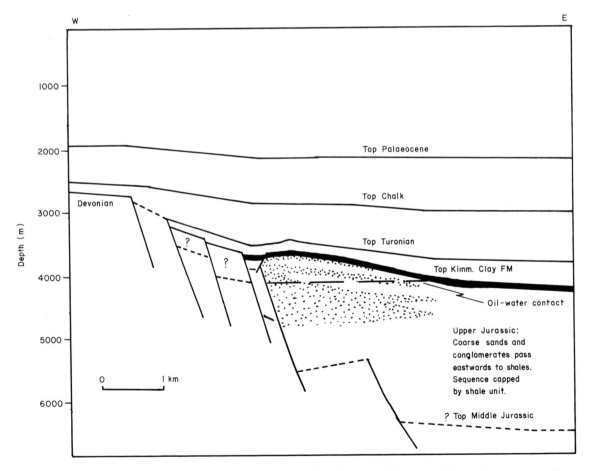

Fig. 6.11. Structural cross-section through the South Brae Field (U.K. Block 16/17), South Viking Graben. (Modified from Harms *et al.*, 1981)

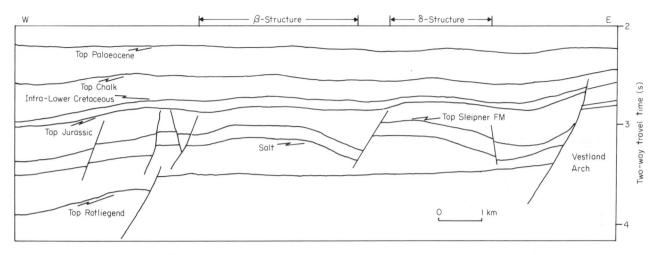

Fig. 6.12. Seismic profile through the Sleipner Field Complex (Norwegian Block 15/19), South Viking Graben. (After Larsen and Jaarvik, 1981)

terial on an anoxic sea-floor rather than high algal productivity is emphasised as the origin of the organic-rich Kimmeridge Clay Formation shales. A stratified water column, with a vertically migrating O_2-H_2S interface is invoked to account for the formation of organic-rich shales and associated coccolith limestone bands in the southern England succession. Tyson *et al.* (1979) consider that high stands of sea-level during the late Jurassic favoured the production of an anoxic bottom layer by increasing water depths in the basins and further separating the oxygenated surface waters from the sea-bottom. In addition, anoxia was also promoted by restricted water circulation from adjacent oceans. A carbonate platform in southern Europe may have contributed to poor circulation between the Tethys ocean and the basins receiving Kimmeridge Clay sediments on the broad shelf to the north.

6.9 A note on the Jurassic-Cretaceous boundary

The identification of a 'late Cimmerian' or 'base Cretaceous' unconformity defining the top of the Jurassic is commonplace in North Sea literature, but increasingly refined palynological age determinations (Rawson and Riley, 1982) have confirmed the view of Whitbread (1975) that such an interpretation is oversimplified (see Johnson, 1975 and Fyfe *et al.*, 1981 for further discussion of the 'unconformity').

An unconformity between Jurassic and Cretaceous strata may occur over marginal and some intra-basinal 'highs', but over much of the North Sea, the system boundary is a conformable one and, furthermore, frequently lies within the organic-rich shales of the Kimmeridge Clay Formation and its equivalents. Therefore, within the main North Sea grabens the 'hot' shale source rocks commonly extend into the Ryazanian. Where an unconformable system boundary can be traced into thicker sections off-structure, the single hiatus is often found to be the result of a number of minor discontinuities within the Volgian-Ryazanian

interval converging up-dip. Vail and Todd (1981) and Rawson and Riley (1982) attribute many of the discontinuities to changes in sea-level rather than to tectonic events within the basin. Further confusion can arise from the misinterpretation of a condensed sedimentary sequence as an unconformity.

Within the main North Sea grabens, the boundary between the Kimmeridge Clay Formation and the overlying Cromer Knoll Group (Lower Cretaceous) is almost invariably sharp and characterised by a pronounced facies change, from organic-rich shale up to calcareous mudstone, limestone, or locally sandstone (Deegan and Scull, 1977). On geophysical well logs the boundary is associated with an upward increase in seismic velocity and density and usually gives rise to a strong seismic event with good lateral continuity.

The termination of Kimmeridge Clay Formation deposition throughout the North Sea grabens during the Ryazanian is interpreted by Rawson and Riley (1982) as an isochronous environmental change, marked by the flushing-out of the anaerobic basins and destruction of the stratified, water column. This change coincided with a late Ryazanian marine transgression.

6.10 Acknowledgements

The author is indebted to many colleagues within the British Geological Survey for their contributions to his understanding of the Jurassic of the North Sea. Special thanks are due to Mr J.M. Dean and Dr C.E. Deegan for fruitful discussions and invaluable advice. The author also wishes to acknowledge the U.K. Department of Energy, Petroleum Engineering Division who fund the offshore work of the Survey and made this contribution possible. This chapter is published by permission of the Director, British Geological Survey (N.E.R.C.).

6.11 References

Albright, W.A., Turner, W.L. and Williamson, K.R. (1980) Ninian Field, U.K. sector, North Sea. In: Halbouty, M.T.

(Ed.) *Giant oil and gas fields of the Decade: 1968-1978*, pp. 173-194. *Mem. Am. Ass. Petrol. Geol.* **30**, Tulsa, Oklahoma.

Al Kasim, F., Rønnevik, H.C. and Ulleberg, K. (1975) Review of the Jurassic offshore Norway. In: *Jurassic Northern North Sea Symposium, Stavanger*, Norw. Petrol. Soc. pp. 3/1-18.

Bailey, C.C., Price, I. and Spencer, A.M. (1981) The Ula Oil Field, Block 7/12, Norway. In: *Norwegian Symposium on Exploration*, Norw. Petrol. Soc. pp. 18/1-26.

Barnard, P.C. and Cooper, B.S. (1981) Oils and source rocks of the North Sea area. In: Illing, L.V. and Hobson, C.D. (Eds.) *Petroleum geology of the Continental Shelf of North-west Europe* Heyden & Son, London, pp. 169-175.

Barnard, P. and Cooper, B.S. (1983) A review of geochemical data related to the Northwest European gas province. In: *Petroleum Geochemistry and Exploration of Europe*, pp. 19-34. Spec. Publ. geol. Soc. Lond. 12. Blackwell Scientific Publications, Oxford. Brooks, J. (Ed.)

Berridge, N.G. and Ivimey-Cook, H.C. (1967) The geology of a Geological Survey borehole at Lossiemouth, Morayshire. *Bull. geol. Surv. G.B.* **27**, 155-169.

Bertelsen, F. (1978) The Upper Triassic-Lower Jurassic Vinding and Gassum Formations of the Norwegian-Danish Basin. *Geol. Surv. Denmark Series B*, **3**, p. 26.

Best, G., Kockel, F. and Schöneich, H. (1983) Geological history of the southern Horn Graben. In: Kaasschieter, J.P.H. and Reijers, T.J.A. (Eds.) *Petroleum geology of the southeastern North Sea and adjacent onshore areas. The Hague, 1982 Geol. Mijnbouw*, **62**, 25-33.

Bjørlykke, K., Dypvik, H. and Finstad, K.G. (1975) The Kimmeridge shale, its composition and radioactivity. In: *Jurassic Northern North Sea Symposium, Stavanger*, Norw. Petrol. Soc. pp. 12/1-20.

Blanche, J.B. and Whitaker, J.H. McD. (1978) Diagenesis of part of the Brent Sand Formation (Middle Jurassic) of the northern North Sea Basin. *J. geol. Soc. Lond.* **135**, 73-82.

Bodenhausen, J.W.A. and Ott, W.F. (1981) Habitat of the Rijswijk Oil Province, onshore, The Netherlands. In: Illing, L.V. and Hodson, G.D. (Eds.) *Petroleum Geology of the Continental Shelf of North-West Europe* Heyden & Son, London. pp. 301-309.

Bowen, J.M. (1975) The Brent oilfield. In: Woodland, A.W. (Ed.) *Petroleum and the Continental Shelf of North-West Europe, Vol. 1 : Geology* Applied Science Publishers, Barking. pp. 353-362.

Brekke, T., Pegrum, R.M. and Watts, P.B. (1981) First exploration results in Block 31/2, offshore Norway. In: *Norwegian Symposium on Exploration*, Norw. Petrol. Soc. pp. 16/1-34.

Brooks, J.R.V. (1977) Exploration status of the Mesozoic of the U.K. Northern North Sea. In: *Mesozoic Northern North Sea Symposium, Oslo*, Norw. Petrol. Soc. pp. 2/1-28.

Budding, M.C. and Illing, H.F. (1981) A reservoir geological model of the Brent Sands in Southern Cormorant. In: Illing, L.V. and Hodson, G.D. (Eds.) *Petroleum Geology of the Continental Shelf of North-West Europe*, Heyden & Son, London. pp. 326-334.

Callomon, J.H. (1975) Jurassic ammonites from the northern North Sea. *Norsk geol. Tiddskr.* **55**, 373-386.

Callomon, J.H. (1979) Marine boreal Bathonian fossils from the northern North Sea and their palaeogeographical significance. *Proc. Geol. Ass.* **90**, 163-169.

Casey, R. (1973) The ammonite succession at the Jurassic-Cretaceous boundary in eastern England. In: Casey, R. and Rawson, P.F. (Eds.) *The Boreal Lower Cretaceous*, Geol. J. Spec. Issue 5. pp. 193-266.

Chauvin, A.L. and Valachi, L.Z. (1980) Sedimentology of the Brent and Statfjord Formations of Statfjord Field. In: *The Sedimentation of the North Sea Reservoir Rocks, Geilo*, Norw. Petrol. Soc. pp. 16/1-17.

Chesher, J.A. and Bacon, M. (1975) A deep seismic survey in the Moray Firth. *Rept. Inst. geol. Sci.* 75/11, pp. 13.

Chesher, J.A. and Lawson, D. (1983) The geology of the Moray Firth. Rep. Inst. geol. Sci. 83/5, pp. 32.

Chiarelli, A. and Duffaud, F. (1980) Pressure origin and distribution in Jurassic of Viking Basin (United Kingdom-Norway). *Bull. Am. Ass. Petrol. Geol.* **64**, 1245-1266.

Christie, P.A.F. and Sclater, J.G. (1980) An extensional origin for the Buchan and Witchground Graben in the North Sea. *Nature* **283**, 729-732.

Cope, J.C.W., Getty, T.A., Howarth, M.K., Morton, N. and Torrens, H.S. (1980a) *A correlation of Jurassic rocks in the British Isles. Part 1: Introduction and Lower Jurassic.* Spec. Report geol. Soc. Lond. 14, Blackwell Scientific Publications, Oxford. pp. 73.

Cope, J.C.W., Duff, K.L., Parsons, C.F., Torrens, H.S., Wimbledon, W.A. and Wright, J.K. (1980b) *A correlation of Jurassic rocks in the British Isles. Part 2: Middle and Upper Jurassic.* Spec. Report geol. Soc. Lond. 15, Blackwell Scientific Publications, Oxford, p. 109.

Davies, E.J. and Watts, T.R. (1977) The Murchison oilfield. In: *Mesozoic Northern North Sea Symposium, Oslo*, Norw. Petrol. Soc. pp. 15/1-24.

Day, G.A., Cooper, B.A., Andersen, C., Burgers, W.F.J., Rønnevik, H.C. and Schöneich, H. (1981) Regional seismic structure maps of the North Sea. In: Illing, L.V. and Hobson, G.D. (Eds.) *Petroleum Geology of the Continental Shelf of North-West Europe*, Heyden & Son, London. pp. 76-84.

De'Ath, N.G. and Schuyleman, S.F. (1981) The geology of the Magnus oilfield. In: Illing, L.V. and Hobson, G.D. (Eds.) *Petroleum Geology of the Continental Shelf of North-West Europe*, Heyden & Son, London. pp. 342-351.

Deegan, C.E. and Scull, B.J. (1977) (Compilers) A standard lithostratigraphic nomenclature for the Central and Northern North Sea. *Rept. Inst. geol. Sci.* 77/25, pp. 36.

Donato, J.A. and Tully, M.C. (1981) A regional interpretation of North Sea gravity data. In: Illing, L.V. and Hobson, G.D. (Eds.) *Petroleum geology of the Continental Shelf of North-West Europe*, Heyden & Son, London. pp. 65-75.

Dypvik, H. and Vollset, J. (1980) Deltaic sedimentation during the Jurassic in the Norwegian-Danish Basin (North Sea). *Geol. Mijnbouw.* **59**, 25-32.

Eynon, G. (1981) Basin development and sedimentation in the Middle Jurassic of the northern North Sea. In: Illing, L.V. and Hobson, G.D. (Eds.) *Petroleum Geology of the Continental Shelf of North-West Europe*, Heyden & Son, London. pp. 196-204.

Fall, H.G., Gibb, F.G.F. and Kanaris-Sotiriou, R. (1982) Jurassic volcanic rocks of the northern North Sea. *J. geol. Soc. Lond.* **139**, 277-292.

Fisher, M.J. and Miles, J.A. (1983) Kerogen types, organic maturation and hydrocarbon occurrences in the Moray Firth and South Viking Graben, North Sea Basin. In: Brooks, J. (Ed.) *Petroleum geochemistry and exploration of Europe.* Spec. Publ. geol. Soc. Lond. 12. Blackwell Scientific Publications, Oxford. pp. 195-202.

Furnes, H., Elvsborg, A. and Malm, O.A. (1982) Lower and Middle Jurassic alkaline magmatism in the Egersund Sub-basin, North Sea. *Marine Geology* **46**, 53-69.

Fyfe, J.A., Abbotts, I. and Crosby, A. (1981) The subcrop of the mid-Mesozoic unconformity in the U.K. area. In: Illing, L.V. and Hobson, G.D. (Eds.) *Petroleum Geology of the Continental Shelf of North-West Europe*, Heyden & Son, London. pp. 236-244.

Gallois, R.W. (1976) Coccolith blooms in the Kimmeridge Clay and origin of North Sea Oil. *Nature* **259**, 473-475.

Gibbs, A.D. (1983a) (Abstract) Secondary detachment above basement faults in North Sea: Clyde Field growth fault. *Bull. Am. Ass. Petrol. Geol.* **67**, 469-470.

Gibbs, A.D. (1983b) Balanced cross-section construction

from seişmic sections in areas of extensional tectonics. *J. Stuct. Geol.* **5**, 153-160.

Glennie, K.W. and Boegner, P.L.E. (1981) Sole Pit inversion tectonics. In: Illing, L.V. and Hobson, G.D. (Eds.) *Petroleum Geology in the Continental Shelf of North-West Europe*, Heyden & Son, London. pp. 110-120.

Goff, J.C. (1983) Hydrocarbon generation and migration from Jurassic source rocks in the E. Shetland Basin and Viking Graben of the northern North Sea. *J. geol. Soc. Lond.,* **140**, 445-474.

Gray, W.D.T. and Barnes, G. (1981) The Heather oil field. In: Illing, L.V. and Hobson, G.D. (Eds.) *Petroleum Geology of the Continental Shelf of North-West Europe,* Heyden & Son, London, pp. 335-341.

Hallam, A. (1978) Eustatic cycles in the Jurassic. *Palaeogr. Palaeoclim. Palaeoecol.* **23**, 1-32.

Hallam, A. (1981) A revised sea-level curve for the early Jurassic. *J. geol. Soc. Lond.* **138**, 735-743.

Hallet, D. (1981) Refinement of the geological model of the Thistle field. In: Illing, L.V. and Hobson, G.D. (Eds.) *Petroleum Geology of the Continental Shelf of North-West Europe*, Heyden & Son, London. pp. 315-325.

Halstead, P.H. (1975) Northern North Sea faulting. In: *Jurassic Northern North Sea Symposium, Stavanger,* Norw. Petrol. Soc. pp. 10/1-38.

Hamar, G.P., Fjaeran, T. and Hesjedal, A. (1983) Jurassic stratigraphy and tectonics of the south-southeastern Norwegian offshore. In: Kaasschieter, J.P.H and Reijers, T.J.A. (Eds.) *Petroleum geology of the southeastern North Sea and the adjacent onshore areas, The Hague, 1982, Geol. Mijnbouw.* **62**, 103-114.

Hamar, G.P., Jakobsson, K.H., Ormaasen, D.E. and Skarpnes, O. (1980) Tectonic development of the North Sea north of the Central Highs. In: *The Sedimentation of the North Sea Reservoir Rocks, Geilo,* Norw. Petrol. Soc. pp. 3/1-11.

Hancock, N.J. and Fisher, M.J. (1981) Middle Jurassic North Sea deltas with particular reference to Yorkshire. In: Illing, L.V. and Hobson, G.D. (Eds.) *Petroleum Geology of the Continental Shelf of North-West Europe*, Heyden & Son, London. pp. 186-195.

Hancock, N.J. and Taylor, A.M. (1978) Clay mineral diagenesis and oil migration in the Middle Jurassic Brent Sand Formation. *J. geol. Soc. Lond.* **135**, 69-72.

Hansen, J.M. and Mikkelsen, N. (1983) Hydrocarbon geological aspects of subsidence curves: interpretations based on released wells in the Danish Central Graben. *Bull. geol. Soc. Denmark* **31**, 159-169.

Harms, J.C. and McMichael, W.J. (1982) Sedimentology of the Brae Oilfield area, North Sea. *J. Petrol. Geol.* **5**, 437-439.

Harms, J.C., Tackenberg, P., Pickles, E. and Pollock, R.E. (1981) The Brae oilfield area. In: Illing, L.V. and Hobson, G.D. (Eds.) *Petroleum Geology of the Continental Shelf of North-West Europe*, Heyden & Son, London. pp. 352-357.

Hay, J.T.C. (1978) Structural development in the Northern North Sea. *J. Petrol. Geol.* **1**, 65-77.

Hazeu, G.J.A. (1981) 34/10 Delta structure, geological evaluation and appraisal. In: *Norwegian Symposium on Exploration*, Norw. Petrol. Soc. pp. 13/1-36.

Hemingway, J.E. (1974) Jurassic. In: Rayner, D.H. and Hemingway, J.E. (Eds.) *The Geology and Mineral Resources of Yorkshire*, Yorks. geol. Soc. pp. 161-223.

Heritier, F.E., Lossel, P. and Wathne, E. (1981) The Frigg gas field. In: Illing, L.V. and Hobson, G.D. (Eds.) *Petroleum Geology of the Continental Shelf of North-West Europe*, Heyden & Son, London. pp. 380-391.

Hesjedal, A. and Hamar, G.P. (1983) Lower Cretaceous stratigraphy and tectonics of the south-southeastern Norwegian offshore. In: Kaasschieter, J.P.H. and Reijers,

T.J.A. (Eds.) *Petroleum geology of the southeastern North Sea and adjacent onshore areas, The Hague, 1982 Geol. Mijnbouw* **62**, 135-144.

Heybroek, P. (1974) (Compiler) Explanation to tectonic maps of The Netherlands. *Geol. Mijnbouw* **53**, 43-50.

Hodson, G.M. (1975) Some aspects of the geology of the Middle Jurassic in the Northern North Sea with particular reference to electro-physical logs. In: *Jurassic Northern North Sea Symposium, Stavanger*, Norw. Petrol. Soc. pp. 16/1-39.

Howitt, F., Aston, E.R. and Jacqué, M. (1975) The occurrence of Jurassic volcanics in the North Sea. In: Woodland, A.W. (Ed.) *Petroleum and the Continental Shelf of North-West Europe, Vol. 1 : Geology*, Applied Science Publishers, Barking. pp. 379-388.

Irwin, H. (1979) On an environmental model for the type Kimmeridge Clay. *Nature* **279**, 819.

Jackson, J. and McKenzie, D. (1983) The geometrical evolution of normal fault systems. *J. Struct. Geol.* **5**, 471-482.

Johnson, R.J. (1975) The base of the Cretaceous: a discussion. In: Woodland, A.W. (Ed.) *Petroleum and the Continental Shelf of North-West Europe, Vol. 1 : Geology*, Applied Science Publishers, Barking. pp. 389-402.

Johnson, R.J. and Dingwall, R.G. (1981) The Caledonides: their influence on the stratigraphy of the North Sea. In: Illing, L.V. and Hobson, G.D. (Eds.) *Petroleum Geology of the Continental Shelf of North-West Europe*, Heyden & Son, London. pp. 85-97.

Kent, P.E. (1975) The tectonic development of Great Britain and the surrounding seas. In: Woodland, A.W. (Ed.) *Petroleum and the Continental Shelf of North-West Europe, Vol. 1 : Geology*, Applied Science Publishers, Barking. pp. 3-28.

Kent, P.E. (1980) Subsidence and uplift in East Yorkshire and Lincolnshire: a double inversion. *Proc. Yorks, geol. Soc.* **42**, 505-524.

Kirk, R.H. (1980) Statfjord Field: a North Sea giant. In: Halbouty, M.T. (Ed.) *Giant Oil and Gas Fields of the Decade: 1968-1978, Mem. Am. Ass. Petrol. Geol.* 30, Tulsa, Oklahoma. pp. 95-116.

Koch, J.-O. (1983) Sedimentology of Middle and Upper Jurassic sandstone reservoirs of Denmark. In: Kaasschieter, J.P.H. and Reijers, T.J.A. (Eds.) *Petroleum geology of the southeastern North Sea and the adjacent onshore areas, The Hague, 1982, Geol. Mijnbouw* **62**, 115-129.

Lam, K. and Porter, R. (1977) The distribution of palynomorphs in the Jurassic rocks of the Brora outlier, NE Scotland. *J. geol. Soc. Lond.* **134**, 45-55.

Larsen, R.M. and Jaarvik, L.J. (1981) The geology of the Sleipner Field complex. In: *Norwegian Symposium on Exploration*, Norw. Petrol. Soc. pp. 15/1-31.

Larsen, V., Aasheim, S.M. and Masset, J.M. (1981) 30/6-Alpha structure, a field case study in the Silver Block. In: *Norwegian Symposium on Exploration*, Norw. Petrol. Soc. pp. 14/1-34.

Lee, G.W. (1925) *The geology of the country around Golspie, Sutherlandshire: Mesozoic rocks of East Sutherland and Ross. Mem. geol. Surv. G.B.*

Leeder, M.R. (1983) Lithospheric stretching and North Sea Jurassic clastic sourcelands. *Nature* **305**, 510-514.

Lindberg, P., Riise, R. and Fertl, W.H. (1980) Occurrence and distribution of overpressures in the Northern North Sea area. *Proc. Soc. Petroleum Engineers Conference, Dallas*, paper 9339.

Linsley, P.N. (1972) *The stratigraphy and sedimentology of the Kimmeridgian deposits of Sutherland, Scotland.* Ph.D. Thesis, University of London (unpubl.).

Linsley, P.N., Potter, H.C., McNab, G. and Racher, D. (1980) The Beatrice Field, Inner Moray Firth, U.K. North Sea. In: Halbouty, M.T. (Ed.) *Giant Oil and Gas Fields of the Decade: 1968-1978. Mem. Am. Ass. Petrol. Geol.* 30, Tulsa, Oklahoma. pp. 117-130.

Maher, C.E. (1980) Piper Oil Field. In: Halbouty, M.T. (Ed.) *Giant Oil and Gas Fields of the Decade: 1968-1978. Mem. Am. Ass. Petrol. Geol.* 30, Tulsa, Oklahoma. pp. 131-172.

Malm, O.A., Furnes, H. and Bjørlykke, K. (1979) Volcaniclastics of Middle Jurassic age in the Statfjord oilfield of the North Sea. *N. Jb. Geol. Paläont. Mh.* **10**, 607-618.

Marcum, B.L., Al-Hussainy, R., Adams, G.E., Croft, M. and Block, M.L. (1978) Development of the Beryl 'A' Field. *Proc. European Offshore Petroleum Conference, London,* 319-321.

McKenzie, D. (1978) Some remarks on the development of sedimentary basins. *Earth Planet. Sci. Letts.* **40**, 25-32.

McQuillin, R., Donato, J.A. and Tulstrup, J. (1982) Dextral displacement of the Great Glen Fault as a factor in the development of the Inner Moray Firth Basin. *Earth Planet. Sci. Letts.* **60**, 127-139.

Michelsen, O. (1978) Stratigraphy and distribution of Jurassic deposits of the Norwegian-Danish Basin. *Geol. Surv. Denmark. Series B,* **2**, p. 28.

Michelsen, O. (1982) (Ed.) Geology of the Danish Central Graben. *Geol. Surv. Denmark. Series B,* **8**, p. 133.

Morton, A.C. and Humphreys, B. (1983) The petrology of the Middle Jurassic sandstones from the Murchison Field, North Sea. *J. Petrol. Geol.* **5**, 245-260.

Nadir, F.T. and Hay, J.T.C. (1978) Geological and reservoir modelling of the Thistle Field. *Proc. European Offshore Petroleum Conference, London,* pp. 233-237.

Nederlandse Aardolie Maatschappij, B.V. and Rijks Geologische Dienst (1980) Stratigraphic nomenclature of The Netherlands. *Verh. Kon. Ned. Geol. Mijnbk. Gen.* **32**, 1-77.

Neves, R. and Selley, R.C. (1975) A review of the Jurassic rocks of north-east Scotland. In: *Jurassic Northern North-Sea Symposium, Stavanger,* Norw. Petrol. Soc. pp. 5/1-19.

Oele, J.A., Hol, A.C.P.J. and Tiemens, J. (1981) Some Rotliegend gas fields of the K and L Blocks, Netherlands offshore (1968-1978)—a case history. In: Illing, L.V. and Hobson, G.D. (Eds.) *Petroleum Geology of the Continental Shelf of North-West Europe,* Heyden & Son, London. pp. 289-300.

Ofstad, K. (1983) The southernmost part of the Norwegian section of the Central Trough. *N.P.D. paper 32,* Norwegian Petroleum Directorate. p.40.

Olsen, J.C. (1983) The structural outline of the Horn Graben. In: Kaasschieter, J.P.H. and Reijers, T.J.A. (Eds.) *Petroleum geology of the southeastern North Sea and the adjacent onshore areas, The Hague, 1982, Geol. Mijnbouw* **62**, 47-50.

Olsen, R.C. and Strass, I.F. (1982) The Norwegian-Danish Basin. *N.P.D. paper 31,* Norwegian Petroleum Directorate p.76.

Parry, C.C., Whitley, P.K.J. and Simpson, R.D.H. (1981) Integration of palynological and sedimentological methods in facies analysis of the Brent Formation. In: Illing, L.V. and Hobson, G.D. (Eds.) *Petroleum Geology of the Continental Shelf of North-West Europe,* Heyden & Son, London. pp. 205-215.

Pearson, M.J. and Watkins, D. (1983) Organofacies and early maturation effects in Upper Jurassic sediments from the Inner Moray Firth Basin, North Sea. In: Brooks, J. (Ed.) *Petroleum Geochemistry and Exploration of Europe.* Spec. Publ. geol. Soc. Lond., 12. Blackwell Scientific Publications, Oxford. pp. 147-160.

Proctor, C.V. (1980) Distribution of Middle Jurassic facies in the East Shetlands Basin and their control on reservoir capability. In: *The Sedimentation of the North Sea Reservoir Rocks, Geilo.* Norw. Petrol. Soc. pp. 15/1-22.

Rawson, P.F., Curry, D., Dilley, F.C., Hancock, J.M.,

Kennedy, W.J., Neale, J.W., Wood, C.J. and Worssam, B.C. (1978) A correlation of Cretaceous rocks in the British Isles. Spec. Report geol. Soc. Lond. 9, Blackwell Scientific Publications, Oxford. p. 70.

Rawson, P.F. and Riley, L.A. (1982) Latest Jurassic-early Cretaceous events and the 'Late Cimmerian Unconformity' in the North Sea area. *Bull. Am. Ass. Petrol. Geol.* **66**, 2628-2648.

Rhys, G.H. (1974) (Compiler) A proposed standard lithostratigraphic nomenclature for the southern North Sea and an outline structural nomenclature for the whole of the (U.K.) North Sea. *Rept. Inst. geol. Sci.* 74/8, p. 14.

Ridd, M.F. (1981) Petroleum geology west of the Shetlands. In: Illing, L.V. and Hobson, G.D. (Eds.) *Petroleum Geology of the Continental Shelf of North-West Europe,* Heyden & Son, London. pp. 414-425.

Rijswijk, J.J. van, Robottom, D.J., Sprakes, C.W. and James, G. (1980) The Dunlin Field, a review of field development and reservoir performance to date. Proc. European Offshore Petroleum Conference, London. pp. 217-222.

Riley, L.A. (1977) Stage nomenclature at the Jurassic-Cretaceous boundary, North Sea Basin. In: *Mesozoic Northern North Sea Symposium, Oslo.* Norw. Petrol. Soc. pp. 4/1-11.

Roe, S.L. and Steel, R.J. (1981) Fan delta development at the Triassic-Jurassic Boundary, Tampen Spur, Northern North Sea. Abstracts Int. Ass. Sediment. 2nd European Regional Meeting, Bologna, p. 168.

Rønnevik, H.C., van den Bosch, W. and Bandlien, E.H. (1975) A proposed nomenclature for the main structural features in the Norwegian North Sea. In: *Jurassic Northern North Sea Symposium, Stavanger,* Norw. Petrol. Soc. pp. 18/1-19.

Roos, B.M. and Smits, B.J. (1983) Rotliegend and Main Buntsandstein gas fields in block K/13—a case history. In: Kaasschieter, J.P.H. and Reijers, T.J.A. (Eds.) *Petroleum geology of the southeastern North Sea and the adjacent onshore areas, The Hague, 1982, Geol. Mijnbouw* **62**, 75-82.

Simpson, R.D.H. and Whitley, P.K.J. (1981) Geological input to reservoir simulation of the Brent Formation. In: Illing, L.V. and Hobson, G.D. (Eds.) *Petroleum Geology of the Continental Shelf of North-West Europe,* Heyden & Son, London. pp. 310-314.

Skarpnes, O., Hamar, G.P., Jacobsson, K.H. and Ormaasen, D.E. (1980) Regional Jurassic setting of the North Sea north of the Central Highs. In: *The Sedimentation of the North Sea Reservoir Rocks, Geilo.* Norw. Petrol. Soc. pp. 13/1-8.

Skjerven, J., Rijs, F. and Kalheim, J.E. (1983) Late Palaeozoic to Early Cenozoic structural development of the south-southeastern Norwegian North Sea. In: Kaasschieter, J.P.H. and Reijers, T.J.A. (Eds.) *Petroleum geology of the southeastern North Sea and adjacent onshore areas, The Hague, 1982, Geol. Mijnbouw* **62**, 35-45.

Sommer, F. (1978) Diagenesis of Jurassic sandstones in the Viking Graben. *J. geol. Soc. Lond.* **135**, 63-68.

Staalduinen, C.J. van, Adrichem Boogaert, H.A. van, Bless, M.J.M., Doppert, J.W. Chr., Harsveldt, H.M., Montfrans, H.M. van, Oele, E., Wermuth, R.A. and Zagwijn, W.H. (1979) The geology of the Netherlands. *Med. Rijks Geol. Dienst* **31-2**, 11-49.

Stille, H. (1924) Grundfragen der Vergleichenden Tektonik, Borntraeger, Berlin, p. 443.

Stow, D.A.V., Bishop, C.D. and Mills, S.J. (1982) Sedimentology of the Brae Oil Field, North Sea: fan models and controls, *J. Petrol. Geol.* **5**, 129-148.

Surlyk, F. (1978) Submarine fan sedimentation along fault scarps on tilted fault blocks (Jurassic-Cretaceous boundary, East Greenland). *Bull. Grønlands geol. Unders.* **128**, p. 108.

Sykes, R.M. (1975) The stratigraphy of the Callovian and Oxfordian stages (Middle-Upper Jurassic) in northern Scotland. *Scott. J. Geol.* **11**, 51-78.

Threlfall, W.F. (1981) Structural framework of the central and northern North Sea. In: Illing, L.V. and Hobson, G.D. (Eds.) *Petroleum Geology of the Continental Shelf of North-West Europe*, Heyden & Son, London. pp. 98-103.

Tyson, R.V., Wilson, R.C.L. and Downie, C. (1979) A stratified water column environmental model for the type Kimmeridge Clay. *Nature* **277**, 377-380.

Vail, P.R. and Rodd, R.G. (1981) Northern North Sea Jurassic unconformities, chronostratigraphy and sea-level changes from seismic stratigraphy. In: Illing, L.V. and Hobson, G.D. (Eds.) *Petroleum Geology of the Continental Shelf of North-West Europe*, Heyden & Son, London. pp. 216-235.

Whitbread, D.R. (1975) Geology and petroleum possibilities west of the United Kingdom. In: Woodland, A.W. (Ed.) *Petroleum and the Continental Shelf of North-West Europe, Vol. 1 : Geology*, Applied Science Publishers, London. pp. 45-60.

Whiteman, A.J., Rees, G., Naylor, D. and Pegrum, R.M. (1975) North Sea troughs and plate tectonics. In: Whiteman, A.J., Roberts, D. and Sellevolle, M.A. (Eds.) *Petroleum geology and geology of the North Sea and NE Atlantic continental margin, Bergen. Norg. geol. Unders.* **316**, 137-162.

Williams, J.J., and Conner, D.C. and Peterson, K.E. (1975) Piper oilfield, North Sea: fault-block structure with Upper Jurassic beach/bar reservoir sands. *Bull. Am. Ass. Petrol. Geol.* **59**, 1581-1601.

Wood, R. and Barton, P. (1983) Crustal thinning and subsidence in the North Sea. *Nature* **302**, 134-136.

Woodhall, D. and Knox, R.W.O.B. (1979) Mesozoic volcanism in the northern North Sea and adjacent areas. *Bull. geol. Surv. G.B.* **70**, 34-56.

Ziegler, P.A. (1975) North Sea basin history in the tectonic framework of North-Western Europe. In: Woodland, A.W. (Ed.) *Petroleum and the Continental Shelf of North-West Europe, Vol. 1 : Geology*, Applied Science Publishers, London. pp. 131-150.

Ziegler, P.A. (1980) Northwest European Basin: geology and hydrocarbon provinces. In: Miall, A.D. (Ed.) *Facts and Principles of World Petroleum Occurrence. Can. Soc. Petrol. Geol. Mem.* 6. pp. 672-706.

Ziegler, P.A. (1981) Evolution of sedimentary basins in North-West Europe. In: Illing, L.V. and Hobson, G.D. (Eds.) *Petroleum Geology of the Continental Shelf of North-West Europe*, Heyden & Son, London. pp. 3-42.

Ziegler, P.A. (1982a) Faulting and graben formation in western and central Europe. *Phil. Trans. R. Soc. Lond.* A, **305**, 113-143.

Ziegler, P.A. (1982b) Geological atlas of Western and Central Europe, Shell Int. Petrol. Maats. B.V., Distributed by Elsevier, Amsterdam. p. 130.

Chapter 7 Cretaceous

JAKE M. HANCOCK

7.1 General

In the North Sea area, the Cretaceous spans the change from the tectonically-controlled sedimentation of the Early Cretaceous, with its extensive areas of land, to the much quieter fully marine conditions of regional subsidence centred over the axial graben system, which characterises the Late Cretaceous and Tertiary.

7.1.1 Tectonic setting

The tectonic setting of the Cretaceous was mostly a continuation of that established during the Jurassic. Ancient massifs remained stable, passively awaiting submergence by the Cretaceous sea. The fault-controlled basins, both large and medium, subsided further. One of the times of accelerated movement was during the Ryazanian, the earliest age of the period; these movements are called Late Cimmerian. In some areas the movements were largely along pre-existing lines of faulting (e.g. Moray Firth Basin), but in the Viking Graben new faults were also initiated, particularly outside the original margins of the trough, which had the effect of broadening the depositional basin (Fig. 7.1a). This arrangement does not occur everywhere, for in the Piper field, between the Halibut Horst and Fladen Ground Spur, faulting migrated basinwards (Fig. 7.2).

In some basins the rate of subsidence across these faults was faster than the rate of supply of clastics, and in such areas the water deepened. In parts of the Central Graben this overdeepening continued even during the weaker movements of the later Late Cretaceous,

although as Ziegler (1978) has remarked, tectonics in that basin are complicated by strong salt diapirism.

Whilst control by marginal faults (taphrogenism) is an obvious feature of most of these Cretaceous basins, there was additionally a true downwarping on a major scale within some basins (e.g. Egersund). Some geologists believe that in addition to the deepening of the basins, there was an uplift of the margins, either by faulting or by tilting. In practice, except where fault movement is very large as in figure 2.1, it is difficult to disentangle such possible movements from eustatic changes of sea-level. Ziegler (1978) has remarked that major rifting phases seem to coincide with sharp falls of sea-level.

Not all structural changes during the Cretaceous were matters of broad downwarping and extensional faulting. Some areas that had accumulated a thick sequence of Jurassic sediments now reversed the process and were uplifted. Such uplifts are commonly elongated and are known as inversion axes. Most of them are Cretaceous-Tertiary events, but a few are Jurassic, (e.g. Pompeck's Swell, Voigt 1963; Kemper 1974). The change from basin to inversion axis was usually episodic and there is no general pattern of timing. On some (e.g. Central Graben), the uplift was completed before the late Palaeocene (Heybroek, 1975), whilst in others the main uplift was mid-Tertiary, as in the Weald of south-east England. In the Sole Pit inversion (Fig. 7.1b; see also 3.1b) the uplift shifted laterally south-westwards over a long time as elegantly shown by Glennie and Boegner (1981), the trough of deposition moving in front; following the Late Cretaceous phase of inversion, erosion cut deep into the Jurassic.

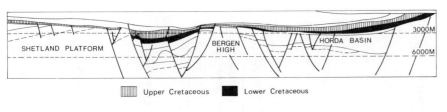

Fig. 7.1(a). Typical E-W cross section in the northern North Sea, around latitude 59 N. (Based on section supplied by Shell Expro)

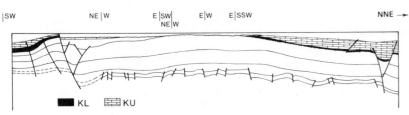

Fig. 7.1(b). Typical section across the Dowsing Fault and Sole Pit Inversion. (Based on section supplied by Shell Expro)

133

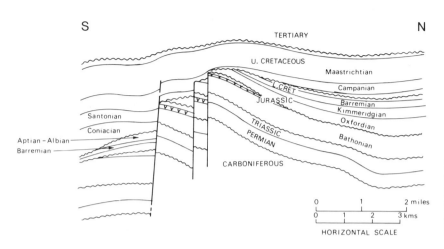

Fig. 7.2 North-south section across the Piper field to show the style of the structural and stratigraphical relationships in the northern North Sea. (Based on Williams, Conner and Peterson 1975.)

In the southern Central Graben the inversion axis was exactly over the maximum Jurassic thickness (Heybroek, 1975).

7.1.2 Sea-levels

Sea-level in north-west Europe at the start of the Cretaceous was relatively low, and much of the North Sea region was land. In a series of basins, often elongated, sedimentation continued from the Jurassic. Through the sea-level oscillations during the Early Cretaceous, the trend of sea level was upwards (Fig. 7.3). During this time the facies in the North Sea were controlled by the usual factors of tectonic setting, climate and availability of source materials. Some time during the Aptian or Albian, sea-level was probably about the same as it is to-day (Vail *et al.*, 1977; Hancock and Kauffman, 1979), and land in the North Sea

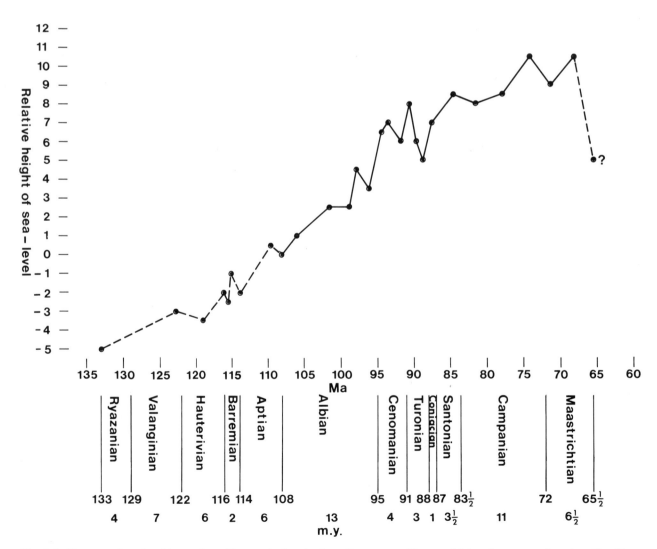

Fig. 7.3. Changes of sea-level in northern Europe during the Late Cretaceous. The pre-Albian changes are based on Kauffman (1977) and are generalised eustatic; the post-Aptian changes are after Hancock and Kauffman, 1979.

was reduced to an archipelago. Sea-level continued to rise during most of the Late Cretaceous, reaching a maximum in the Late Campanian-Middle Maastrichtian, when only the Scottish and Norwegian Highlands continued to be land (Fig. 7.7). These Late Cretaceous sea-levels were exceptionally high, and a normal oceanic pelagic chalk facies was carried on to large areas of the continental shelf; only in the Viking Graben and on the northern part of the Shetland Platform do fine-grained clastics predominate.

7.1.3 Climate and area of land

During the Early Cretaceous there was plenty of land available for erosion, as can be seen from Fig. 7.7, even when inversion axes are discounted. This changed markedly in the Late Cretaceous as sea-level rose (Figs. 7.3 and 7.7). Around the North Sea area, the only ancient land unsubmerged was in the highlands of Scotland, Norway and Greenland. The Upper Cretaceous clastics of the Viking Graben were probably derived from Greenland (Hancock and Scholle, 1975); the silty slipper-clay facies of the Shetland Group can be matched with very late Cretaceous sediments of the Kangerdlugssuaq Group in central east Greenland (Soper *et al.*, 1975). Along inversion axes the uplift in some areas was sufficiently great to produce land. The most prominent occurrences are around the Central Netherlands Inversion, where there are local developments of sandy chalk in various stages from the Turonian to the Maastrichtian. Near the Sole Pit inversion east of Lowestoft there is a sandstone, possibly Turonian, within the Chalk.

There is Early Cretaceous fossil evidence, both onshore and in the North Sea, of a plentiful vegetation (e.g. Hughes, 1976). The considerable quantity of Lower Cretaceous detritus shows that the climate was also sufficiently seasonal to allow erosion of land areas. The scarcity of land during the Late Cretaceous makes it more difficult to determine the climate, but there are indications that away from Greenland it was non-seasonal and possibly arid (Hancock, 1975).

7.1.4 Limitations inherent in the available data

1. All wells are drilled where structures are potentially capable of trapping oil or gas, and thus are generally over structural highs where the succession will be incomplete and hence atypical. This is less of a problem than it used to be because: (a) there are many more wells, particularly ones seeking structures in pre-Cretaceous formations, and (b) it has become easier to make refined interpretations of geophysical traverses. Nevertheless, there is more Lower Cretaceous present than would appear to be the case from published reconstructions.

2. In regions affected by movements in Zechstein salt (for limits see Taylor, this volume, Fig. 4.12) there will be local anomalies that may run contrary to regional trends. Figure 7.4 shows examples of these, including sudden lateral disappearances of individual stages by salt-induced faulting. For discussion see Christian, 1969; Heybroek, 1975; Krey and Marschall, 1975; Taylor, this volume, section 4.8.

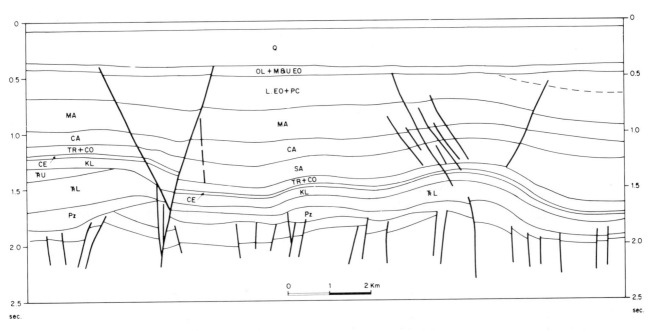

Fig. 7.4. Late Jurassic-Early Cretaceous (commonly called Late Kimmerian) truncation of older Mesozoic strata and the differential movements of Permian Zechstein salt (Pz) during the Late Cretaceous and Early Tertiary in the southern North Sea. Traced from a seismic line. Formation and stage boundaries controlled by well data. Note the apparent complete absence of Santonian Chalk (SA) in the left-hand part of the section and its considerable thickness in the right-hand part.

7.1.5 Difficulties of interpretation

1. The objectives of most North Sea wells are in pre-Cretaceous rocks; the Cretaceous was merely an overburden. Especially in older wells, few details of the Cretaceous were recorded beyond that seen on wireline logs (Gamma Ray, Sonic) and samples were rarely taken for palaeontological dating.

2. Especially in the early days of North Sea exploration, several companies have used different standards for dating, so that the thicknesses of Cretaceous sequences in wells of adjoining fields may be more similar than the published records suggest.

3. The boundary between the Jurassic and Cretaceous can be difficult to fix in reasonably well exposed onshore areas, and in the North Sea it can be much more difficult to define. Some published wells provide misleading data on this. The boundary is commonly placed at a marked change in lithology, partly for convenience, especially in correlating with seismic reflectors, and partly in the mistaken belief that this will correspond with a 'Late Cimmerian regional unconformity'. Birkelund *et al.* (1983) have now shown that in the Danish sector of the Central Graben, for example, the lithological change from the Kimmeridge Clay to the more calcareous Valhall Formation lies within the Valanginian (but see also Table 7.1 on p. 147). Isopachs for the Lower Cretaceous should be regarded as trends rather than absolute figures.

4. In many places (but not as many as is commonly thought) there is a stratigraphic break between the Jurassic and the Cretaceous. This is almost always ascribed to 'Late Cimmerian movements', whereas in many places it was no more than the effect of a marked fall in sea-level that removed the sea from much of the region (Rawson and Riley, 1982). Similarly, over inversion axes (see below) it may be very difficult to distinguish breaks due to tectonic uplift from those caused by eustatic falls in sea-level.

5. There is a common tendency to standardise all tectonic movements into the scheme of Stille (1924). This assumes that all stratigraphic breaks will be synchronous throughout the region, coupled with the practice of using only Stille's broad classification. The so-called 'Cimmerian movements' may be of any age from Late Triassic to Late Aptian. It is tempting to use this tectonic-stratigraphic nomenclature because it allows one to suggest a date for an unconformity without a biostratigraphic dating of the formations above and below. As our knowledge of North Sea geology has increased, these Cimmerian movements have been subdivided (as Stille himself did). In 1975 Ziegler referred to early, mid and late movements; by 1978 he recognised six divisions of the Cimmerian phase, and his Austrian phase corresponds with the Late Cimmerian of some other authors. Whereas earth movements may have been in pulses rather than evenly continuous, detailed studies of individual regions show that not all tectonic movements fall into Stille's phases (see Heybroek, 1974; Hancock and Scholle, 1975).

7.1.6 Influence of Cretaceous tectonics on hydrocarbon accumulations

The inversion movements of the Late Cretaceous are significant for hydrocarbon accumulations in several ways:

1. Traps were sometimes formed. This is well shown in the Rijswijk Oil Province in the Netherlands (Bodenhausen and Ott, 1981) where both domal and fault-controlled traps were developed during the Santonian-Campanian (the so-called Sub-Hercynian phase) and the Maastrichtian-Palaeocene (the so-called Laramide phase).

2. Hydrocarbons may be lost. This probably happened to some of the Rotliegend gas in the inversion of the Broad Fourteens Basin of the Dutch offshore (Oele *et al.*, 1981). In the northern part of the Dutch Rijswijk Province, the hydrocarbons derived from Toarcian source rocks were similarly lost by deep erosion accompanying the uplift of the Central Netherlands Inversion (Bodenhausen and Ott, 1981).

3. The downwarp that preceded inversion sometimes carried source rocks to depths that were hot enough for hydrocarbon generation, and reservoir sediments to depths where diagenesis modified porosities and permeabilities. The gas in the Rotliegend reservoirs is derived from the Upper Carboniferous, principally from coals. Eames (1975) expressed the opinion that in the Southern North Sea Basin, the necessary cooking temperature for hydrocarbon expulsion was from igneous intrusions; there is, however, no evidence of the necessary intrusions. It now seems more likely that sufficient temperatures were reached during the Turonian-Santonian downwarp and prior to Maastrichtian inversion; in the Sole Pit Trough these coals have been estimated to reach depths of 4,000-4,800 m (Glennie and Boegner, 1981). The Rotliegend, having been buried to nearly the same depth, has lost half its porosity, partly by compaction and partly by overgrowths on the quartz and feldspar grains and by the growth of authigenic illite and chlorite (Glennie *et al.*, 1978).

7.2 Lower Cretaceous

7.2.1 Palaeogeography

The general distribution of the Lower Cretaceous is shown in Fig. 7.5. Most of the areas now occupied by inversion axes would have been submerged for at least part of the Early Cretaceous, but there was an archipelago of changing proportions throughout that time span. At the beginning of the period the sea occupied only basinal regions, and even of these the Weald and those in the Netherlands were accumulating freshwater sediments. The North Sea region was cut off from the ocean in southern Europe by a land connection running from Ireland, through Wales and central England via the London-Brabant Platform across to the Rhenish Massif and thence to Bohemia. Marine connections were to the east through Germany and Poland, and northwards via the Viking Graben into the early Atlantic. As a result the faunas are northern in character and difficult to correlate with Tethyan regions (Kemper *et al.*, 1981).

The northern and southern seas connected across the western slopes of the London-Brabant Platform during the Late Aptian transgression. The transgressions of the Late Aptian, and more particularly of the

Middle and Late Albian (see Fig. 7.3) more than doubled the submerged area of the North Sea. Only Aptian-Albian sediments are likely to be present in the area outside the basins; for example, in that part of the U.K. sector between the Mid North Sea High and the Sole Pit inversion.

7.2.2 Moray Firth Basin, South Halibut Basin, Witch Ground Graben

The U.K. sector east of the northern Scottish Highlands is a fault-controlled complex of basins and horsts (sometimes referred to collectively as the Moray Firth Basin). They were almost isolated from the Central Graben and only just connected with the Viking Graben via the Witch Ground Graben and Fisher Bank Basin.

The best known of these is the Moray Firth Basin which is itself about 150 km across. Its landward boundaries coincided approximately with the faults of the

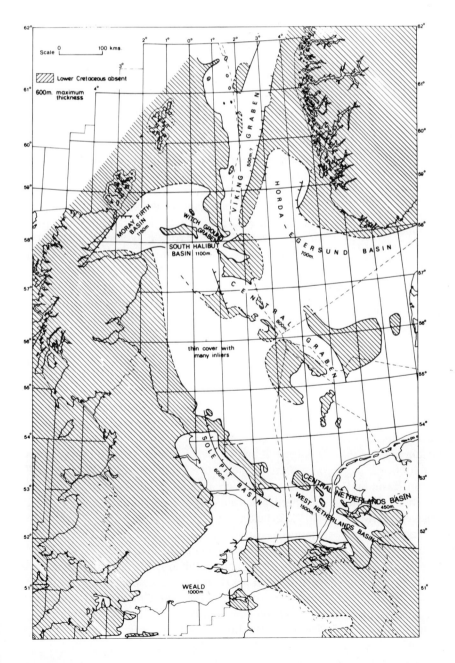

Fig. 7.5. Distribution of the Lower Cretaceous and its principal basins.

present day shore lines and their extension out to sea; that is, the Lossiemouth-Banff coast on the south and the Golspie-Helmsdale coast on the north-west. Within this area, there is a 'main basin' south-east of the line of the Great Glen Fault, and a 'secondary basin' off the Lossiemouth coast, the two being separated by a minor horst, the 'central ridge' of Chesher (1977). Jurassic sediments outcrop within recesses of the Firth and nearer to the shore, but the Lower Cretaceous thickens rapidly offshore to c.650-700 m of pre-Aptian alone. At the foot of fault scarps there are conglomerates and turbidites deposited as submarine fans (Sellwood, 1979). In the centre of the basin there are some 1150 m of shaly sandstones and sandstones, the thickest Lower Cretaceous arenaceous sequence in the North Sea.

Similar block-faulted structures and sediments are to be found in the South Halibut Basin (Woodhall and Knox 1979, Fig. 6D). Many of these faults were active during the Late Cimmerian movements, and in many parts of these three basins there is a seismic break at the top of the Kimmeridgian or Volgian; but even where this is not obvious there is usually a change in colour (and sometimes in lithology) into Cretaceous grey-green to red-brown, often pyritic, sandstone or shaly sandstone. Over local highs there is an obvious break at the base; for example, in the Beatrice Field, Valanginian gravelly and pyritic deep-water channel sands rest directly on Kimmeridgian shales (Linsely *et al.*, 1980). In the deeper parts of these basins there is a complete succession of all the stages of the Lower Cretaceous.

Most of the sands in these basins contain too much clay to be good reservoirs, but at the southern end of the north-east flank of the Witch Ground Graben, south of the Fladen Ground Spur, a Lower Cretaceous reservoir contains gas condensate. The sand-clay alternations result from turbidites deposited by currents from the north-west. They have commonly been assigned to the Barremian-Aptian, but are probably Lower Aptian to Albian (Rawson and Riley, 1982). Here again, the thicknesses of Lower Cretaceous are related to upfaulted blocks of Jurassic in the graben (Brooks, 1977).

Faulting in the general region of the Moray Firth and associated basins seems to have continued into mid-Cretaceous times, for the extent and size of the mid-Cretaceous breaks could hardly be accounted for only by erosion during the mid-Turonian regression (see details in Burnhill and Ramsay, 1981).

At the eastern ends of the South Halibut Basin and Witch Ground Graben there are pale tuffs in the Aptian-Albian, possibly from a volcanic centre at the southern end of the Viking Graben (Woodhall and Knox 1979, Fig. 2).

7.2.3 Viking Graben

The Lower Cretaceous component of this basin is rather sharply defined by bounding faults. Early Cretaceous downwarping of the basin was accompanied by upfaulting on the east and west: the Late Cimmerian movements (Ziegler, 1978). Within the basin, deposition was probably continuous from the Jurassic, and hundreds of metres (increasing northwards) of shales were deposited (relatively deepwater facies according to Ziegler). Few published wells have penetrated this series.

There are subsidiary basins in the much faulted western flanks of the graben, which are separated by elongated areas without any Lower Cretaceous. Within these smaller basins on the eastern flank of the Shetland Platform, however, there are always breaks between the Jurassic and Cretaceous, and it may be that there are no pre-Barremian sediments preserved. Relatively thin Aptian-Albian marls or limestones commonly start the Cretaceous. On the margin of the Shetland Platform there are up to 50 m of sandy sediments.

7.2.4 Horda-Egersund Basin

At its northern end this basin may have connected at times with the Viking Graben across the Vestland High. South-eastwards and eastwards it passes into the Danish-Polish Trough. The centre of thick deposition seems to have been similar to that in the Jurassic Egersund Basin, about 150 km south-west of Stavanger, where thicknesses of 500-700 m are common. Sedimentation was continuous from the Jurassic in most of the central region, but in some areas the junction is unconformable, possibly as the result of salt diapirism. The sequence is made of grey clays or shales, that are sometimes pyritic and glauconitic, with a little limestone and dolomite. Thin sandstones are rare and show no regular pattern, except near the Scandinavian border of the Danish-Polish Trough, where Ziegler (1982) reports they become a major component of the Lower Cretaceous.

The flanks of the basin, although gradational, also show a break with the Jurassic, and extend as far as the Jurassic onto the Vestland High. The lowest part of the Cretaceous in some places here is non-marine with coal.

7.2.5 Central Graben

As in the Egersund Basin, the Lower Cretaceous is 500-700 m thick, but occasionally exceeds 800 m in the Dutch sector. Again most of the succession is grey shales, but marls are also prominent in some areas, and parts of the succession contain a little sand. Sedimentation was continuous from the Jurassic in the northern and central parts of the graben (e.g. Riise, 1977; Birkelund *et al.*, 1983), but in the Dutch sector there is a distinct break, and the earliest Cretaceous is often sandy Valanginian (Heybroek, 1975). The Dutch part of the Central Graben is also distinguished by a further total break in the Upper Barremian, whilst inversion uplift very late in the Cretaceous has allowed the total erosion of the Lower Cretaceous over large areas (see Heybroek, 1975, Fig. 7).

At times during the Mesozoic the Central Graben connected south-eastwards with the Central Netherlands Basin, but during part of the Early Cretaceous they were separated by the Texel-Ijsselmeer High. North of this high lies the little Vlieland Basin in which the Valanginian sands at the base of the marine succession form the gas reservoirs of the Zuidwal, Leeuwarden and Harlingen fields (Cottençon *et al.*, 1975). As usual, the Upper Carboniferous is believed to be the source, and the traps are formed between Zechstein anhydrite and Hauterivian clays.

Within the Zuidwal field, and partly forming the trap structure, is a pre-Valanginian volcanic neck containing a trachyte and phonolite explosion breccia (Cottençon *et al.*, 1975). There have been suggestions that it might be Cretaceous, partly because of the need to find a source for the widespread trachytic tuffs in the Aptian of north-west Europe (Jeans *et al.*, 1977, 1982; Gaida *et al.*, 1978; Zimmerle, 1979), but there are several $^{40}Ar/^{39}Ar$ ages of 144-145 Ma that are definitely pre-Cretaceous (Dixon *et al.*, 1981). The minor phonolitic intrusions in the Permian on the flank of the Mid North Sea High at 138 ± 4 Ma may be truly Cretaceous.

7.2.6 Broad Fourteens-Central Netherlands Basin

In the onshore Central Netherlands Basin, the Jurassic-Cretaceous boundary lies within non-marine sediments (Hageman and Hooykaas, 1980; Kemper, 1974). Eastwards, the basin is barely separated from the Lower Saxony Basin in Germany by the East Netherlands Triassic High. Just within the Lower Saxony Basin, straddling the Netherlands-German border, lies the Schoonebeek oil field, the largest onshore oil accumulation in western Europe (Troost, 1981). The reservoir is formed by a regressive coastal barrier sequence of the Bentheim Sandstone (Valanginian). In spite of porosities of up to 32% and original oil saturations as high as 95%, the high viscosity of the crude oil has made recovery increasingly difficult (Troost, 1981).

7.2.7 West Netherlands Basin

Only a bare outline can be given here of the considerable work on the Lower Cretaceous of the Netherlands (e.g. Haanstra, 1963; Heybroek, 1974; Hageman and Hooykaas, 1980). The general succession is similar to that in the Central Netherlands Basin. The two basins are now separated by the Central Netherlands Inversion, but even during the Early Cretaceous there was a partial barrier in the form of the Zandvoort Ridge and Maasbommel High (Heybroek, 1974, Fig. 5). Erosion following inversion uplift has removed much of the original Lower Cretaceous.

The Jurassic-Cretaceous boundary is contained within the non-marine Delfland Group. The Valanginian transgression brought the sea back into the basin, although probably a little later than in the Central Netherlands Basin. There was a further extension of the sea on the flanks of the Zandvoort Ridge during the Aptian. Throughout the Early Cretaceous the clastic material was derived from the London-Brabant Platform to the south.

On the southern side of the basin there is a succession of east-west trending bodies of sandstone (e.g. Vlieland Sandstone, Rijswijk Sandstone), but northwards, almost the whole of the Valanginian, Hauterivian and Barremian is made up of the Vlieland Shale, which is grey and slightly marly.

Glauconite appears in the Upper Barremian De Lier Sand-Shale member, whilst the Lower Albian Holland Greensand is so rich in glauconite as to be bright green. Most of the Aptian-Albian is represented by the Holland Shales and Marls, which are usually micaceous, silty and slightly glauconitic, but northwards they become more argillaceous.

Some of the sandstones are sufficiently clean to form reservoirs, notably the Rijswijk, Berkel, Ijsselmonde and De Lier Sands, together forming the Rijswijk Province (excellent description by Bodenhausen and Ott, 1981). Most of the oil and gas is derived from the Toarcian Posidonia Shale, with some contribution from the basal Cretaceous Delfland Group.

7.2.8 Sole Pit Basin

The Lower Cretaceous of this basin is best known onshore where it outcrops in Yorkshire (Neale, 1974), Lincolnshire (Swinnerton and Kent, 1976; Casey, 1974; Gallois, 1975) and Norfolk (Casey and Gallois, 1973); there is also much information in Casey (1971) and Rawson *et al.* (1978). An example of the offshore succession is given in Fig. 7.6. The inversion tectonics of the Sole Pit High have been described by Glennie (1981) and their extension onshore by Kent (1980); a section across the inversion axis is shown in Fig. 7.1b.

The interpretation of the stages of the Lower Cretaceous succession in Fig. 7.6 is taken from the published well record, and is probably not correct at the base. The log is probably representative of the stratigraphic subdivisions applied, with the support of minimal faunal dating, to rock units that had no known hydrocarbon-bearing reservoirs at the time of drilling (1968). The Spilsby Sandstone probably extends down into the Volgian-Ryazanian. Even the onshore succession in southern Lincolnshire is largely complete across the Volgian-Ryazanian (i.e. Jurassic-Cretaceous) boundary within the Spilsby Sandstone (Casey, 1974), and the same is true in the central part of the basin between Norfolk and the Sole Pit inversion (Glennie, 1981). For north-east of the Sole Pit area, see Crittenden (1982).

Most of the succession is represented by the Speeton Clay, grey to black pyritic clays with occasional cementstone bands, but onshore, nearer the western borders of the basin, there is a more complex alternation of sandstones, ironstones, clays and limestones. The facies becomes geographically more uniform in the Albian, most of which is represented by the Red

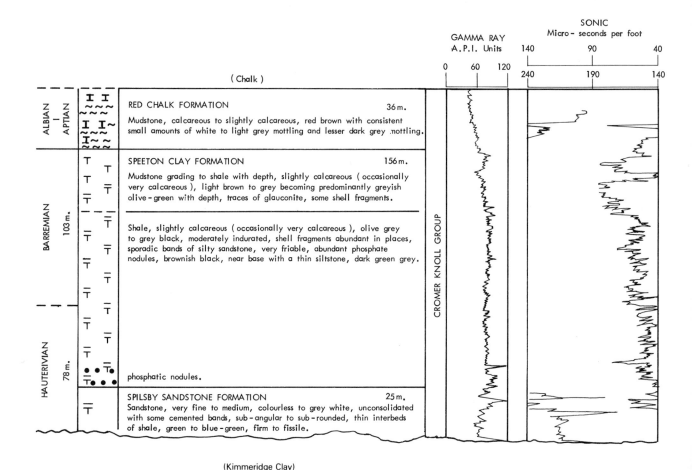

Fig. 7.6. Type section of the Lower Cretaceous Cromer Knoll Group in the southern North Sea. Burmah 48/22-2.

Chalk, a condensed ferruginous marlstone with some of the features of the Tethyan Ammonitico Rosso (Jeans, 1973; Sellwood, 1979).

7.3 Upper Cretaceous

7.3.1 Palaeogeography

During the Late Cretaceous the stable massifs and highs within the North Sea (London-Brabant Platform, the Mid North Sea High, the Ringkøbing-Fyn High, Shetland Platform and Vestland High) were passive except for some faulting marginal to the Viking Graben, and to a much lesser degree in the Central Graben. Most of the London-Brabant Platform, Mid North Sea High and Ringkøbing-Fyn High were submerged during the Albian, and this submergence was completed in the Campanian, when the sea also spread over the Shetland Platform, including the Fladen Ground Spur and Halibut Horst (Fig. 7.8).

Thus the only land in the North Sea region was in the areas now occupied by: (a) Norway; (b) the Utsira High, a small area of the former Vestland High; (c) the Highlands of Scotland, including a number of islands over the East Shetland Platform during the earlier part of the period; (d) a small area near the southern end of the Sole Pit Inversion for a short time; (e) much of the Central Netherlands Inversion from some time in the Turonian onwards.

Almost all the North Sea was basinal in a facies sense, with thicknesses away from massifs of typically 400-500 m when complete (Fig. 7.8). Within the broad basinal regions were a number of trenches with much greater maximum thicknesses: Viking Graben (more than 1,800 m in the far north); Central Graben (more than 1,200 m, 1,350 m if Danian is included; the South Halibut Basin and Witch Ground Graben (each 1,000 m) were structurally not connected to the Central Graben, although on the map they lie as a north-westerly extension of it; Central Netherlands Basin, north of the Central Netherlands Inversion (more than 1,200 m); West Netherlands Basin, south of the Central Netherlands Inversion (more than 1,600 m); Sole Pit Basin, west of the Early Cretaceous Sole Pit Basin (1,000 m). The greatest trench of all is outside the North Sea: this is the Danish-Polish Trough with thicknesses up to nearly 2,000 m.

Much of the Upper Cretaceous of the Moray Firth Basin was removed by erosion during the Tertiary, but it never continued its great downwarping of the Jurassic and Early Cretaceous. The Horda-Egersund Basin had largely ceased to be an area of major downwarp; the maximum thicknesses seldom exceed 500 m.

The onshore Upper Cretaceous Chalk accumulated at depths of 100 to around 600 m (Scholle, 1974; Kennedy and Garrison, 1975; Hancock, 1976). In parts of the trenches of the North Sea greater depths prevailed, possibly reaching 1,000 m (Hancock and Scholle,

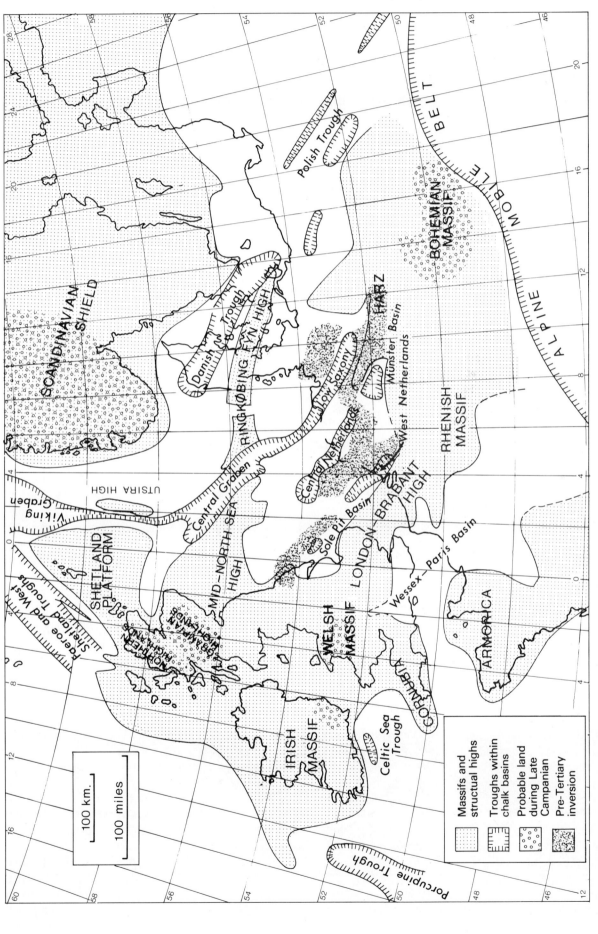

Fig. 7.7. Geological setting of the Upper Cretaceous in north-west Europe. Based on Hancock (1976) with some modifications from Ziegler (1982).

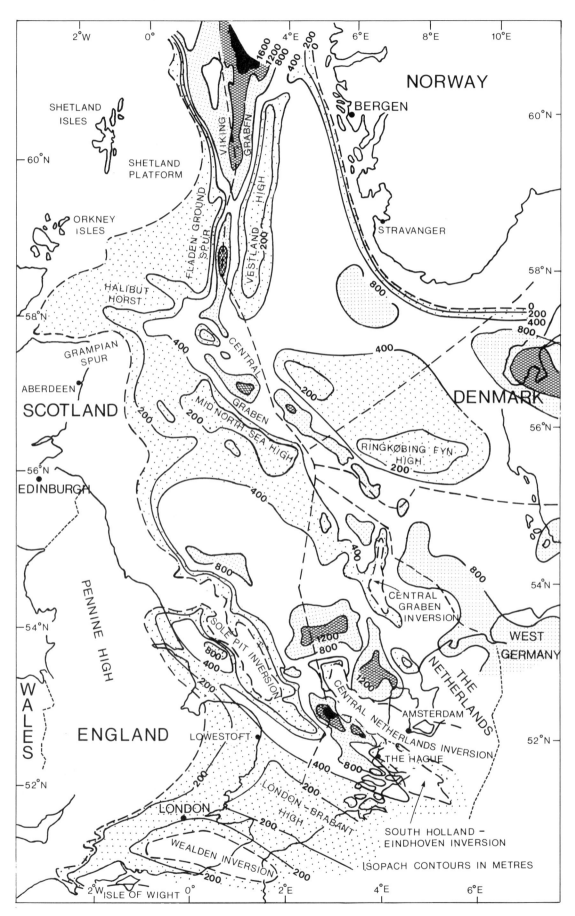

Fig. 7.8. Isopachs of the Chalk Group (Upper Cretaceous plus Lower Palaeocene) and Shetland Group in the region of the North Sea. Based on Hancock and Scholle (1975) with modifications from Hardman (1982), Ziegler (1982) and elsewhere.

1975). Much of the chalk in these trenches was probably deposited by mass-flows from the flanks of the trench.

7.3.2 Distribution of lithologies in the Upper Cretaceous

Most of the Upper Cretaceous of the southern North Sea is chalk (Fig. 7.9) and this lithology often continues into the Lower Palaeocene (so-called Danian or Ekofisk Formation), particularly in the Central Graben (Fig. 7.10). North of latitude 57° a clastic component enters and chalk decreases in proportion. In the Viking Graben and over the Shetland Platform north of 59½° almost the whole succession is clastic (the Shetland Group), but even at the most northerly explored limits of the Graben there are still a few metres of chalk in the Maastrichtian (Fig. 7.11).

In the Central Graben, the clastic component is mainly of pre-Santonian age and is clay mixed with the chalk, although occasionally there are traces of sand, and there can be scattered dolomite crystals in the chalk. In the southern part of the Viking Graben it is the middle part of the succession that contains the least chalk, but because of an unequal representation of the different stages, this development of claystone, sometimes pyritic, or marl with limestone streaks, occurs in the Coniacian-Lower Campanian. The underlying Cenomanian-Turonian, thin if present at all, is of marly chalk, as is the overlying Upper Campanian-Maastrichtian. All changes, both vertically and laterally, are gradational. At the northern end of the Graben nearly the whole succession is clastic; the clay has become silty, and in the Santonian-Campanian there are streaks of quartz sand. The Shetland Group is a slipper-clay facies, dark grey to black laminated silty shale similar to the Gault of southern England and much of the Pierre Shale of western Kansas (Hancock, 1975), and is possibly a continuation of the Kangerdlugssuaq Group of east Greenland.

There is a minor clastic component through much of the Upper Cretaceous of the West Netherlands Basin. The Cenomanian Texel Chalk is marly and at the margins of the basin the Texel Greensand would seem to be a true greensand facies, calcareous glauconitic sandstone. The Texel Chalk is capped by an equivalent of the Plenus Marl, and there is also a widespread distribution of marl within the Coniacian-Santonian Chalk. All stages from the Turonian onwards can contain sand near the Central Netherlands Inversion.

The dominance of chalk as a facies in the Upper Cretaceous of the North Sea as a whole, and the fact that it forms the reservoirs of the Dan and Ekofisk fields, makes it advisable to examine its lithology more closely (see also Table 7.1).

7.3.3 Chalk

Chalk is a very fine-grained limestone (micrite of Folk's classification) that is largely an accumulation of the skeletons of planktonic marine algae of the class Haptophyceae (the golden-brown algae). These algae include calcispheres, thoracospheres and *Nannoconus*. Other organisms are collectively subordinate, but include planktonic and benthonic foraminifera, fragments of bivalves (notably *Inoceramus*), echinoderm plates and bryozoa. Not all Haptophyceae have hard skeletons, but those that do, secrete calcite in little tablet-shaped plates around ½-1 μm across. The plates are arranged in rings called coccoliths, and in the living alga some 7-20 coccoliths are grouped into a globular coccosphere about 10-13 μm in diameter, embedded in a pliable membrane. Complete coccospheres are rare in the Chalk, but complete coccoliths are quite common, though still subordinate to single plates. One could define chalk as a coccolithic limestone.

Such small particles would take years to sink but for the fact that golden-brown algae are the staple diet of copepods. These little crustacea, only a few mm long, are the most abundant multicellular animals in the sea. Their faecal pellets take only a day or so to drop to the bottom of a shelf sea; each pellet contains tens of thousands of coccoliths.

Fundamental to the understanding of chalks is that coccoliths are secreted as low-magnesian calcite. As such they are stable at surface temperatures and pressures, whereas in almost all other carbonate sediments high-magnesian calcite and/or aragonite predominates, and both of these are unstable at surface temperatures and pressures. Most carbonate sediments start to undergo diagenesis as soon as the skeletal particles are no longer protected by organic membranes. Chalk, however, can remain unchanged almost indefinitely, and secondary porosity does not develop.

Even before diagenesis, the coccolithic plates do not have a completely random distribution, but are partly grouped into irregular clusters, each of which may contain up to perhaps 50 plates, but equally may contain only 3 or 4. The effect is to produce many relatively large pores, 6 μm or more across, which would be absent without the clustering. Chalk, as it accumulates, can have a porosity as high as 70%, but nearly half of this is lost by dewatering during the first tens to hundreds of metres of burial (i.e. to porosities of 35-47%; Scholle, 1977). These are typical of values obtained for onshore white-chalk (Hancock, 1963; Scholle, 1974; Carter and Mallard, 1974). Unfortunately, this high porosity is normally accompanied by very low matrix permeabilities (2-13 mD, typically 6-8 mD), because of the minute intergranular pore diameters, which are mostly in the range of 0.1-1 μm (Harper and Shaw, 1974; Price *et al.*, 1976; Scholle, 1977). Onshore chalks often have a high mass permeability from joints, fractures and passages enlarged by solution, but without these, chalk often acts as a seal rather than a reservoir (Scholle, 1977).

Chalk is not immune to early diagenesis (Kennedy and Garrison, 1975; Hancock, 1976), but it is later diagenesis that is often a problem, and this can be induced by the weight of overburden, by high heat-

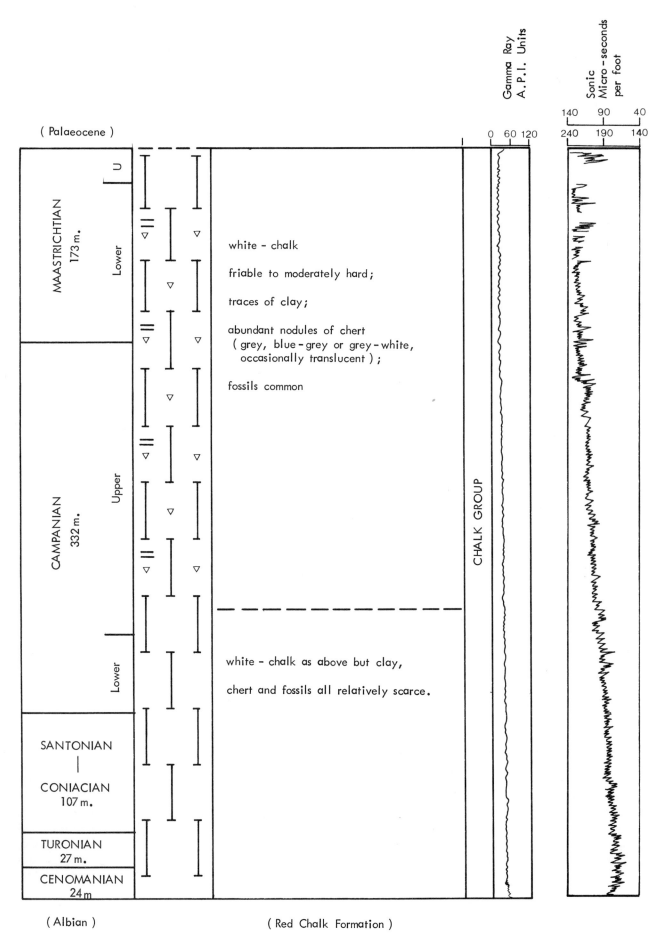

Fig. 7.9. Type section of the Upper Cretaceous Chalk Group in the southern North Sea. Shell-Esso 49/24-1.

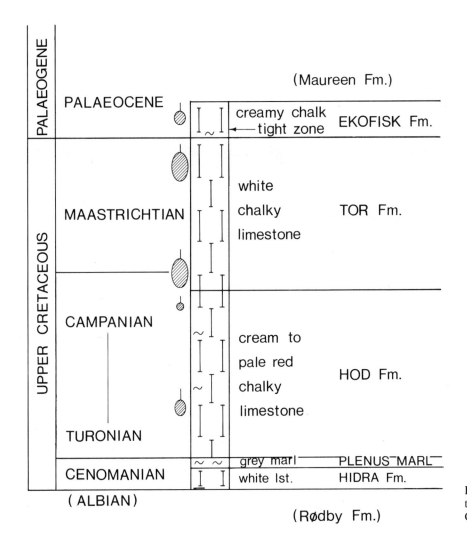

Fig. 7.10. Generalised succession of the Chalk Group in the Central Graben.

flow or by local tectonic stresses (Scholle, 1974; Neugebauer, 1973, 1974; Hancock, 1976). Any one of these can cause: (a) smaller crystals to go into solution and be precipitated as overgrowths on larger coccolithic plates; (b) spot welding; and (c) microspar infill. Given exceptional depths of accumulation, such changes can occur without any of the three factors mentioned above, but this may not be relevant to the North Sea where burial diagenesis is probably the most important factor, although heat-flow cannot be ignored (Oxburgh and Andrews-Speed, 1981), and tectonic stresses have played a role in places (Hancock and Scholle, 1975). Overburden weight produces a proportionate decrease in porosity and matrix permeability; at 1,500-2,000 m, porosities are generally 15-30%, permeabilities 0.1-1 mD; at depths of 2,700-3,300 m the porosities are 2-25% and permeabilities are 0-0.5 mD (Hancock and Scholle, 1975). When the porosity is reduced to 20%, matrix permeabilities are usually less than 0.5 mD (Scholle, 1977a). Considering that in the Dan Field there is an overburden of 1,765 m of Cenozoic (Childs and Reed, 1975), 2,600 m in the Hod field (Hardman and Kennedy, 1980), and 3,200 m in the Ekofisk field (Heur, 1980), it is surprising, at first, that there are any productive chalk reservoirs at all.

The actual porosities in productive reservoirs are much higher: in the Dan Field most of the chalk has porosities of 25-30%, and at two levels it is 30-40% (Childs and Reed, 1975); the average in the Ekofisk field is 30% and can be as high as 40% (Scholle, 1977b); in the Hod field it is over 30% (Hardman and Kennedy, 1980). Scanning electron micrographs show that these are original porosities, which have been preserved by the early entry of the oil itself (late Eocene to early Miocene). In the Hod field, this has involved entry pressures equivalent to a theoretical differential elevation of only 50 m to give 80% saturation in the Maastrichtian chalk (Hardman and Kennedy, 1980, Fig. 3; Hardman, 1982). The emplacement of oil displaces the pore water without which diagenesis is difficult.

Relatively soon after saturation with oil, the chalk was over-pressured, possibly by a combination of retardation of further water expulsion by the deposition of overlying Cenozoic shale and early generation of biogenic gas (Scholle, 1977b). In the Ekofisk area (Fig. 7.12) there are pore-fluid pressures of 7,100 psi at a depth of 3,000 m, where the normal hydrostatic pressure should be 4,300 psi (Harper and Shaw, 1974). Such overpressures are restricted to the Central Graben and do not extend beyond the boundary faults (Scholle, 1977b). The effect of these overpressures is that the oil itself supports about two-thirds of the vast weight of Cenozoic overburden, instead of this being imposed

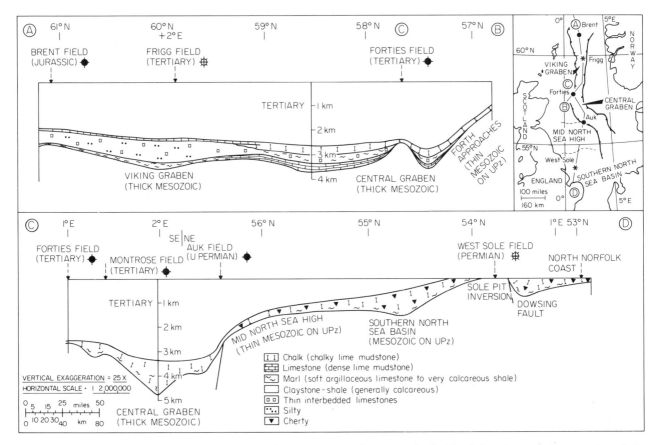

Fig. 7.11. North-south sections showing the lithologies of the Upper Cretaceous in the North Sea; Lower Palaeocene in a chalk facies is included. (From Hancock and Scholle, 1975.)

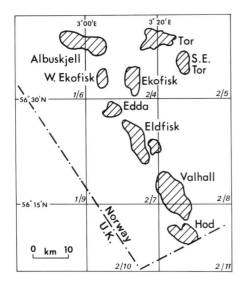

Fig. 7.12. Oilfields in Chalk reservoirs in the Norwegian Central Graben.

on plate-to-plate contacts in the chalk, thus further protecting the original texture and porosity.

Even the presence of oil was not sufficient to protect the chalk from some patchy diagenesis in stratigraphically older reservoirs (e.g. in the Turonian of the Hod Field), possibly because the damage had already been done before oil entered the reservoir. In addition, solution diagenesis associated with fracturing may be marked, as in the Maastrichtian of the Hod Field (Hardman and Kennedy, 1980).

Although, at first sight, the development of a high oil saturation in such fine grained material as chalk may seem surprising, it is readily explained by the laws governing capillary pressure in porous materials. These show that at depths of about 3,000 m and temperatures of around 200°F (94°C), a pressure of only a few pounds per square inch is required to displace water in the matrix of the chalk with oil. Under these conditions, the effect of the very fine pore-size distribution of chalk is outweighed by the extent to which high temperature, pressure and gas/oil ratio reduce the interfacial tension of the water-oil system. This process is, of course, helped by the high gas/oil ratio in these Chalk fields (up to 1000 to 2000 scf/barrel), by the low

Table 7.1. Approximate correlation of Cretaceous formations in the region of the North Sea.

Region	Lower PALAEOCENE	MAASTRICHTIAN	CAMPANIAN	SANTONIAN	CONIACIAN	TURONIAN	CENOMANIAN	ALBIAN	APTIAN	BARREMIAN	HAUTERIVIAN	VALANGINIAN	RYAZANIAN
South England			Upper Chalk	Upper Chalk			Middle Chalk / Lower Chalk	U.G.S. / Gault	Lower Greensand	Weald Clay	Weald Clay	Hastings Beds	Purbeck Beds
West Netherlands		Chalk	Chalk	Chalk	Chalk	Chalk	Texel Marlstone	Upper Holland Marl / Holland G'sand	Lower Holland Marl	De Lier / Ijssel-monde / Berkel S'dst. / Rijswijk S'dst. (Vlieland Shale)	Vlieland Shale	Delfland Group	Delfland Group
Central Graben Danish		Chalk 6	Chalk 5	Chalk 4	Chalk 3	Chalk 2	Chalk 1	Rødby Formation	Valhall Formation	Valhall Formation	Valhall Formation	Valhall Formation	Valhall Formation
Central Graben Norwegian	Ekofisk Chalk	Ekofisk Chalk	Tor Chalk	Hod Chalk	Hod Chalk	Hod Chalk	Plenus Marl / Hidra Chalk	Rødby Formation / Sola Formation	Valhall Formation	Valhall Formation	Valhall Formation	Mandal Formation	Mandal Formation
Viking Graben	Shetland Group	Shetland Group	Shetland Group	Shetland Group	Shetland Group	Shetland Group	Shetland Group	Cromer Knoll Group	Cromer Knoll Group	Cromer Knoll Group	Cromer Knoll Group	Cromer Knoll Group	Cromer Knoll Group
West–Central North Sea	Ekofisk Chalk	Ekofisk Chalk	Tor Chalk	Flounder Chalk	Flounder Chalk	Herring Chalk	Plenus Marl / Hidra Chalk	Valhall Formation	Valhall Formation / Devil's Hole Sandstone	Devil's Hole Sandstone	Devil's Hole Sandstone	Devil's Hole Sandstone	Devil's Hole Sandstone
Sole Pit Basin			Chalk	Chalk	Chalk	Chalk	Plenus Marl / Lower Chalk	Red Chalk	Speeton Clay	Speeton Clay	Speeton Clay	Speeton Clay	Spilsby Sandstone
Eastern England onshore			Flamborough Chalk	Burnham Chalk	Burnham Chalk	Welton Chalk	Lower Chalk / Plenus Marl	Red Chalk / Carstone	Sutterby Marl	Fulletby Beds / Tealby Beds	Claxby Ironstone	Upper Spilsby Sandstone	Spilsby Sandstone / Upper Spilsby Sandstone

viscosity of the oil (36°API; Harper and Shaw, 1974) and by the fracturing of the chalk at an early stage in its history (Watts, 1983).

All the fields are on structural highs resulting from diapirism in Zechstein salt and/or Jurassic shales. Flowage started during the Early Cretaceous, in some places initiated by faulting (Hardman and Eynon, 1978). The movements occurred in pulses whilst the chalk was accumulating, the strongest in the Hod field being at the end of the Coniacian, at the end of the Maastrichtian and at the end of the Early Palaeocene (Hardman and Kennedy, 1980). These movements fractured the chalk and increased the overall permeability to about 12 mD (Scholle, 1977). Earlier ideas of a Palaeogene origin for the oil are now discounted and it is now generally believed to have come from the Upper Jurassic Kimmeridge Shales (Barnard and Cooper, 1981).

A second factor to be considered is that many of these chalks are not simple accumulations of coccolithic material, but are winnowed chalks with repeated hardgrounds, or are chalks that have been redeposited as slumps, as debris flows, or as turbidites shortly after this original accumulation (Kennedy, 1980; Watts *et al.*, 1980). Winnowing of chalks will improve their permeability, but all forms of redeposition will enhance the existing trends that depend on the amount of clay present; clean chalks that have been redeposited have a looser packing, and hence higher porosity, and may well have a higher permeability, but argillaceous chalks will then have still lower permeabilities. A plastic slump with a relatively high clay content, such as the melange described by Hancock and Scholle (1975), would become a seal. This is the situation in much of the Ekofisk 'tight zone'. It is now widely agreed that many of the good reservoirs are in redeposited chalk (Hardman, 1982; Nygaard *et al.*, 1983). Such allochthonous chalks have been known onshore for many years: as debris flows (Voigt, 1962), as turbidites (Voigt and Häntzschel, 1964), as carbonate banks (Kennedy and Juignet, 1974; Hancock, 1976), whilst slumped chalks are known in many places, such as Ireland (Reid in Hancock, 1976), southern England (Gale, 1980), northern France (Kennedy and Juignet, 1974), Denmark (Svendsen in Håkansson *et al.*, 1974) and Rügen (Steinich, 1972).

In addition to the oil-fields in the Norwegian and Danish sectors, there is a gas field at Harlingen in Friesland, northern Netherlands, where the productive reservoir is in the topmost layers of the chalk (Bosch, 1983). Fracturing is weak compared with that in the oil-fields, but again the chalk has been protected from diagenesis by overpressures and by the pressure of the gas itself.

7.4 Acknowledgements

I am grateful to K.W. Glennie and his colleagues of Shell U.K. Exploration and Production Ltd. for much information and helpful advice in the preparation of this paper. Discussions with Drs W.J. Kennedy and P.F. Rawson are much appreciated.

7.5 References

Barnard, P.C. and Cooper, B.S. (1981) Oils and source rocks of the North Sea area. In: Illing and Hobson q.v. 169-175.

Birkelund, T., Clausen, C.K., Hansen, H.N. and Holm, L. (1983) The *Hectoroceras kochi* Zone (Ryazanian) in the North Sea Central Graben and remarks on the Late Cimmerian Unconformity. *Danm. geol. Unders.*, Årbog 1982, 53-72.

Bodenhausen, J.W.A. and Ott, W.F. (1981) Habitat of the Rijswijk oil province, onshore, The Netherlands. In: Illing and Hobson q.v. 301-309.

Bosch, W.J. van den (1983) The Harlingen Field, the only gas field in the Upper Cretaceous Chalk of The Netherlands. In: Kaasschieter and Reijers q.v. 145-156.

Brooks, J.R.V. (1977) Exploration status of the Mesozoic of the U.K. northern North Sea. *Mesozoic northern North Sea Symposium, Oslo*, paper 2, Norsk Petroleumsforening. 28 pp.

Carter, P.G. and Mallard, D.J. (1974) A study of the strength, compressibility, and density trends within the chalk of south-east England. *Q.J. engng. Geol.* 7, 43-55.

Casey, R. (1971) Facies, faunas and tectonics in late Jurassic-early Cretaceous Britain. In: Middlemiss, F.A., Rawson, P.F. and Newall, G. (Eds.) *Faunal Provinces in Space and Time.* Geol. J. Spec. Issue 4, 153-168.

Casey, R. (1974) The ammonite succession at the Jurassic-Cretaceous boundary in eastern England. In: Casey, R. and Rawson, P.F. (Eds.) *The boreal Lower Cretaceous*, Geol. J. spec. issue 5, (mis-dated 1973), pp. 193-266.

Casey, R. and Gallois, R.W. (1973) The Sandringham Sands of Norfolk. *Proc. Yorks. geol. Soc.* 40, 1-22.

Chesher, J.A. (1977) A review of the offshore geology of the Moray Firth. In: Gill, G. (Ed.) *The Moray Firth Area Geological Studies*, Inverness Field Club, Inverness. pp. 60-71.

Childs, F.B. and Reed, P.E. (1975) Geology of the Dan field and the Danish North Sea. In: Woodland, A.W. q.v. 429-438.

Christian, H.E. Jr. (1969) Some observations on the initiation of salt structures of the southern British North Sea. In: Hepple, P. (Ed.) *The exploration for petroleum in Europe and north Africa.* Institute of Petroleum, London. pp. 231-248.

Cottençon, A., Parant, B. and Flacelière, G. (1975) Lower Cretaceous gas-fields in Holland. In: Woodland, A.W. q.v. 403-412.

Crittenden, S. (1982) Lower Cretaceous lithostratigraphy NE of the Sole Pit area in the UK southern North Sea. *J. Petrol. Geol.* 5, 191-202.

Dixon, J.E., Fitton, J.G. and Frost, R.T.C. (1981) The tectonic significance of post-Carboniferous igneous activity in the North Sea Basin. In: Illing and Hobson q.v. 121-137.

Gaids, K.-H., Kemper, E. and Zimmerle, W. (1978) Das Oberapt von Sarstedt und seine Tuffe. Geol Jb. (A) 45, 43-123.

Gale, A.S. (1980) Penecontemporaneous folding, sedimentation and erosion in Campanian Chalk near Portsmouth, England. Sedimentology 27, 137-151.

Gallois, R.W. (1975) A borehole section across the Barremian-Aptian boundary (Lower Cretaceous) at Skegness, Lincolnshire. *Proc. Yorks. geol. Soc.* 40, 499-503.

Glennie, K.W. and Boegner, P.L.E. (1981) Sole Pit inversion tectonics. In: Illing and Hobson q.v. 110-120.

Glennie, K.W., Mudd, G.S. and Nagtegaal, P.J.C. (1978) Depositional environment and diagenesis of Permian Rotliegendes sandstones in Leman Bank and Sole Pit areas of the U.K. southern North Sea. *J. geol. Soc. Lond.* **135**, 25-34.

Haanstra, U. (1963) A review of Mesozoic geological history in The Netherlands. *Geol. Mijnbouw* **21**, 35-57.

Hageman, B.P. and Hooykaas, H. (1980) *Stratigraphic nomenclature of The Netherlands.* Koninklijk Nederlands Geologisch Mijnbouwkundig Genootschap, The Hague, 77 pp. & 36 enclosures.

Håkansson, E., Bromley, R. and Perch Nielsen, K. (1974) Maastrichtian chalk of north-west Europe—a pelagic-shelf sediment. In: Hsü, K.J. and Jenkyn, H.C. (Eds.) *Pelagic sediments: on land and under the sea.* Spec. Publ. int. Ass. Sediment. 1, pp. 211-233.

Hancock, J.M. (1975) The sequence of facies in the Upper Cretaceous of northern Europe compared with that in the western interior. In: Caldwell, W.G.F. (Ed.) *Cretaceous system in the Western Interior of North America.* Spec. Pap. Geol. Ass. Canada 13, pp. 83-118.

Hancock, J.M. (1976) The petrology of the Chalk. *Proc. Geol. Ass.* **86**, (for 1975), 499-535.

Hancock, J.M. and Kauffman, E.G. (1979) The great transgressions of the Late Cretaceous. *J. geol. Soc. Lond.* **136**, 175-186.

Hancock, J.M. and Scholle, P.A. (1975) Chalk of the North Sea. In: Woodland, A.W. q.v. 413-427.

Hardman, R.F.P. (1982) Chalk reservoirs of the North Sea. *Bull. geol. Soc. Denmark* **30**, 119-137.

Hardman, R.F.P. and Eynon, G. (1978) Valhall Field—a structural/stratigraphic trap. *NPF Mesozoic Northern North Sea Symposium*, Oslo 1977. MNNSS/14, 33 pp.

Hardman, R.F.P. and Kennedy, W.J. (1980) Chalk reservoirs of the Hod fields, Norway. *The sedimentation of the North Sea reservoir rocks.* Norwegian Petroleum Society (NPF), Geilo.

Harper, M.L. and Shaw, B.E. (1974) Cretaceous-Tertiary carbonate reservoirs in the North Sea. *Offshore North Sea Technology Conference, Stavanger, Norway*, G-IV/4.

Hesjedal, A. and Hamar, G.P. (1983) Lower Cretaceous stratigraphy and tectonics of the south southeastern Norwegian offshore. In: Kaasschieter and Reijers q.v. 135-144.

Heur, M.D' (1980) Chalk reservoir of the West Ekofisk field. *The sedimentation of the North Sea reservoir rocks.* Norwegian Petroleum Society (NPF), Geilo.

Heybroek, P. (1974) Explanation to tectonic maps of The Netherlands. *Geol. Mijnbouw* **53**, 43-50 & 2 maps.

Heybroek, P. (1975) On the structure of the Dutch part of the Central North Sea Graben. In: Woodland, A.W. q.v. 339-351.

Hughes, N.F. (1976) Plant succession in the English Wealden strata. *Proc. Geol. Ass.* **86** (for 1975), 439-455.

Illing, L.V. and Hobson, G.D. (1981) (Eds.) *Petroleum geology of the continental shelf of north-west Europe* Heyden, London. 521 & xvii pp. & maps.

Jeans, C.V. (1973) The Market Weighton structure; tectonics, sedimentation and diagenesis during the Cretaceous. *Proc. Yorks. geol. Soc.* **39**, 409-444.

Jeans, C.V., Merriman, R.J. and Mitchell, J.G. (1977) Origin of Middle Jurassic and Lower Cretaceous fuller's earths in England. *Clay Miner.* **12**, 11-44.

Jeans, C.V., Merriman, R.J. , Mitchell, J.G. and Bland, D.J. (1982) Volcanic clays in the Cretaceous of southern England and Northern Ireland. *Clay Miner.* **17**, 105-156.

Kaasschieter, J.P.H. and Reijers, T.J.A. (1983) (Eds.) *Petroleum geology of the southeastern North Sea and the adjacent onshore areas.* Geologie en Mijnbouw, special edition, 62, 239 pp.

Kemper, E. (1974) The Valanginian and Hauterivian stages in north-west Germany. In: Casey, R. and Rawson, P.F. (Eds.) The boreal Lower Cretaceous, Geol. J. spec. Issue 5, 327-344 (mis-dated 1973).

Kemper, E., Rawson, P.F. and Thieuloy, J.-P. (1981) Ammonites of tethyan ancestry in the early Lower Cretaceous of north-west Europe. *Palaeontology* **24**, 251-311.

Kennedy, W.J. (1980) Aspects of chalk sedimentation in the southern Norwegian offshore. *The sedimentation of the North Sea reservoir rocks.* Norwegian Petroleum Society (NPF), Geilo.

Kennedy, W.J. and Garrison, R.E. (1975) Morphology and genesis of nodular chalks and hardgrounds in the Upper Cretaceous of southern England. *Sedimentology* **22**, 311-386.

Kent, P. (1975) Closing address. In: Woodland, A.W. q.v. 493-494.

Kent, P.E. (1980) Subsidence and uplift in East Yorkshire and Lincolnshire: a double inversion. *Proc. Yorks. geol. Soc.* **42**, 505-524.

Krey, T. and Marschall, R. (1975) Undershooting salt-domes in the North Sea. In: Woodland, A.W. q.v. 265-273.

Linsley, P.N., Potter, H.C. McNab, G. and Racher, D. (1980) The Beatrice field, inner Moray Firth, U.K. North Sea. *Am. Ass. Petrol. Geol. Mem.* **30**, 117-129.

Michelsen, O. (1982) (Ed.) Geology of the Danish Central Graben. *Danm. geol. Unders. ser. B*, **8**, 1-333.

Neale, J.W. (1974) Cretaceous. In: Rayner, D.H. and Hemingway, J.E. (Eds.) *The geology and mineral resources of Yorkshire.* Yorkshire Geol. Soc. (Occasional Publ. 2). pp. 225-243.

Neugebauer, J. (1973). The diagenetic problem of chalk. *Neues Jb. Geol. Paläont. Abh.* **143**, 223-245.

Neugebauer, J. (1974) Some aspects of cementation in chalk. In: Hsü, K.J. and Jenkyns, H.C. (Eds.) *Pelagic sediments on land and under the sea.* Spec. Publ. int. Ass. Sediment. 1, pp. 149-176.

Nygaard, E., Lieberkind, K. and Frykman, P. (1983) Sedimentology and reservoir parameters of the Chalk Group in the Danish Central Graben. In: Kaasschieter and Reijers q.v. 177-190.

Oele, J.A., Hol, A.C.P.J. and Tiemens, J. (1981) Some Rotliegend gas fields of the K and L blocks, Netherlands offshore (1968-1978)—a case history. In: Illing and Hobson q.v. 289-300.

Oxburgh, E.G. and Andrews-Speed, C.P. (1981) Temperature, thermal gradients and heat flow in the southwestern North Sea. In: Illing and Hobson q.v. 141-151.

Price, M., Bird, M.J. and Foster, S.S.D. (1976) *Chalk pore-size measurements and their significance.* Water Services 80, pp. 596-600.

Rawson, P.F., Curry, D., Dilley, F.C., Hancock, J.M., Kennedy, W.J., Neale, J.W., Wood, C.J. and Worssam, B.C. (1978) *A correlation of Cretaceous rocks in the British Isles.* Specl. Rep. geol. Soc. Lond. 9, 70 pp.

Rawson, P.F. and Riley, L.A. (1982) Latest Jurassic-Early Cretaceous events and the 'Late Cimmerian Unconformity' in North Sea area. *Bull. Am. Ass. Petrol. Geol.* **66**, 2628-2648.

Riise, R. (1977) *Lithology wells 2/8-1 and 2/11-1.* Norwegian Petroleum Directorate, Stavanger. 24 pp. & logs.

Scholle, P.A. (1974) Diagenesis of Upper Cretaceous chalks from England, Northern Ireland, and the North Sea. In: Hsü, K.J. and Jenkyns, H.C. (Eds.) *Pelagic sediments: on land and under the sea.* Spec. Publ. int. Ass. Sediment 1, 177-210.

Scholle, P.A. (1977a) Chalk diagenesis and its relation to petroleum exploration—oil from chalks, a modern miracle? *Bull. Am. Ass. Petrol. Geol.* **61**, 982-1009.

Scholle, P.A. (1977b) *Current oil and gas production from North American Upper Cretaceous chalks.* Circ. U.S. geol. Surv. 767, 51 pp.

Skovbro, B. (1983) Depositional conditions during Chalk sedimentation in the Ekofisk area Norwegian North Sea. In: Kaasschieter and Reijers q.v. 169-175.

Soper, N.J., Higgins, A.C., Downie, C., Matthews, D.W. and Brown, P.E. (1975) Late Cretaceous-early Tertiary stratigraphy of the Kangerdlugssuaq area, east Greenland, and the age of opening of the north-east Atlantic. *J. geol. Soc. Lond.* **132** (for 1976), 85-104.

Steinich, G. (1972) Endogene tektonik in der Unter-Maastricht workommen auf Jasmund (Rügen). *Beih, Z. Geol.* **20**, 1-205.

Stille, H. (1924) *Grundfragen der vergleichenden Tektonik.* Borntraeger, Berlin.

Swinnerton, H.H. and Kent, P.E. (1976) *The geology of Lincolnshire.* Lincolnshire Naturalists' Union, Lincoln. 130 pp.

Troost, P.J.P.M. (1981) Schoonebeek oil field: the RW-2E steam injection project. *Geol. Mijnbouw* **60**, 531-539.

Vail, P.R., Mitchum, R.M. Jr. and Thomson, S. III (1977) Seismic stratigraphy and global changes of sea level, part 4, Global cycles of relative changes of sea level. *Mem. Am. Ass. Petrol. Geol.* **26**, 83-97.

Voigt, E. (1962) Frühdiagenetische deformation der turonen Plänerkalke bei Halle/Westf. *Mitt. geol. StaatInst. Hamburg.* **31**, 146-275.

Voigt, E. (1963) Über Randtröge vor Schollenrändern und ihre Bedeutung im Gebiet der Mitteleuropäischen Senke und angrenzender Gebiete. *Z. dt. geol. Ges.* **114** (for 1962), 378-418.

Voigt, E. and Häntzschel, W. (1964) Gradierte schichtung in der Oberkreide Westfalens. *Fortschr. Geol. Rheinld. u. Westf.* **7**, 495-548.

Watts, N.L. (1983) Microfractures in chalks of Albuskjell Field, Norwegian sector, North Sea: possible origin and distribution. *Bull. Am. Ass. Petrol. Geol.* **67**, 201-234.

Watts, N.L., Lapré, J.F., Schijndel-Goester, F.S. van and Ford, A. (1980) Upper Cretaceous and lower Tertiary chalks of the Albuskjell area, North Sea: deposition in a slope and base-of-slope environment. *Geology* **8**, 217-221.

Woodhall, D. and Knox, R.W. O'B. (1979) Mesozoic volcanism in the northern North Sea and adjacent areas. *Bull geol. Surv. Gt. Br.* **70**, 34-56.

Woodland, A.W. (1975) (Ed.) *Petroleum and the continental shelf of north-west Europe.* 1. Geology. Applied Science Publishers 501 pp.

Ziegler, P.A. (1975) North Sea basin history in the tectonic framework of north-western Europe. In: Woodland, A.W. q.v. 131-149.

Ziegler, P.A. (1978) North-western Europe: tectonics and basin development. *Geol. Mijnbouw* **57**, 589-626 & 4 maps.

Ziegler, P.A. (1982) Geological atlas of western and central Europe. Elsevier, Amsterdam. 130 pp., 40 pls.

Zimmerle, W. (1979) Lower Cretaceous tuffs in north-west Germany and their geotectonic significance. In: Wiedmann, J. (Ed.) Aspekte der Kreide Europas 385-402, Int. Un. geol. Sci (A) 6.

Chapter 8   Cenozoic

J.P.B. LOVELL

8.1 Introduction

The first oilfield to be found in the U.K. sector of the North Sea was discovered in December 1969, in Palaeocene sandstones of the Montrose Field (initial recoverable reserves c. 14×10^6 m³, or 90 million barrels) (Fig. 8.1). Less than a year later a much larger (initial recoverable reserves c. 320×10^6 m³ or 2000 million barrels) Palaeocene sandstone oil reservoir was discovered in the Forties Field, 50 km to the north-west; in 1971 the Frigg Field, one of the world's largest offshore gas-fields (recoverable reserves c. 200×10^9 m³ or 7×10^{12} ft³), was found on the border of the British and Norwegian sectors in lower Tertiary sandstones. By now it was clear that in the northern North Sea the lower Cenozoic was a prime target for exploration; it

is on these lower Cenozoic rocks that attention is concentrated here.

In this chapter, a brief outline is given of the facts (structural, stratigraphical, palaeontological, sedimentological), upon which are based the palaeogeographical interpretations that follow. The implications for hydrocarbon exploration and production are discussed briefly in concluding sections.

8.2 Structure

The main structural features of the northern North Sea are shown in Figure 8.2. A map of the present-day structural contours at the top of the Chalk (Fig. 8.3, see also Fig. 8.1) shows that the thick (up to 3,500 m) Cenozoic sequence above fills a basin with an approxi-

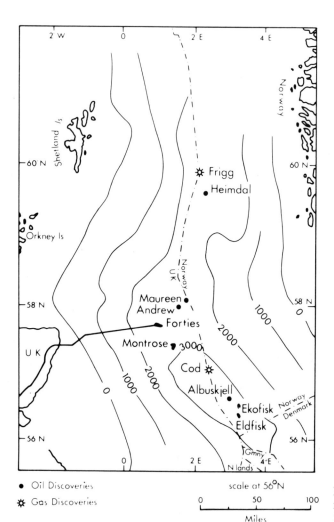

• Oil Discoveries
☼ Gas Discoveries

scale at 56°N

0   50   100

Miles

Fig. 8.1. North Sea hydrocarbon discoveries in Tertiary rocks, and base-Tertiary structural contours (after Walmsley, 1975). (Structural contours in metres below sea level.)

151

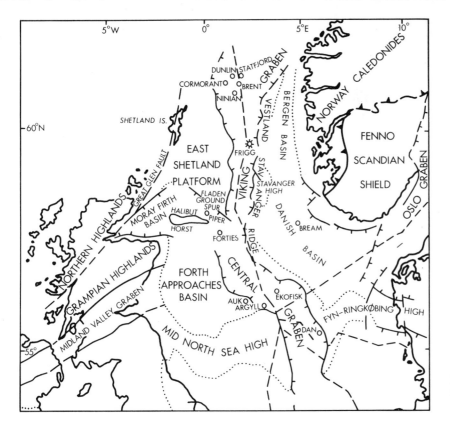

Fig. 8.2. Main structural features of the northern North Sea (after Kent, 1975).

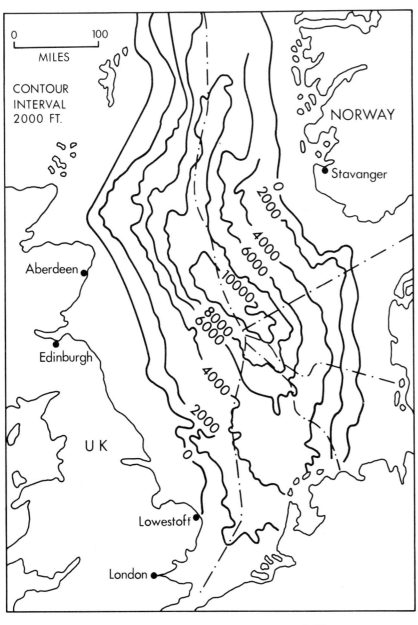

Fig. 8.3. Top-Chalk structural contours (after Parker, 1975).

152

mately north-south axis that follows the trend of the Mesozoic rift systems. The Cenozoic is largely unfaulted, and only slightly deformed as, for example, at the Forties Field where a broad dome provides closure for the reservoir (Fig. 8.4). Drape, compaction and late movement are involved in the structural development of fields like Forties, Frigg and Montrose; salt movement has controlled the tectonic evolution of Ekofisk (Blair, 1975; Byrd, 1975).

8.3 Stratigraphy

The proposed standard lithostratigraphical nomenclature for the Tertiary of the central and northern North Sea is shown in Figure 8.5. This compilation by Deegan and Scull (1977) has been used as a basis for more detailed work; for example, Mudge and Bliss (1983) use the scheme for their studies of the Palaeocene of the northern North Sea, and Sutter (1980) identifies

some of the smaller sand bodies as formations. The Deegan and Scull nomenclature will be followed here, and the character of some of the more important divisions will be discussed below. Their relationship is indicated in Figure 8.6. The rocks are mainly mudstones; subordinate sandstones are found, especially near the base of the sequence.

The stratigraphy of the Tertiary mudstones of the southern North Sea is summarised in Table 8.1, a compilation of Rhys (1974). South of 55°N, halokinesis in the Zechstein has given rise to local variations in thickness in the Tertiary; the major regional variations in thickness are related to widespread unconformities within the Tertiary that affect all subdivisions (see, for example, the discussion of early Palaeogene stratigraphy by Knox *et al.*, 1981).

Throughout the North Sea palaeontological stratigraphy is supported by correlation of lower Tertiary tuffs. There were two main phases of pyroclastic sedi-

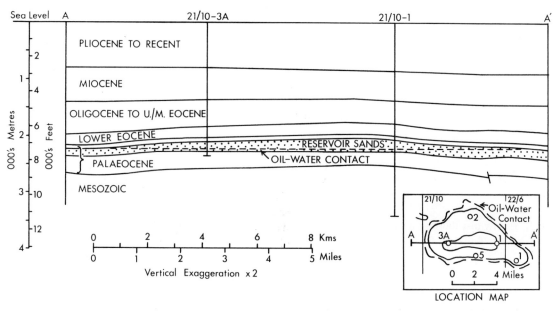

Fig. 8.4. Structural section east-west across Forties Field (after Walmsley, 1975).

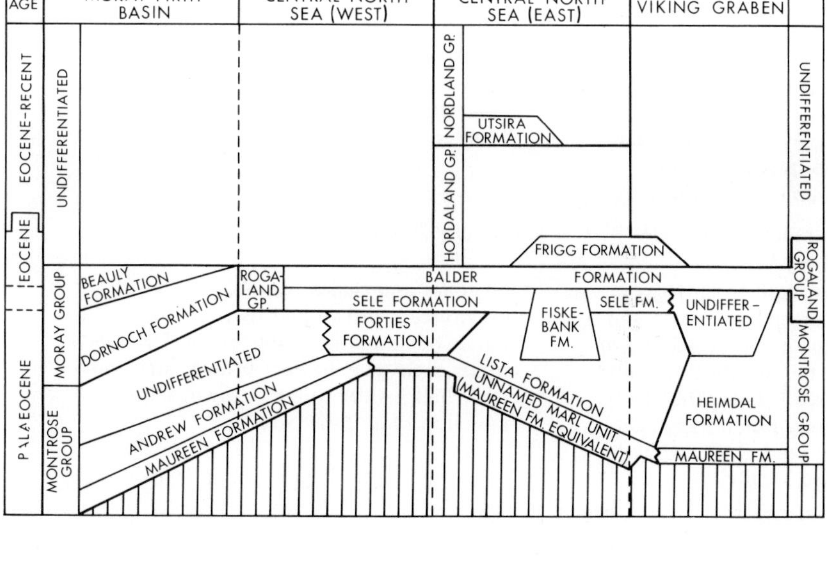

Fig. 8.5. Tertiary lithostratigraphical nomenclature (after Deegan and Scull, 1977).

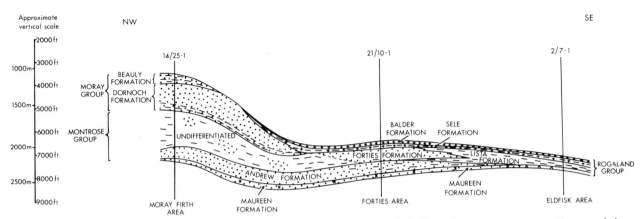

Fig. 8.6. Distribution of the Moray, Montrose and Rogaland Groups and their formations in the area south-east of the Halibut Horst (after Deegan and Scull, 1977).

Table 8.1. Tertiary strata in southern North Sea Basin (after Rhys, 1974).

Epoch	South of 55° North	55° North to 56° North
Pliocene	Commonly up to 200ft but up to 750ft in local basins. Clay, grey, with local developments of sand, light grey, generally in the upper part. Lignites common in the upper part.	Thickness increases to about 1000ft in the north. Clay, grey to grey-green, with local developments of sand, light grey. Some lignite in the upper part. One occurrence of limestone recorded.
Miocene	Patchy distribution, absent in many wells; where present it is generally up to 200ft thick with isolated provings of 600 to 900ft. Clays, mudstones and shales, grey to grey-brown, with traces of sand, white to grey; isolated occurrences of lignite.	Fairly consistent distribution. Maximum recorded thickness 2000ft. Clays, mudstones and shales, grey, green-grey, brown-grey. Thin beds of limestone, light brown, in the north.
Oligocene	Patchy distribution, absent in many wells; where present it is generally up to 300ft thick. Clays and mudstones, light grey to grey and dark green-grey, with minor developments of sand.	Fairly consistent distribution. Maximum recorded thickness 1000ft. Mudstones and shales, light to dark grey, green-grey and brown-grey, thin beds of limestone, grey to green-grey, in the north.
Eocene	Commonly 500-800ft thick, up to 1200ft in places. Mudstones, grey, green-grey and brown-grey with some red-brown colouration in the lower part, with local thin developments of sand, light grey to light brown, occasional thin limestones, grey, brown and red-grey. One thick development of sand recorded in the north. At or near the base, volcanic tuff, light to dark grey and brown-grey, isolated occurrences in south, more common in north; up to 100ft thick.	Up to 2250ft encountered in the north. Mudstones and shales, grey to grey-green and grey-brown, red-brown in parts and particularly at the base, with thin bands of limestone in the north. One isolated occurrence of thick sands, light grey to yellow. Volcanic tuff, common near base, grey, dark grey, brown, brown-red; up to 170ft with shale intercalations.
Palaeocene	Up to 600ft in the extreme south, 200ft-400ft commonly proved. Mudstones and shales, light to dark grey, variable shades of green and brown and, particularly towards the base, red-brown, with variable amounts of sandstone, grey to grey-green and red-brown. DANIAN mudstones, light grey, limestones, light grey to light brown, and chalk, white to light grey, developed locally.	Generally 500-600ft thick. Mudstones and shales, variable shades of grey, green, brown and red, with some thin bands of limestone, light brown. DANIAN marls, light grey to light brown, limestones and chalk, white to grey-brown, more common than in the south.

mentation, one dated c.58-57 Ma, the other c.55-52 Ma (Knox and Morton, 1983). A subphase of the later period forms the main tuff zone of Jacqué and Thouvenin (1975), a widespread seismic marker (part of the Balder Formation, see Figs. 8.5 and 8.6).

Quaternary deposits are summarised in Caston (1977), from which the isopachyte map in Figure 8.7 is taken. Over 1 km thickness of Quaternary sediment (including much glacial material) is found in places in the central North Sea. More detailed descriptions are given in a series of reports of the Institute of Geological Sciences (I.G.S.; e.g. Owens, 1977). Recent sediments have received a good deal of attention (e.g. McCave, 1971; Owens, 1980) in studies that bear on geotechnical work in locating wells, platforms and pipelines, and on geophysical work where there is power-loss in areas of sand-waves.

North Sea stratigraphy may be set in a regional context by consulting Curry *et al.* (1978) on the Tertiary, and Mitchell *et al.* (1973) on the Quaternary. The association of Tertiary tuffs and sediments in the North

Sea serves to bring together two formerly widely separated areas of research: the early Tertiary sediments of south-east England, and the Hebridean igneous province.

8.4 Palaeontology

A tenfold benthonic and planktonic foraminiferal subdivision of the Cenozoic in the British and Norwegian sectors of the North Sea has been established by Berggren and Gradstein (1981). This is linked through dinoflagellates with the standard European Palaeogene subdivision; the zonation based on foraminifera is applicable to a deep-water facies not known from the classical onshore section. In the Palaeogene deposits of the Central and Viking Grabens there is an abundant agglutinated benthonic foraminiferal assemblage and a poor calcareous fauna. Berggren and Gradstein suggest that the agglutinated assemblage established itself in the central North Sea trough as relatively rapid subsidence took place. Minimum water depths of

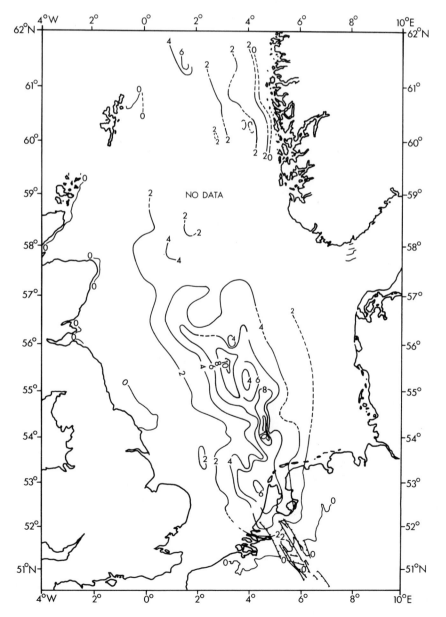

Fig. 8.7. Isopachyte map of total Quaternary thickness in North Sea (after Caston, 1977). Contour interval 100 m.

300-500 m for the Palaeogene clastic sequence of the Central and Viking Grabens are indicated, though palaeobathymetry is not thought to be a key factor controlling the presence of the agglutinated fauna: relatively rapid deposition of organic-rich fine-grained terrigenous sediments with restricted bottom-water circulation is held to favour the assemblage.

King (1983) provides a detailed Cenozoic micropalaeontological biostratigraphy of the North Sea, with some zonal boundaries comparable to those of Berggren and Gradstein (1981). King recognises 17 bethonic zones and 16 planktonic zones, with three major depth-related microfaunal biofacies from Late Palaeocene to Late Miocene. These are: inner sublittoral, outer sublittoral-epibathyal, and the *Rhabdaminna* biofacies (which is characterised by a foraminiferal fauna of agglutinating taxa). Harland *et al.* (1978) have considered the micropalaeontological evidence bearing on recent oceanographic history in late Quaternary deposits in the north-central North Sea. They suggest that the biostratigraphy can be used as climatostratigraphy, with links to North Atlantic circulation, in particular the positions of the climatically important North Atlantic Current and Polar Front.

8.5. Sedimentology

8.5.1 Petrography of Palaeocene sandstones in central North Sea

Provenance

Four major sandstone units within the Palaeocene of the central North Sea are recognised by Morton (1979) and Knox *et al.* (1981). Two units that are mainly con-

fined to the west of the area have assemblages of heavy minerals indicating derivation from the Scottish Highlands; two units in eastern parts of the area have heavy mineral assemblages indicating provenance from pre-existing sandstones on the East Shetland Platform. This work offers detailed support to the earlier suggestions of Scottish provenance for the Palaeocene Forties Formation sandstones (Thomas *et al.*, 1974) and the Palaeocene sandstones of the Montrose Field (Fowler, 1975). Later work by Morton (1982) on lower Tertiary sandstones in the Viking Graben confirms the importance of the Orkney-Shetland area as a source, with metamorphic basement exposed to the north, and probable Late Jurassic sediments exposed in the south.

Diagenesis

The composition and diagenesis of Forties Field sandstones are discussed by Pagan (1980) in relation to facies type (Figs. 8.8, 8.9 and 8.10). She shows the significance of clay authigenesis in controlling poroperm characteristics within a given facies (see the 'muddy' turbidites in Fig. 8.10), as well as the broad control over porosity exercised by the different facies themselves (compare the turbidites and massive sandstones in Fig. 8.10). More recent work by BP has concentrated on pore-lining diagenetic phases of possible significance to programmes of enhanced oil recovery.

Porosity

A broad range of Palaeocene sandstone facies is considered by Selley (1978) in constructing porosity gradients for these rocks: (a) grainflow facies of structureless

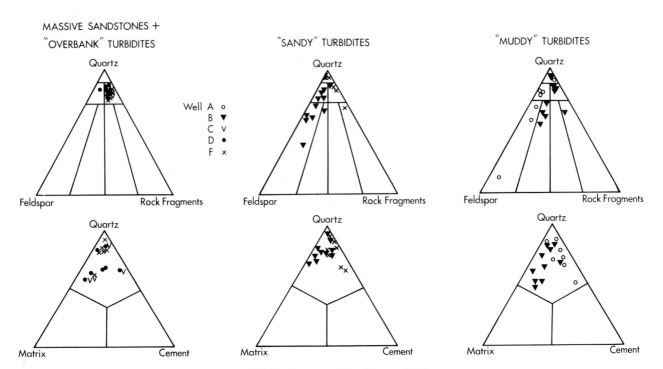

Fig. 8.8. Compositional triangles for Forties Field facies type (after Pagan, 1980).

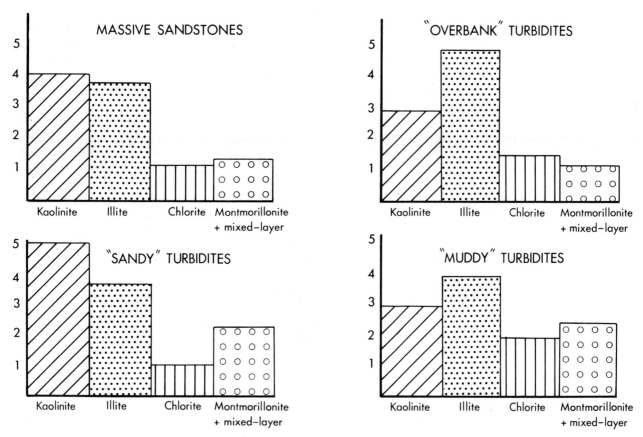

Fig. 8.9. Average abundances of clay minerals for Forties Field facies types: shown as parts-in-ten (after Pagan, 1980).

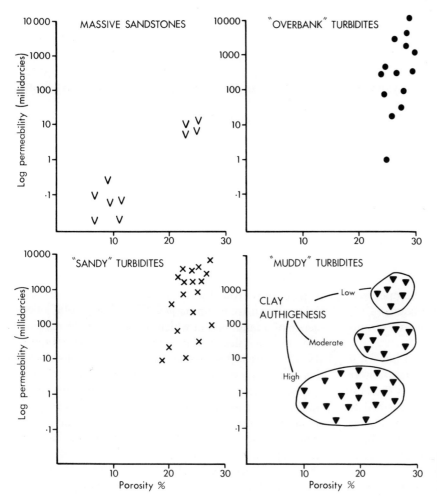

Fig. 8.10. Poro-perm relationships for Forties Field facies types (after Pagan, 1980).

sandstones with porosities at 2300 m of 20-30%; (b) turbidite facies, with porosities at 2300 m of 17.5-22.5%; (c) glauconitic sandstones with porosities at 2100 m of 25-35%. The major producing facies are (a) and (b), as at the Forties, Andrew, Montrose and Cod Fields (Selley, Fig. 9.1). Selley suggests a porosity gradient for the Palaeocene sandstones of 2.9% porosity-loss per 1000 feet (9.5% per 1000 m) of burial, on the strenth of the published data from Forties and Montrose (Walmsley, 1975; Fowler, 1975). For the Forties Formation, later data provided by Carman and Young (1981) show a mean porosity of c. 25% for both sandy turbidites and massive sandstones, and of 17.5% for muddy turbidites, at a depth of c.2200 m. The mean permeability of the massive sandstones is 639 md compared with 418 md for the sandy turbidites and only 8.6. md for muddy turbidites (see next section for identification of facies types).

8.5.2 Facies distribution and interpretation

Discussion here is confined mainly to the lower Tertiary of the central and northern North Sea, in which there is particular hydrocarbon interest.

An initial broad facies division of the lower Tertiary of the central North Sea defines large topset, foreset and bottomset intervals (Parker, 1975). Thick composite sandstones with interbedded lignites form the topset units; mudstones with thin sandstones the foreset units; graded sandstones and mudstones the bottomset units. This is interpreted as: *topset*—shallow-marine to coastal-plain deposits, *foreset*—the near-source equivalent of bottomset, *bottomset*—turbidites deposited in deeper water (forming the reservoirs of fields such as Forties and Montrose). Subsequent work has been largely a refinement of Parker's scheme, with some consideration of Fowler's (1975) alternative hypothesis of shallower-water deltaic deposition. Fowler's conclusion is based on petrographical studies in the Montrose Field of cores through massive sandstones that Parker believes are turbidite facies.

The main division of the lower Tertiary into a younger shallow-water and an older deep-water facies is recognised in the stratigraphical nomenclature of Deegan and Scull (1977), who, broadly, identify 'shelf-deltaic' sediments as the Moray Group and 'submarine-fan' sediments as the underlying Montrose Group (Figs. 8.5 and 8.6). These two divisions have been mapped seismically by Rochow (1981), who also recognises discontinuities with the Moray and Montrose Groups that may be identified as formations.

Development drilling in the Forties Field has led to a more detailed sedimentological interpretation by Carman and Young (1981) of the four main facies of the Palaeocene Forties Formation first described by Thomas *et al.* (1974): *Facies A*—a consistent alternation of sandstone and mudstone (sandy turbidites), *Facies B*—predominantly sandstone with minor pebbly sandstone (massive sandstones), *Facies C*—predominantly grey kaolinitic mudstone with some coarser beds (muddy turbidites), *Facies D*—burrowed waxy green mudstones, rich in montmorillonite. Carman and Young believe that the Forties Formation was deposited in the middle to lower parts of a submarine fan, with transport of sediment from a main northwesterly direction.

The Lower Eocene sandstones forming the gas reservoir in the Frigg Field are also interpreted as submarine-fan deposits (Héritier *et al.*, 1979, 1981), though Morton (1982) suggests that they are shallow-marine. These overlie Palaeocene fan-complexes of sediments brought in from the west (see below). They are sealed by Middle Eocene marine mudstones; these are overlain by later Eocene and Oligocene mudstones that pass up in turn into a predominantly sandy Miocene and Pliocene section described by Héritier *et al.* (1979, p. 2003) as having a 'pronounced continental character' in well 25/1-1. Elsewhere in the central and northern North Sea the younger Palaeogene rocks, and the Neogene sequence, are predominantly fine-grained sediments (Deegan and Scull, 1977).

The Palaeocene sandstones of the Balder area on the eastern margin of the Viking Graben are held to provide a further example of a submarine-fan reservoir with a mudstone seal (Sarg and Skjold, 1982). Interpretation of the regional pattern of Palaeocene sandstone sedimentation in the Viking Graben is still under review (Morton, 1982). Mudge and Bliss (1983) suggest that the middle Palaeocene sandstones in the northern North Sea were deposited in a dominantly nearshore shallow-marine setting.

The lowest Tertiary in places consists of chalk (Byrd, 1975; Childs and Reed, 1975; Hancock and Scholle, 1975); in the Albuskjell area in the Central Graben, Maastrichtian-Danian chalks contain rocks that Watts *et al.* (1980) interpret as large-scale allochthonous chalk units emplaced by slumping, and débris and turbidity flows in a slope and base-of-slope setting. These rocks provide a link between the contrasting palaeogeographies of the Late Cretaceous and the early Tertiary, which are discussed in the following section.

8.6 Palaeogeography

Figures 8.11 and 8.12 show the great changes that took place in the palaeogeography of the area of the British Isles from Late Cretaceous to early Tertiary, from almost total submergence, to almost total emergence of a landmass in which the shape of the present-day British Isles can be seen quite clearly (Lovell, 1977). There is good evidence for these changes in the lower Tertiary of the North Sea, in particular the influx of non-carbonate sediment from the uplifted mainland in the west, and the widespread tuffs. The ashes result from contemporaneous igneous activity in the Hebridean province and elsewhere, that was associated with a particularly active phase in the opening of the North Atlantic. More details of the regional picture of early Tertiary palaeogeography will emerge as data are

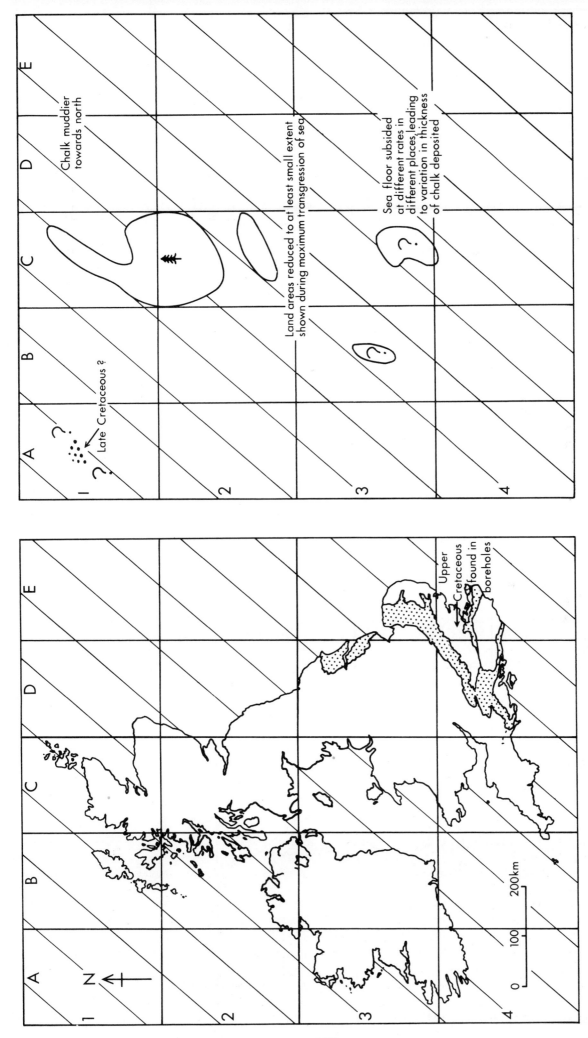

Fig. 8.11. Upper Cretaceous outcrop and Late Cretaceous palaeogeography (after Lovell, 1977).

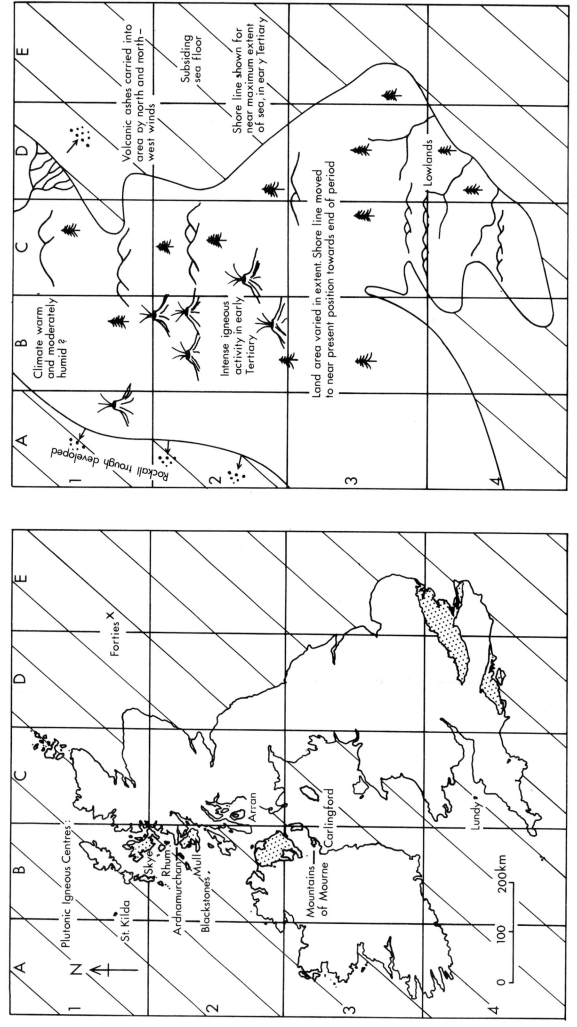

Fig. 8.12. Tertiary outcrop and palaeogeography (after Lovell, 1977).

released from the exploration drilling west of the Shetland Islands, where thick Cenozoic sedimentary sequences with general eastern provenance and associated early Tertiary volcanics are to be expected.

W.H. Ziegler's (1975) version of Palaeogene palaeogeography for the North Sea as a whole is shown in Figure 8.13. A more detailed picture of the palaeogeographical evolution of the northern and central areas has been given by Sutter (1980) (Figs. 8.14, 8.15, 8.16 and 8.17). Sutter's version of Palaeocene palaeogeography follows quite closely the early interpretation of Parker (1975) (Fig. 8.18). It is also in broad agreement with that of Héritier *et al.* (1979) for the Frigg area (Fig. 8.19), and the conclusions reached from seismic stratigraphy by Rochow (1981) (Figs. 8.20, 8.21, 8.22 and 8.23). Rochow's maps show the relationship between the shallow-water facies of the Moray Group and the deeper-water facies of the earlier Montrose Group. This broad Palaeocene pattern, of supply

of sand from the north-west through an area of shelf-deltaic deposition to submarine fans in the Viking and Central Grabens, is one within which a good deal of detail may be recognised (Deegan and Scull, 1977; Sutter, 1980; Rochow, 1981; Morton, 1982; Mudge and Bliss, 1983). Some of the detail that is of economic significance is considered in the next section.

8.7 Facies and reservoir geology

By 1975, Parker was already able to claim that, with data from 77 wells plus extensive seismic coverage, the North Sea Palaeocene provided 'one of the better controlled examples of an ancient deep-water fan'. By the late 1970s production drilling in the Forties Formation had given an even more detailed picture (Hill and Wood, 1980; Carman and Young, 1981). Facies identified from cores are recognised on petrophysical logs; these results are combined with information on pressure-

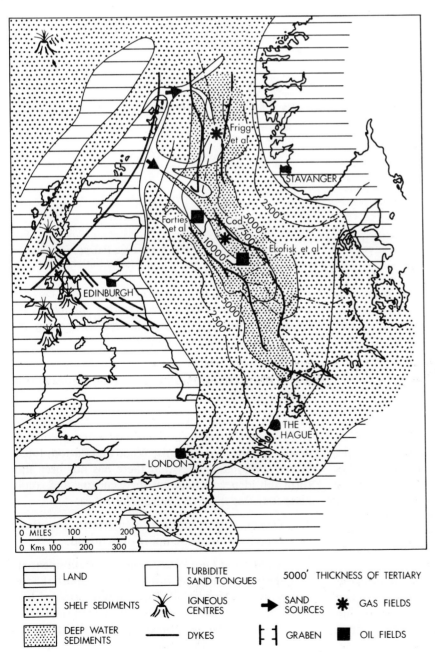

LAND	TURBIDITE SAND TONGUES	5000' THICKNESS OF TERTIARY
SHELF SEDIMENTS	IGNEOUS CENTRES	SAND SOURCES	GAS FIELDS
DEEP WATER SEDIMENTS	DYKES	GRABEN	OIL FIELDS

Fig. 8.13. Palaeogene palaeogeography (after Ziegler, 1975).

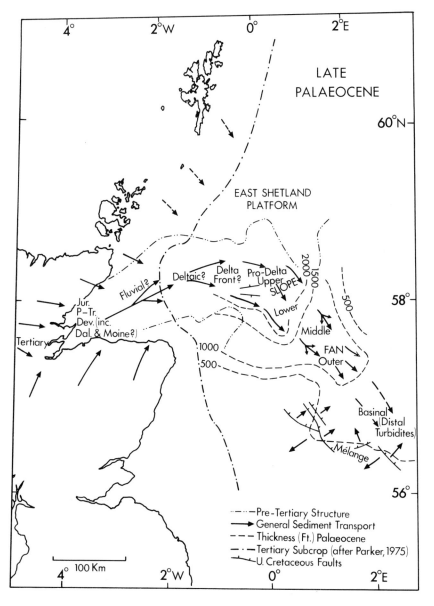

Fig. 8.14. Late Palaeocene palaeo-geography (after Sutter, 1980).

decline and production-logging data to map out the complex submarine-fan sandstone-geometry, which in turn provides a guide to reservoir performance and further drilling. The formation is divided into a lower Shale Member and an upper Sandstone Member. The Shale Member is interpreted as predominantly basin-plain and lower-fan deposits. The Sandstone Member contains predominantly mid-fan deposits, and can itself be subdivided into a Main Sand Unit (progradational-lobe and channel sequences), the Charlie Shale Unit ('abandonment' thin-bedded turbidites on top of the Main Sand Unit), the Charlie Sand Unit (thick mid-fan channel), and the Upper Shale Unit ('abandonment' mudstones marking the end of fan sedimentation in the Forties area). Charlie Sand Unit was recognised as being a separate sandstone body because the pressure decline within it was initially more rapid than the decline within the Main Sand Unit itself.

Thick channel sandstones have also been mapped out in detail for the lobate Frigg fan-complex, to aid in gas production (Héritier *et al.*, 1979). The relationship of these channel sandstones to levée and interchannel deposits leads to suggestions of a range of depositional

processes, including grain-flow and contour currents (but cf. Morton, 1982). The resulting submarine fan depositional topography is enhanced by draping and differential compaction of sands.

8.8 Cenozoic subsidence and oil generation and migration

Mesozoic subsidence was dominated by major faulting and rifting, Cenozoic subsidence by a broad synclinal downwarp (see, for example, Wood and Barton, 1983). Sutter (1980) concludes that there has not been uniform subsidence throughout the Cenozoic and recognises four stages in the central North Sea:

1. *Danian*. Slow subsidence of North Sea Basin coupled with gentle uplift of Scotland.

2. *Middle and Late Palaeocene*. Rapid subsidence in the basin combined with large-scale uplift of Scotland (see Morton (1979) for a discussion of four phases of Palaeocene subsidence, with sand deposition alternating between the Central Graben and the Moray Firth Basin).

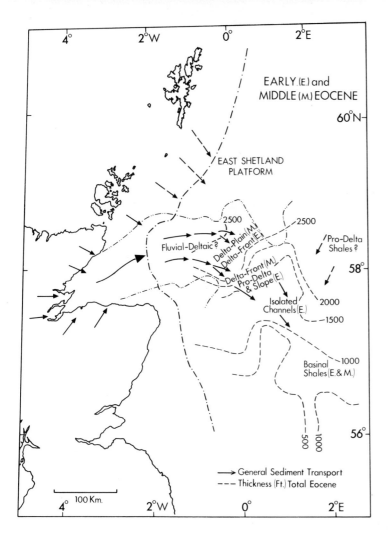

Fig. 8.15. Early and Middle Eocene palaeogeography (after Sutter, 1980).

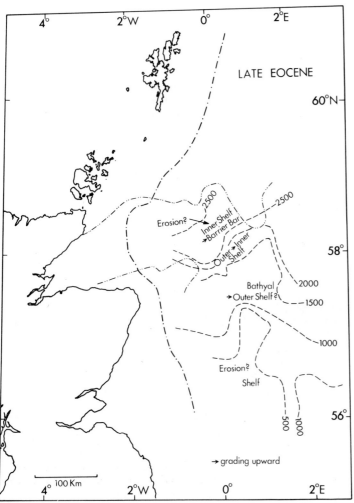

Fig. 8.16. Late Eocene palaeogeography (after Sutter, 1980), Isopachytes for total Eocene (feet).

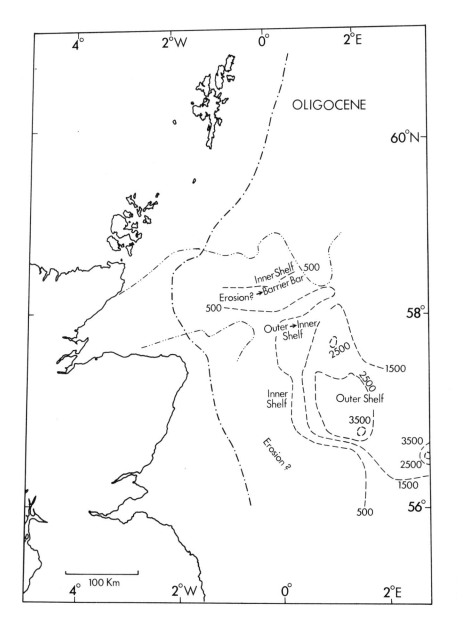

Fig. 8.17. Oligocene palaeogeography (after Sutter, 1980). Isopachytes in feet.

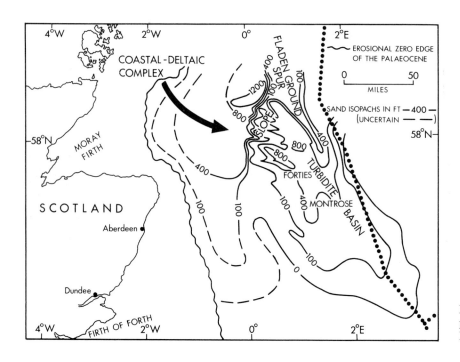

Fig. 8.18. Palaeocene sandstone isopachytes and palaeogeography (after Parker, 1975).

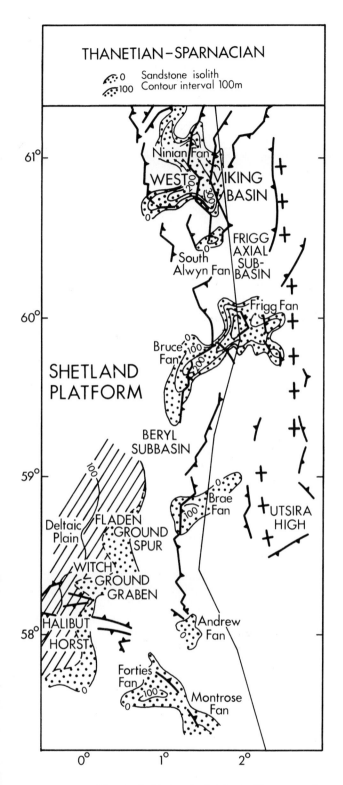

Fig. 8.19. Late Palaeocene palaeogeography, Frigg Fan area (after Héritier *et al.*, 1979).

3. *Eocene.* Slow subsidence in basin, with prograding sedimentation from the margins to the centre.

4. *Oligocene to present.* Water depths decreased throughout the Oligocene, by the end of which the basin was at its smallest extent. Sedimentation rates have increased since the Oligocene, so there must have been an overall if irregular increase in rate of subsidence since that time (cf. Fig. 2.1). (Maximum average Quaternary sedimentation rates in the North Sea are 0.3 to 0.5 mm a year, up to ten times as high as figures for the Tertiary (Caston, 1977; Zagwijn and Doppert, 1978).)

Associated with this considerable subsidence (Fig. 8.3) were maximum northern North Sea Basin palaeo-temperatures (Cooper *et al.*, 1975). There is general conformity in the palaeotemperature profiles; in particular there are no apparent sudden increases from low temperatures to higher temperatures at unconformities. Cooper *et al.* therefore conclude that the palaeotemperatures recorded all belong to one geothermal event, the development and filling of the Tertiary basins. They also note that in many places, palaeotemperature-gradients have higher values than those found at the present day. It should be added

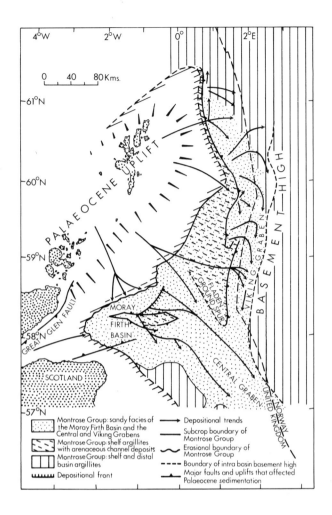

Fig. 8.20. Palaeogeology and palaeogeography of Montrose Group (after Rochow, 1981).

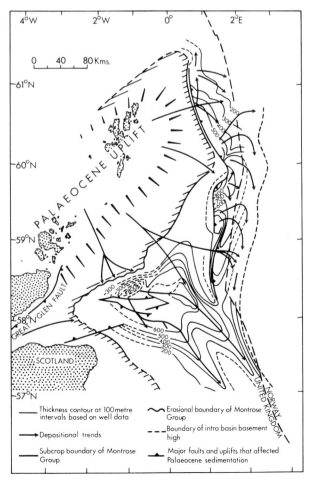

Fig. 8.21. Isopachytes and palaeogeography of Montrose Group (after Rochow, 1981).

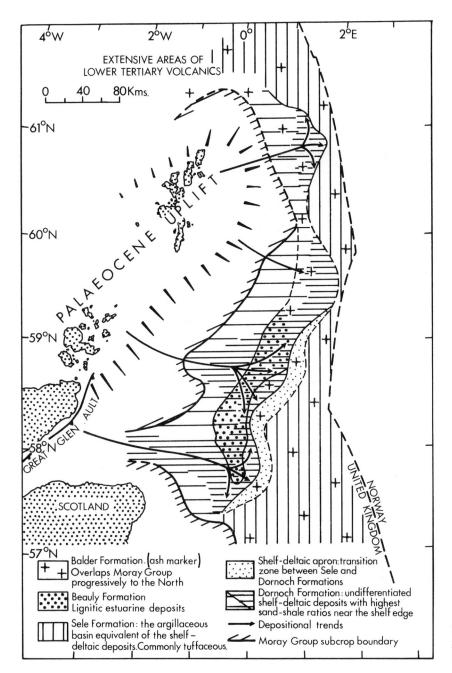

Legend:

+ + Balder Formation (ash marker)
Overlaps Moray Group
progressively to the North

• Beauly Formation
Lignitic estuarine deposits

Sele Formation: the argillaceous
basin equivalent of the shelf –
deltaic deposits. Commonly tuffaceous.

Shelf-deltaic apron: transition
zone between Sele and
Dornoch Formations

Dornoch Formation: undifferentiated
shelf-deltaic deposits with highest
sand-shale ratios near the shelf edge

→ Depositional trends

≪ Moray Group subcrop boundary

Fig. 8.22. Palaeogeology and palaeogeography of Moray and Rogaland Groups (after Rochow, 1981).

that some stratigraphic horizons are experiencing their highest temperatures now. Barnard and Cooper (1981) conclude that in the North Sea as a whole, oil generation and migration took place during this Tertiary geothermal event: with continuing burial, this process continues to the present day. The richest and most extensive source rock is the Kimmeridge Clay, from which, Barnard and Cooper suggest, most North Sea oils are derived. Héritier *et al.* (1979) consider that an exception is the oil underlying the gas at Frigg, which appears to have been generated at an early stage in deeply buried Lower and Middle Jurassic source rocks (but compare with Goff, 1983).

The Frigg gas itself is believed to result from a late stage of alteration of the Jurassic source rocks, which show vitrinite reflectance values of 1.5 to 1.8 that are characteristic of gas-zone diagenesis. Lower Tertiary and Cretaceous potential source rocks in the area show a low degree of diagenesis; a vitrinite reflectance value of 1.0 is reached only at a depth of 4000 m in the deep Frigg wells.

The structural relationship between Jurassic source rocks and Tertiary reservoirs for the Forties and Frigg areas is shown in Brown, Figure 6.11 (this volume).

Heavy-oil discoveries have been made in early Tertiary sandstones at several localities in the Viking Graben; the Alpha and Bressay structures are examples.

8.9 Conclusions

It was the prospect of finding hydrocarbons in sandstones in the thick Tertiary sequence that led to initial interest in the northern North Sea Basin (Kent, 1975). Subsequent detailed work on those sandstones during exploration and development has made a contribution

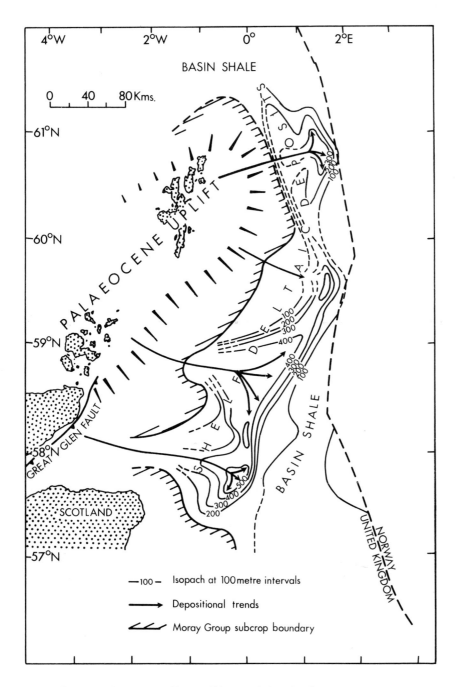

Fig. 8.23. Isopachytes and palaeo-geography of Moray and Rogaland Groups (after Rochow, 1981).

to petroleum geology extending well beyond the North Sea itself.

Current work in the North Sea on a range of scales, from basin to pore-throat, will lead to more technical progress. For example, further exploration of the early Tertiary sandstones throughout the basin will require understanding of detailed tectonic and eustatic controls of turbidite sedimentation. In development work, quantification of description and prediction of both sandstone-body geometry and pore morphology is demanded by attempts to enhance recovery of oil from fields such as Forties.

8.10 Acknowledgements

I thank many of those whose work is cited for helpful discussions, and the Chairman and Board of Directors of The British Petroleum Company plc for permission to publish.

8.11 References

Barnard, P.C. and Cooper, B.S. (1981) Oils and source rocks of the North Sea area. In: Illing, L.V. and Hobson, G.D. (Eds.) q.v. 169-175.

Berggren, W.A. and Gradstein, F.M. (1981) Agglutinated benthonic foraminiferal assemblages in the Palaeogene of the central North Sea: their biostratigraphic and depositional significance. In: Illing, L.V. and Hobson, G.D. (Eds.) q.v. 282-285.

Blair, D.G. (1975) Structural styles in North Sea oil and gas fields. In: Woodland, A.W. (Ed.) q.v. 327-328.

Byrd, W.D. (1975) Geology of the Ekofisk Field, offshore Norway. In: Woodland, A.W. (Ed.) q.v. 439-445.

Carman, G.J. and Young, R. (1981) Reservoir geology of the Forties Oilfield. In: Illing, L.V. and Hobson, G.D. (Eds.) q.v. 371-379.

Catson, V.N.D. (1977) Quaternary deposits of the central North Sea.
  1. A new isopachyte map of the Quaternary of the North Sea.

2. The Quaternary deposits of the Forties Field, northern North Sea. *Rep. Inst. geol. Sci. London* 77/11, 21 p.

Childs, F.P. and Reed, P.E.C. (1975) Geology of the Dan Field and the Danish North Sea. In: Woodland, A.W. (Ed.) q.v. 429-438.

Cooper, B.S., Coleman, S.H., Barnard, P.C. and Butterworth, J.S. (1975) Palaeotemperatures in the Northern North Sea Basin. In: Woodland, A.W. (Ed.) q.v. 487-492.

Curry, D., Adams, G.C., Boulter, M.C., Dilley, F.C., Eames, F.E., Funnell, B.M. and Wells, M.K. (1978) A correlation of Tertiary rocks in the British Isles. *Spec. Rep. geol. Soc. London 12*, 72 p.

Deegan, C.E. and Scull, B.J. (compilers) (1977) A proposed standard lithostratigraphic nomenclature for the central and northern North Sea. *Rep. Inst. geol. Sci. London 77/25; Bull. Norw. Petrol. Direct. 1*, 36 p.

Fowler, C. (1975) The geology of the Montrose Field. In: Woodland, A.W. (Ed.) q.v. 467-476.

Goff, J.C. (1983) Hydrocarbon generation and migration from Jurassic source rocks in East Shetland Basin and Viking Graben of the northern North Sea. *J. geol. Soc. London 140*, 445-474.

Hancock, J.M. and Scholle, P.A. (1975) Chalk of the North Sea. In: Woodland, A.W. (Ed.) q.v. 413-427.

Harland, R., Gregory, D.M., Hughes, M.J. and Wilkinson, I.P. (1978) A late Quaternary bio- and climatostratigraphy for marine sediments in the north-central part of the North Sea. *Boreas* 7, 91-96.

Héritier, F.E., Lossel, P. and Wathne, E. (1979) Frigg Field—large submarine-fan trap in lower Eocene rocks of North Sea Viking Graben. *Bull. Am. Assoc. Petrol. Geol. 63*, 1999-2020.

Héritier, F.E., Lossel, P.J. and Wathne, E. (1981) The Frigg gas field. In: Illing, L.V. and Hobson, G.D. (Eds.) q.v. 380-391.

Hill, P.J. and Wood, G.V. (1980) Geology of Forties Field, U.K. Continental Shelf, North Sea. In: Halbouty, M.T. (Ed.) *Giant oil and gas fields of the decade 1968-1978. Am. Assoc. Petrol. Geol. Memoir 30*, 81-93.

Illing, L.V. and Hobson, G.D. (Eds.) (1981) *Petroleum Geology of the Continental Shelf of North-West Europe.* Heyden & Son, London, 521 p.

Jacqué, M. and Thouvenin, J. (1975) Lower Tertiary tuffs and volcanic activity in the North Sea. In: Woodland, A.W. (Ed.) q.v. p. 455-465.

Kent, P.E. (1975) Review of North Sea Basin development. *J. geol. Soc. London 131*, 435-468.

King, C. (1983) Cainozoic micropalaeontological biostratigraphy of the North Sea. *Rep. Inst. geol. Sci. London 82/7*, 40 p.

Knox, R.W.O'B. and Morton, A.C. (1983) Stratigraphical distribution of early Palaeocene pyroclastic deposits in the North Sea Basin. *Proc. Yorkshire geol. Soc.* 44, 355-363.

Knox, R.W.O'B., Morton, A.C. and Harland, R. (1981) Stratigraphic relationships of Palaeocene sands in the U.K. sector of the central North Sea. In: Illing, L.V. and Hobson, G.D. (Eds.) q.v. 267-281.

Lovell, J.P.B. (1977) *The British Isles through geological time: a northward drift.* George Allen and Unwin, London, 40 p.

McCave, I.N. (1971) Sand waves in the North Sea off the coast of Holland. *Mar. Geol.* 10, 199-225.

Mitchell, G.F., Penny, L.F., Shotton, F.W. and West, R.G. (1973) A correlation of Quaternary deposits in the British Isles. *Spec. Rep. geol. Soc. London* 4, 99 p.

Morton, A.C. (1979) The provenance and distribution of the Palaeocene sands of the central North Sea. *J. Petrol. Geol. Beaconsfield* 2, 11-21.

Morton, A.C. (1982) Lower Tertiary sand development in Viking Graben, North Sea. *Bull. Am. Assoc. Petrol. Geol.* 66, 1542-1559.

Mudge, D.C. and Bliss, G.M. (1983) Stratigraphy and sedimentation of the Palaeocene sands in the northern North Sea. In: Brooks, J. (Ed.). *Petroleum Geochemistry and Exploration of Europe. Spec. Publ. geol. Soc. London.* 12, Geological Society/Blackwell, 95-111.

Owens, R. (1977) Quaternary deposits of the central North Sea, 4. Preliminary report on the superficial sediments of the central North Sea. *Rep. Inst. geol. Sci. London 77/13*, 16 p.

Owens, R. (1980) Holocene sedimentation in the north-western North Sea. In: Nio, S.D., Schuttenhelm, R.T.E. and van Weering, T.C.E. (Eds.) *Holocene marine sedimentation in the North Sea Basin. Spec. Publ. Int. Assoc. Sedimentol. No. 5*, 303-322.

Pagan, M.C.T. (1980) *Diagenesis of the Forties Field sandstones.* M.Phil. thesis, Univ. Edinburgh.

Parker, J.R. (1975) Lower Tertiary sand development in central North Sea. In: Woodland, A.W. (Ed.) q.v. 447-453.

Rhys, G.H. (compiler) (1974) A proposed standard lithostratigraphic nomenclature for the southern North Sea and an outline structural nomenclature for the whole of the (U.K.) North Sea. A report of the joint Oil Industry—Institute of Geological Sciences Committee on North Sea Nomenclature. *Rep. Inst. geol. Sci. London 74/8*, 14 p.

Rochow, K.A. (1981) Seismic stratigraphy of the North Sea 'Palaeocene' deposits. In: Illing, L.V. and Hobson, G.D. (Eds.) q.v. 255-266.

Sarg, J.F. and Skjold, L.J. (1982) Stratigraphic traps in Paleocene sands in the Balder area, North Sea. In: Halbouty, M.T. (Ed.), *The Deliberate Search for the Stratigraphic Trap. Am. Assoc. Petrol. Geol. Memoir 32*, 197-206.

Selley, R.C. (1978) Porosity gradients in North Sea oil-bearing sandstones. *J. geol. Soc. London* 135, 119-132.

Sutter, A.A. (1980) *Palaeogene sediments from the United Kingdom sector of the central North Sea.* Ph.D thesis, Univ. Aberdeen.

Thomas, A.N., Walmsley, P.J. and Jenkins, D.A.L. (1974) Forties Field, North Sea. *Bull. Am. Assoc. Petrol. Geol.* 58, 396-406.

Walmsley, P.J. (1975) The Forties Field. In: Woodland, A.W. (Ed.) q.v. 477-485.

Watts, N.L., Lapré, J.F., Van Schijndel-Goester, F.S. and Ford, A. (1980) Upper Cretaceous and lower Tertiary chalks of the Albuskjell area, North Sea: deposition in a slope and a base-of-slope environment. *Geology* 8, 217-221.

Wood. R. and Barton, P. (1983) Crustal thinning and subsidence in the North Sea. *Nature* 302, 134-136.

Woodland, A.W. (Ed.) (1975) *Petroleum and the Continental Shelf of North-West Europe, Vol. 1 : Geology.* Applied Sci. Publ., Barking. 501 p.

Zagwijn, W.H. and Doppert, J.W.CHR. (1978) Upper Cenozoic of the Southern North Sea Basin: palaeoclimatic and palaeogeographic evolution. *Geol. Mijnbouw* 57, 577-588.

Ziegler, W.H. (1975) Outline of the geological history of the North Sea. In: Woodland, A.W. (Ed.) q.v. 165-190.

Chapter 9

Source Rocks and Hydrocarbons of the North Sea

CHRIS CORNFORD

9.1 Introduction

This chapter has three aims: to review briefly the general characteristics of hydrocarbon source rocks and the generation and migration of oil therefrom; to stratigraphically catalogue the proven and putative source rocks of the North Sea and surrounding areas; and finally to summarise the properties of the reservoired hydrocarbons in the North Sea area. This text does not attempt to cover exploration geochemistry. The interested reader is referred, in the first instance, to Tissot and Welte (1978).

Source rocks can be defined as sediments which are (or were) capable of generating migratable oil or gas. Whether there is sufficient oil or gas to form a commercial accumulation depends largely on the volume and richness of the source rock, its maturity history, the geological framework in which it occurs, and the current economics of exploitation.

A significant abiogenic origin for hydrocarbons (see Gold and Soter, 1982) is not, for a number of serious scientific reasons, considered. The North Sea provides excellent evidence for the association of oil and gas with thick sedimentary sequences, in contrast to the total absence of indigenous hydrocarbons in the metamorphic shield areas. Overwhelming scientific evidence shows that oil and gas derive from the organic remains trapped in, and buried with, sedimentary rocks (Tissot and Welte, 1978).

This chapter is concerned mainly with clastic source rocks (i.e. shales): the concepts used, and values given to boundary conditions, cannot be readily applied to areas such as the Middle East, where chemical sediments (limestones, dolomites, evaporites) dominate. Some common depositional environments for clastic source rocks are summarised in Fig. 9.1.

9.2 Recognition of hydrocarbon source rocks

9.2.1 Introduction

The classical hydrocarbon source rock in a clastic environment is an organic-rich, dark olive grey to black, laminated mudstone. The organic matter in the source rock is broadly termed kerogen if solid or insoluble, bitumen if fluid or solvent extractable, and gas if gaseous. Kerogen decomposes to bitumen and gas as a result of the increased temperature experienced during burial. This process, termed maturation, gives rise to the generation of oil and gas. However, recent integrated studies of hydrocarbon generation and migration have highlighted a number of geological situations (e.g. Tertiary deltas) where hydrocarbons are sourced from greater thicknesses of relatively organic-lean but well drained sediment. As a result, explorationists now increasingly use source-rock volumetrics and drainage in addition to the amount of oil or gas prone kerogen, to determine the source potential of a prospect or basin (e.g. Goff, 1983). The concept of source rock drainage is summarised in Fig. 9.2.

Use of this approach requires the recognition of a number of properties of the source rock:

—Quantity of organic matter (TOC or total organic carbon).
—Type of organic matter (oil or gas prone, or inert).
—Thickness of unit.
—Areal extent and lateral variation.
—Interbedded sands or silts (drainage).
—Regional maturity boundaries.

These are discussed in more detail later.

Organic-rich shales are fairly uncommon in the geological record, since their deposition requires the co-existance of high bioproductivity and high preservation rates (Müller and Suess, 1979). Recent studies suggest that preservation rather than productivity *per se* is the controlling factor. Fig. 9.1 summarises schematically some of the environments favouring the deposition of organic rich sediments. High organic preservation is promoted by high sedimentation rates and reduced oxygen concentrations in the water column; see Fig. 9.1 inset (Demaison and Moore, 1980). Reduced oxygen concentrations are found in stratified lakes and shelf and oceanic basins, in delta swamps, and in oceanic mid-water oxygen minima. Sediments associated with these environments are likely to be rich in organic matter. A well-referenced discussion of these processes *vis à vis* the Cretaceous black shales of the Atlantic is given by Waples (1983).

The organic matter in a sediment can broadly derive from three sources: higher (land) plants (trees, ferns etc.), lower (aquatic) plants (planktonic algae etc.), and bacteria. Volumetrically, animal tissue makes only a minor contribution to kerogen. A land plant input to a sediment generally produces a gas prone, Vitrinitic or Type III kerogen, unless it is altered (e.g. oxidised to charcoal), when it will produce dead carbon, termed

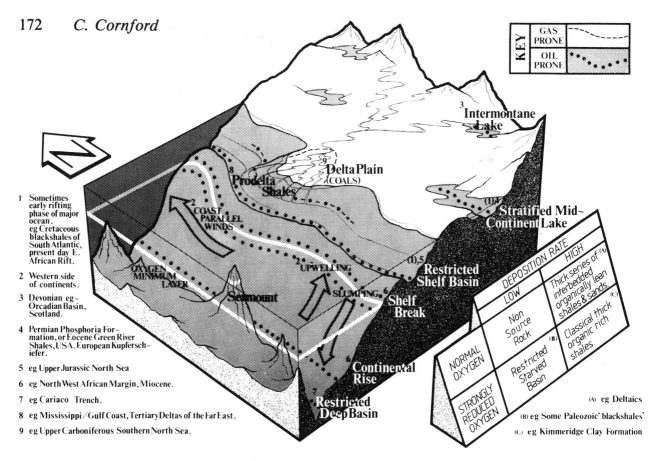

Fig. 9.1. Schematic summary of depositional environments favouring the accumulation of organic-rich oil or gas source-rocks in a clastic regime on a passive continental margin. Note that the gas-prone source-rocks of the North Sea (the Westphalian and Middle Jurassic coals) accumulated in a delta plain environment, and the oil-prone black shales of the Upper Jurassic Kimmeridge Clay Formation accumulated in a restricted, shelf-basin environment.

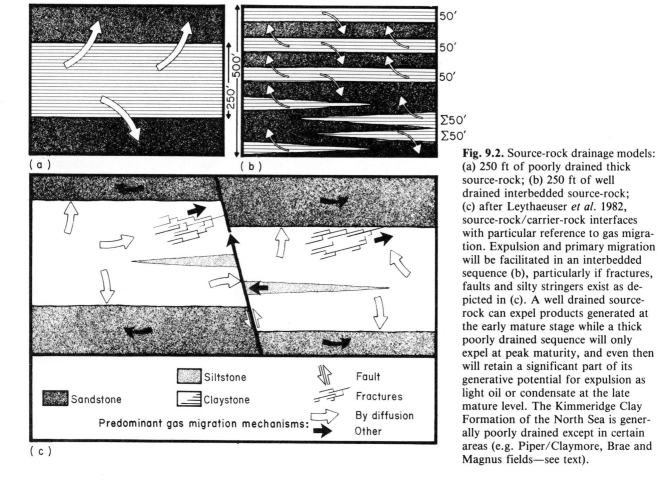

Fig. 9.2. Source-rock drainage models: (a) 250 ft of poorly drained thick source-rock; (b) 250 ft of well drained interbedded source-rock; (c) after Leythaeuser *et al.* 1982, source-rock/carrier-rock interfaces with particular reference to gas migration. Expulsion and primary migration will be facilitated in an interbedded sequence (b), particularly if fractures, faults and silty stringers exist as depicted in (c). A well drained source-rock can expel products generated at the early mature stage while a thick poorly drained sequence will only expel at peak maturity, and even then will retain a significant part of its generative potential for expulsion as light oil or condensate at the late mature level. The Kimmeridge Clay Formation of the North Sea is generally poorly drained except in certain areas (e.g. Piper/Claymore, Brae and Magnus fields—see text).

Inertinite or Type IV kerogen. However, land plant-derived spores, resin or cuticles can, if present in sufficiently high concentrations in the sediment, generate oil or condensates of a characteristic type. This group of land-plant tissues is collectively termed Exinite, and comprises part of the Type II kerogen group. An algal and/or bacterial input has the potential to produce oil and associated condensate and gas upon burial. This is termed a Liptinitic kerogen and falls in the Type I or Type II group.

The definition and use of these terms are summarised in Table 9.1 and Figs. 9.3 and 9.4. Using these concepts we can consider the properties of source rocks in more detail.

9.2.2 Quantity of organic matter

An adequate amount of organic matter (measured as organic carbon) is a necessary prerequisite for a sediment to source oil or gas. The quantity of organic matter required for a sediment to be considered a source rock is, like all attempts to define a multi-parameter system by a single variable, a much disputed point. For a typical poorly drained, thick, homogenous shale (Fig. 9.2a), the following values may be used to rate source rock potential in terms of quantity (but not quality) of kerogen.

<0.5% TOC	very poor
0.5-1.0% TOC	poor
1.0-2.0% TOC	fair
2.0-4.0% TOC	good
4.0-12.0% TOC	very good
>12.0% TOC	oil shale/carbargillite

These ratings should be used to describe the *amount of organic matter* and *not* the hydrocarbon source potential, since this will also depend on kerogen type. In addition to kerogen type, drainage must be considered in an assessment of the amount of organic matter required for a sediment to be considered a source rock: in a well drained sequence of interbedded shales and sands (Fig. 9.2b), 1% might be considered good and 2% very good.

A worldwide compilation of TOC data is shown in Fig. 9.5. This shows that high TOC sediments are found in the Cambro-Ordovician and the Lower Carboniferous, with increasing levels from the Jurassic onwards. Claystones have higher TOC contents than sandstones or carbonates, while the size fractions from a single sample show increasing TOC content with decreasing grain size. In addition, petroliferous basins have higher levels than non-petroliferous basins. This is emphasised by the low mean value for the 7253 Deep Sea Drilling Project analyses of non-exploration-related samples.

9.2.3 Type of organic matter

For application to hydrocarbon exploration, sedimentary organic matter can conveniently be divided, like Gaul, into three parts:

—Oil-prone components
—Gas-prone components
—Inert components (or dead carbon)

Table 9.1. Recognition of oil-prone or gas-prone kerogen using various analytical techniques

Hydrocarbon potential	Organic petrography		Pyrolysis[1] (Rock Eval)		$\delta^{13}C$ (pdb)	H/C atomic[2]	Origin (not depositional environment)
Oil- (→condensate →gas) prone	Liptinite	Algal/ Amorphous (sapropel)	Amorphogen[3]	Type I	< −28 (Freshwater algae −28 to −32)	1.5-2.0	Aquatic (freshwater or marine) algae, often bacterially degraded to yield amorphous material. Terrigenous spore, pollen, cuticle or resin, can also be degraded.
	Exinite		Phyrogen	Type II		1.0-1.5	
Condensate-prone	Vitrinite	Herbaceous	Hyalogen		> −28	0.5-1.0	Terrigenous, ligno-cellulosic tissue, relatively unaltered—present in a particulate or amorphous form.
Gas-prone		Woody		Type III			
Inert (dead carbon)	Inertinite	Coaly	Melanogen	Type IV (Type IIIB)		<0.5	Terrigenous as above but altered by oxidation in soil, during transport, or from forest fires, etc.

[1] Rock Eval is a pyrolysis technique carried out on the whole rock, and yields values of Hydrogen and Oxygen Indices as defined in Fig. 9.4. Type II is often a mixture of Type I plus Type III/IV kerogen.
[2] atomic ratio of the immature kerogen.
[3] Bujak, Barss and Williams, 1977.

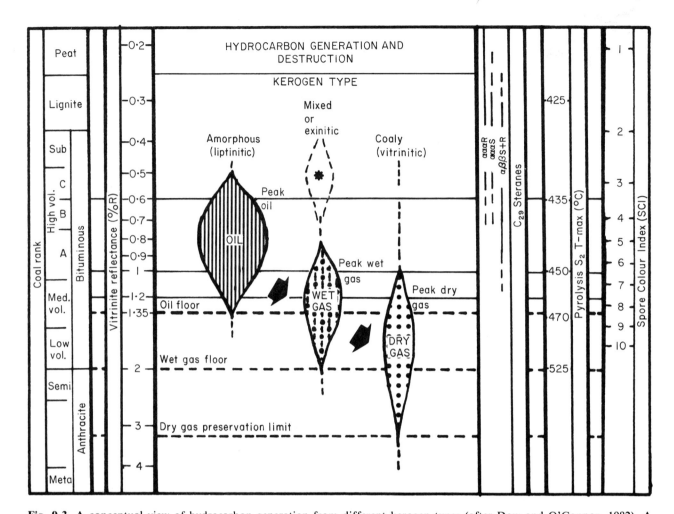

Fig. 9.3. A conceptual view of hydrocarbon generation from different kerogen types (after Dow and O'Connor, 1982). A detailed equivalence of the maturity parameters is not implied since the 'time' factor is not considered—see text and Figs. 9.7 and 9.8 for detailed generation curves. Note that kerogens typically contain a mixture of organic matter types: the pure wet gas-prone kerogen type termed 'mixed' comprises spore/pollen, cuticle, resins and possibly more hydrogen rich vitrinitic debris. The generation of early mature wet-gas/condensate (*) has been described from this type of organic matter at vitrinite reflectance values as low as 0.4%R (e.g. Connan and Cassou, 1977; Snowdon and Powell, 1979, 1982). No generation of early mature condensate has been described in the North Sea area. Sterane stereochemistry designated 5α, 14α, 17α, 20R, etc.

The kerogen of a typical rock will contain a mixture of all three components. When estimating the oil potential of a sediment, only the oil-prone part of the TOC of that sediment should be considered. For example, a 4% TOC sediment with 50% oil-prone kerogen will have 2% oil-prone organic carbon (2% OPOC). If, in addition, it had 25% gas-prone organic matter it would be said to have 1% gas-prone organic carbon (1% GPOC).

In apportioning gas or oil generative capacity to kerogen, it should be remembered that oil-prone kerogen (or reservoired oil itself), will crack to yield both wet and dry gas if buried, and hence heated, sufficiently. The generation of hydrocarbons from the broad categories of kerogen type is summarised in Fig. 9.3.

Kerogen type is generally determined by microscropy (organic petrography), by pyrolysis (e.g. the Rock Eval method) or by elemental (C,H,O) analysis. Confirmatory evidence can be provided by stable carbon isotope ($\delta^{13}C$), light hydrocarbon, or sediment extract analyses. Because, at present, no single technique is totally reliable, a combination of complementary methods is typically used. Table 9.1 shows a comparison of a number of techniques in terms of their ability to categorise oil and gas potential.

The definition of kerogen types I, II, III and IV from Rock Eval pyrolysis using Hydrogen and Oxygen indices is shown in Fig. 9.4. The changes of these indices with maturity are indicated by arrows in this figure. As pyrolysis is a bulk determination, a 'Type II' kerogen can be (and in fact generally is) a mixture of Type I algal material, degraded by bacteria and mixed with terrigenous Type III material (Barnard *et al.*, 1981). A pure Type II kerogen comprises spores, pollen, cuticle etc.

9.2.4 Recognising source rocks from wireline logs

Organic-rich rocks can often be recognised using a normal suite of wireline logs (Fertl, 1976). Meyer and Nederlof (1984) have recently discussed the identification of source-rocks from a statistical evaluation of gamma ray, sonic, density and resistivity log responses, with examples given from the North Sea. Indeed the term 'hot shale' applied to organic-rich

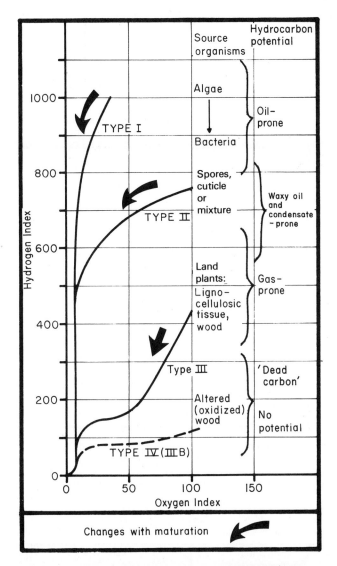

Fig. 9.4. Definition of kerogen Types I, II and III based on a Rock Eval Oxygen Index/Hydrogen Index diagram. Note that Type II kerogen can (occasionally) be a pure spore/pollen/cuticle/resin kerogen, but is generally a mixture of bacterially degraded algal debris of Type I composition, mixed with a terrigenous component of Type III composition. Type IV (sometimes called Type IIIb) comprises altered terrigenous debris (Inertinites).

sections of the Kimmeridge Clay Formation of the North Sea derives from its high natural radio-activity. The log response of a source-rock unit is largely that which would be predicted from its physical properties as summarised in Table 9.2. The use of logs to define source rocks is particularly appropriate for volumetric studies of source-rocks since it allows a continuous monitor of source quality over the whole section. Some conventional analyses are, however, always required in order to calibrate the log responses. Typical log responses through some of the Cretaceous and Jurassic rock intervals of the North Sea are shown in Fig. 9.6.

9.3. Maturation and generation

Kerogen matures during burial, that is, it undergoes physical and chemical changes that result largely from temperature increases related to depth of burial. One

result of these changes is the generation of hydrocarbons—first liquids (oil), and then gas. This process is shown schematically in Fig. 9.3.

Quantitative generation from an oil-prone (Type II) kerogen of the type found in the Kimmeridge Clay Formation of the North Sea is shown in Fig. 9.7. The vertical axis is temperature, which can be equated with depth by using an average geothermal gradient (North Sea average 29°C/km (1.6°F/100 ft), range 22-40°C/km (1.2-2.2°F/100 ft); see Fig. 9.9). North Sea temperature data has been treated in detail by Harper (1971), Cornelius (1975), Cooper *et al.* (1975), Carstens and Finstad (1981), and Toth *et al.* (1983). In Fig. 9.7, generation of oil is given in m^3 oil/km^3 of 1% TOC source rock (bbls/acre ft), and associated gas in terms of m^3 gas at standard temperature and pressure/km^3 of 1% TOC source rock (mcf/acre ft).

The progressive generation of dry gas (methane) from coal and coaly shales can be seen in Fig. 9.8, where maturity is measured in terms of vitrinite reflectance and coal rank.

A number of experimentally determined parameters are used to measure kerogen maturity, including vitrinite reflectance, and the properties of coal in general (Teichmüller and Teichmüller, 1979; Bostick, 1979). Spore or kerogen colour (Smith, 1983) and, to a lesser extent, fluorescence, are used together with the temperature of maximum pyrolysis yield (T_{max}) from the Rock Eval pyrolysis technique, and molecular ratios e.g. sterane/triterpane isomer ratios (Cornford *et al.*, 1983; Mackenzie *et al.*, 1980). No one technique can be applied universally, and no simple equivalence can be drawn up between the various techniques. Vitrinite reflectance is the most widely used technique, while the rapidly developing field of molecular ratios has the important advantage that measurements can be made on both source-rock and reservoired oil, thus allowing the establishment of a direct genetic link (Cornford *et al.*, 1983). These measurements also have the advantage of reflecting maturation changes within the oil-prone material directly, and do not require empirical correlation with the generation process.

An approximate equivalence of some of these maturation parameters is given in Fig. 9.3.

The approximate maturity of a source rock can be predicted from the maximum temperature it has experienced during burial. In continuously subsiding basins such as the North Sea Grabens, the maximum temperature experienced is related to depth of burial via the local geothermal gradient (°C/km or °F/100 ft), see Fig. 9.9.

However, the time that a source rock has been exposed to a certain temperature has a major effect on some maturity parameters. For a given temperature the reflectance of vitrinite is strongly time dependent (Hood *et al.*, 1975; Waples, 1980; Gretener and Curtis, 1982), spore colour is less dependent (Barnard *et al.*, 1981a), whilst some common isomer ratios appear to be minimally effected (Mackenzie and McKenzie, 1983; Cornford *et al.*, 1983). The effect of time on the actual

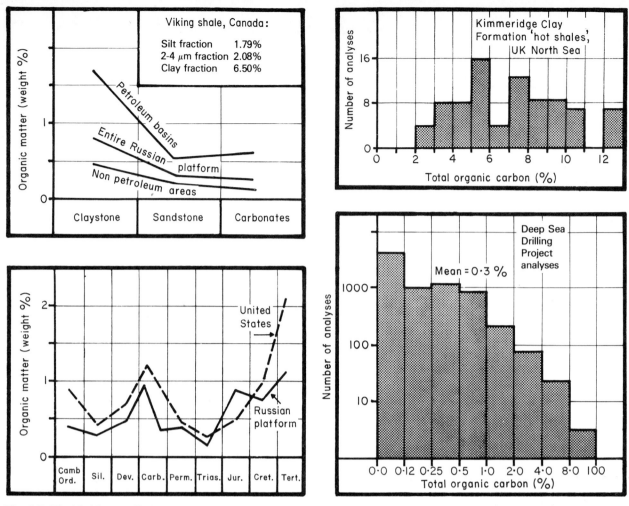

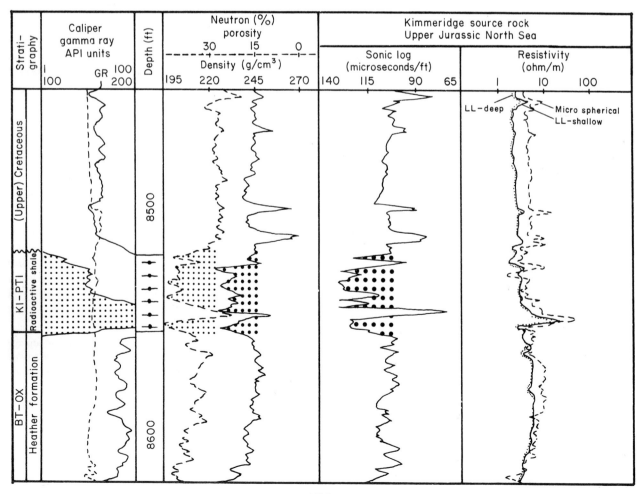

Fig. 9.5. Worldwide compilation of data on Total Organic Carbon (TOC) and organic matter of sediments showing fluctuation with lithology, with grain size within a single sample, and with stratigraphic interval (Bostick, 1974). Histograms are also shown of 7253 analyses from the Deep Sea Drilling Project analyses (McIver, 1975), and the Kimmeridge Clay Formation of the U.K. North Sea. Note that sampling is probably biased towards higher TOC intervals in all data collections.

Table 9.2. Source rock characterisation from wireline log response

Log (Units)	Response — Change with increase in organic matter	Coals	Values in Kimmeridge Clay Formation	Comments
Natural gamma ray (API units)	Generally very[1] high	Low except in uranium rich coals	Typically 100-200 API units	Response mainly due to uranium[6] (Bjørlykke *et al.*, 1975; Berstad and Dypvik, 1982) associated with planktonic matter. Lower response in gas prone shale. Very low response in lacustrine organic rich shales.
Formation density (g/cc)	Decreases[2] [3]	Very low	Typically 2.2-2.3g/cc	Density contrast can be offset by high pyrite contents.
Resistivity ohm.m	Increases	Very high	Immature <5ohm.m Mature >20ohm.m Late mature <2ohm.m (Goff, 1983)	Resistivity low in immature shales due to saline pore water, and high due to oil filled porosity at mature stage. Resistivity falls at late mature stage due to increased pore water content and 'graphitisation' of the organic matter (Meissner, 1978)
Sonic interval velocity (μsec/ft)	Increases[4]	Very high	Variable	Also increases with burial, little affected by pyrite and hole condition.
Neutron (counts/second)	Decreases[5]	Low	—	
Pulsed neutron carbon/oxygen	—	—	—	New logs (Hopkinson *et al.*, 1982): may be able to recognise richness or even, when calibrated, kerogen type in source rocks.

(1) Schmoker (1981) has defined the following relationship between the volume fraction of organic matter (ϕ_0), and Υ, the gamma ray log response in API units for the Devonian shales of the Appalachians: $\phi_0 = \Upsilon_B - \Upsilon/1.378A$, where Υ_B is gamma log response if no organic matter is present and A is the slope of the crossplot of gamma and density logs.

(2) Schmoker (1979) has shown that the volume fraction of the organic matter ϕ_0 is related to log density in the Devonian shales of the Appalachians by $\phi = \rho_B - \rho/1.378$ where ρ_B is the rock density where no organic matter is present.

(3) Tixier and Curtis (1967) have related Fisher Assay pyrolysis yield to log density (ρ) for the Green River oil shales with Type I kerogen as follows: yield = $154.81 - 59.43\rho$ gals/ton ($\times 3.8 \approx$ litres/tonne), while Meyer and Nederlof (1984) have found the following relationship for the Posidoniaschiefer of Germany (generally a Type II kerogen): yield = $1.113 + 63.14\triangle\rho$ gals/ton where $\triangle\rho$ is the density difference between lean and rich shale.

(4), (5) See Fertl (1976) Figures 10.7 and 10.8 respectively for Fisher Assay pyrolysis yields vs. log response.

(6) The gamma ray log response in the Kimmeridge Clay Formation shales of Amoco's NOCS well 2/11-1 derives 61% from Uranium, 33% from Thorium and 6% from Potassium (Bjørlykke *et al.*, 1975).

volumetric generation of oil itself is not clear, but it seems to be relatively small (Price, 1983). The quantitative prediction of maturity from subsidence curves and geothermal gradients is a complex matter: those wishing to take this subject further should refer to the publications listed above.

9.4 Source rock volumetrics and migration

For exploration in frontier areas it is probably sufficient to know that you have a source rock and that it is likely to be mature: major structures will certainly be drilled even without the comfort of this information. For more marginal structures especially in mature exploration areas, however, a comprehensive prospect evaluation requires evidence that there is sufficient volume of mature source rock of an adequate quality to fill the proposed trap with oil, and that the oil was generated at an appropriate time with respect to trap formation. This approach is of great value when rating prospects, that is, deciding which one to drill first, or which acreage to bid for or to relinquish.

Fig. 9.6. *opposite.* Generalised sketch of log motif through the hot shales of the Kimmeridge Clay Formation source-rocks of the North Sea (Meyer and Nederlof, 1984).

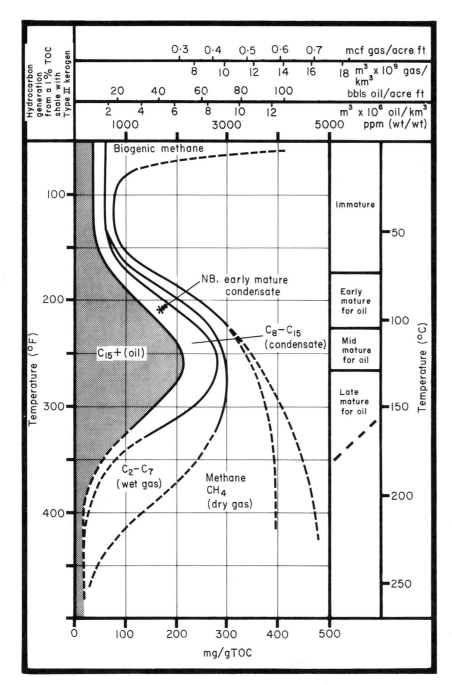

Fig. 9.7. Qualitative generation of oil and condensate: m³ oil/km³ of 1% TOC source-rock (bbls/acre ft); and gas: m³CH₄/km³ of 1% TOC source-rock (mcf/acre ft), as a function of down-hole temperature for oil-prone kerogen. For details see Brooks, Cornford and Archer (in press). Note the possible generation of larger quantities of early mature condensate from spore/pollen/cuticle/resin rich sediments, and expulsion if drainage is good. To convert to a specific hydrocarbon type of density ρ, use ppm wt/wt × (rock bulk density/ρ) = ppm vol/vol. (ρ-oil ≈ 0.8–0.9; ρ-methane = 7.1×10^{-4} at standard temperature and pressure); 1bbl = 5.614 cu ft = 159 litres; 1 acre ft = 1233.5 m³.

The concepts of source-rock volumetric estimation have been discussed by Fuller (1975), Tissot and Welte (1978) and Goff (1983). For a simple approach it consists of four steps:

1 . Establishing the areal extent and thickness of source rock(s) draining into the structure.

2 . Defining those parts of the drainage area that reached maturity after trap/seal formation.

3 . Determine the volume of hydrocarbon (oil or gas) generated per unit volume of source rock.

4 . Estimate the efficiency of (a) expulsion from the source rock; (b) the migration path to the trap, and (c) of the seal on the trap.

Although it is beyond the scope of this paper to deal with this topic in detail, the following comments, which are summarised in Fig. 9.10, outline the most important points.

The effective drainage area of a structure may be bounded by one or all of the following features (Fig. 9.10a):

The immature/early mature, or early mature/mid mature boundary, above which no hydrocarbon generation will have occurred.

The late mature/post mature boundary for the source rock that was attained at the time of trap/seal formation. This will, in a continuously subsiding basin, generally be stratigraphically deeper than the present day late mature/post mature boundary.

Sealing faults.

Erosional or regressional sub-crop of the source-rock unit.

Facies change of the source-rock from, say, oil-prone to gas-prone, or from shale to sandstone.

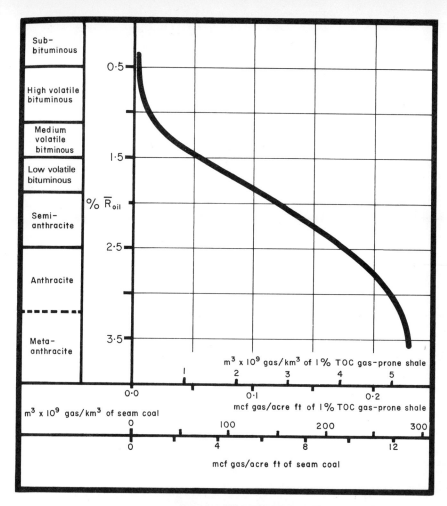

Fig. 9.8. Cumulative methane generation in m³/km³ (millions of cu ft/acre ft) from 1% TOC gas-prone shale (upper horizontal scale) and coal (lower horizontal scale) as a function of maturity measured in terms of coal rank and vitrinite reflectance. Since methane is highly mobile, the stated volumes of gas will not be present in the coal or shale, but will start to migrate out as soon as generated. The curve is based on laboratory data of Juntgen and Karweil (1966) and stochiometric calculations based on the change of hydrogen content with rank. Strictly, the plot can be used only for vitrinite-rich coals and kerogens, for example, the Westphalian of N.W. Europe.

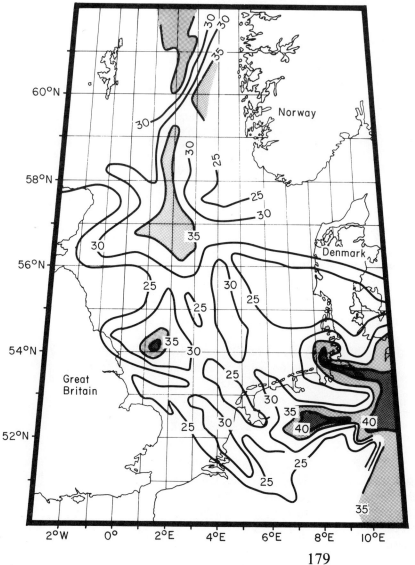

Fig. 9.9. Present day North Sea geothermal gradients (after Cornelius (1975) and Carstens and Finstad (1981); see also Harper, 1971). Considerable errors exist in obtaining geothermal gradients from well-log data—see references above. The gradient plotted is the overall geothermal gradient from sea floor to the lowest measuring point, normally at terminal depth. Geothermal gradients vary consistently with depth, and with lithology: heat flow is the conserved property at any one location (Oxburgh and Andrews-Speed, 1981).

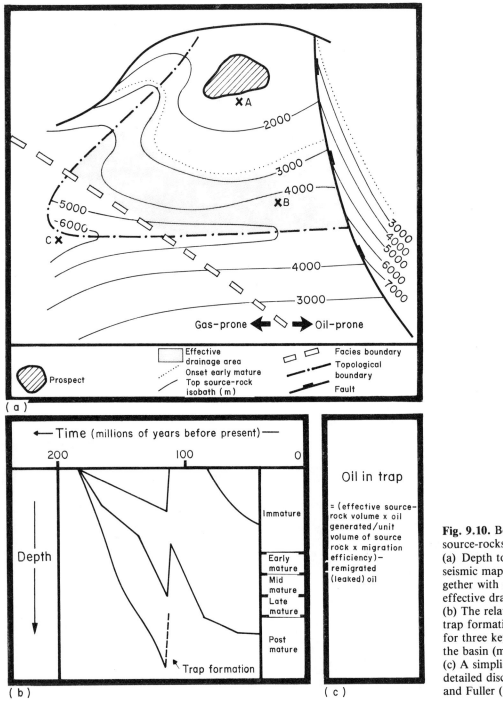

Fig. 9.10. Boundary conditions for source-rocks volumetric calculations. (a) Depth to top source-rock on seismic map showing prospect, together with some boundaries for the effective drainage area. (b) The relationship of generation to trap formation using subsidence plots for three key points A, B and C in the basin (marked with 'x' on map). (c) A simplistic mass balance—for detailed discussion see Goff (1983), and Fuller (1975).

In addition, a maximum migration distance may be a limiting factor, but this is contentious. Certainly, the regional extent of the proposed migration pathway (e.g. a sheet sand) could place limits on the effective drainage area. The thickness of the source-rock unit can be determined from seismic mapping, or from regional compilations. Only that part of the drainage area which reaches maturity after trap formation should be considered. This can be deduced from a subsidence curve for representative parts of the drainage area (Fig. 9.10b). In Fig. 9.10b, part of the source-rock in the basin centre (location A) generated its hydrocarbon prior to trap formation at about 110my bp.

The volume of hydrocarbon generated per unit volume of source-rock can be obtained from Figs. 9.7 and 9.8, given an average value for the total organic carbon content of the unit. If the quantity or type of kerogen varies laterally or vertically, then the drainage area should be compartmentalised, and each compartment treated separately.

Lastly, uncertainties about migration probably contribute the major sources of error in such calculations. Mechanisms of expulsion and migration are currently the topic of animated debate—largely in the absence of sound geological observation (Roberts and Cordell, 1980; Durand, 1983). The movement of light hydrocarbons has, however, been extensively studied (Leythaeuser *et al.*, 1982, and references therein).

The results of these studies show that for oil, expulsion of between 1% and 20% of the total generated, is normal in thick, rich source rocks (e.g. Fig. 9.2a). Expulsion efficiencies can rise as high as 50%, or even 80%, in well-drained, interbedded sand-shale sequences (e.g. Fig. 9.2b). In the North Sea region, the highest oil expulsion efficiencies are to be expected in areas where sand intercalates with the 'hot shales' of the Kimmeridge Clay Formation, as in the Piper, Brae and Magnus fields. Values for migration efficiency derive from mass-balance calculations equating the volume of source-rock with the volume of reservoired oils. In the case of low expulsion values from thick, rich source-rocks, the oil remaining *in situ* will be thermally cracked on further burial and expelled as condensate or wet gas (Fig. 9.7). Late stage gas generated in the basin centre can displace oil from upflank structures, giving rise to accumulations of commercially less desirable gas.

The efficiency of secondary migration can be quite high (e.g. $\sim$ 80%) in cases of simple up-flank migration through a laterally continuous conduit to a structure at the natural focus (Pratsch, 1983) of the migration path. Such a simple secondary migration pathway may exist in the Brent Sands of the East Shetland Basin (see for example the cross-section in Figure 2 of Eynon, 1981). However, long distance or devious migration pathways through low permeability conduits (e.g. ratty sands, faults, fractured chalk, etc.) are likely to be much less efficient. The presence of oil sourced from the Kimmeridge Clay Formation in the Tertiary reservoirs of the North Sea can sometimes pose problems in terms of defining a plausible migration route. In the Ekofisk field faults are believed to have played a major role (Van den Bark and Thomas, 1980). It is even more difficult to explain the presence of biodegraded Jurassic oils in apparently closed, lensoid, Tertiary sand bodies, with no faulting apparent on seismic lines.

Gas expulsion is probably nearly 100% efficient but, as Leythaeuser *et al.* (1982) have pointed out, loss by diffusion during migration and through the reservoir seal can be very significant over geological time. These authors conclude that without continuous topping-up with gas from source-rock kitchen areas, most gas fields will have a relatively limited life. In the North Sea Ekofisk area, Van den Bark and Thomas (1980) have identified a 'gas chimney' on seismic lines, interpreted as the leakage of gas from the Chalk through the Tertiary. Active replenishment is expected in this area (see section 9.5.8). Other examples of 'gas chimneys' defined on seismic lines are given by Nordberg (1981).

In contrast, the gas in some of the gas fields of the Southern North Sea must have been trapped since its generation in the late Cretaceous (Glennie and Boegner, 1981), although in the Broad Fourteens Basin (Oele *et al.*, 1981) and the giant Groningen gasfield (van Wijhe *et al.*, 1980) the gas may still be being 'topped-up' to the present day. Remigration to new reservoirs also has occurred as a result of tectonic inversion of former basinal areas (see Glennie, Chapter 3 of this book). The longevity of the gas fields of the U.K.

Southern North Sea may in large part be due to the excellent seal afforded by the Zechstein halite and anhydrite deposits.

Finally, source-rock studies have now moved into the area of three dimensional modelling of the basic physical and chemical processes (Welte and Yukler, 1981). These large scale computer models are becoming able to predict accurately where oil is generated by thermal modelling, and where it flows by pressure modelling. They do, however, require a large amount of input information. These large quantitative models force the explorationist to view oil generation and accumulation as a unified process, and have greatly benefitted our understanding by focusing attention on a number of poorly understood but critical areas of hydrocarbon genesis.

9.5 Stratigraphic distribution of recognised and possible source rocks

9.5.1 Introduction

This section comprises a review of the Phanerozoic organic-rich sediments of the North Sea and surrounding areas (Fig. 9.11), giving, where possible, the amount (TOC) and type (oil-prone, gas-prone) of the sedimentary organic matter, and its maturity, where known from published sources. In covering the whole stratigraphic column, it should not be forgotten that the Kimmeridge Clay Formation (Upper Jurassic/basal Lower Cretaceous) is overwhelmingly the most important oil source-rock, and the Carboniferous coal measures (Westphalian) and locally the Middle Jurassic coals are the only well established source-rocks for dry gas in the region. Barnard and Cooper (1981) have previously reviewed the hydrocarbon source-rocks of the North Sea, whilst more recently, Thomsen *et al.* (1983) and Rønnevik *et al.* (1983) have summarised data on Danish and Norwegian acreage respectively.

9.5.2 Cambro-Ordovician

Cambro-Ordovician black shales are found on the shelves and basins of the Iapetus Ocean, within and flanking the Caledonides, from the Kukersite oil shales of Estonia in the east (Duncan and Swanson, 1965) to the Appalachians in the west (Islam *et al.*, 1982). Within the area of interest, the Cambro-Ordovician black shales outcrop in the Southern Uplands of Scotland and in Scandinavia, and could source oil and gas. The Alum shale of southern Sweden contains as much as 17.5% TOC (Andersson *et al.*, 1982; Bitterli, 1963), and is oil-prone (4-8% oil yield upon pyrolysis as an oil shale). At least part of the outcrop currently falls within the oil generation window (Bergstrøm, 1980).

In contrast, the Ordovician black shales of the Southern Uplands of Scotland are post-mature for oil and gas generation at Hartfell and Dobb's Linn, and late mature for gas generation at Mountbenger (average uncorrected graptolite reflectance values of 4.45%R,

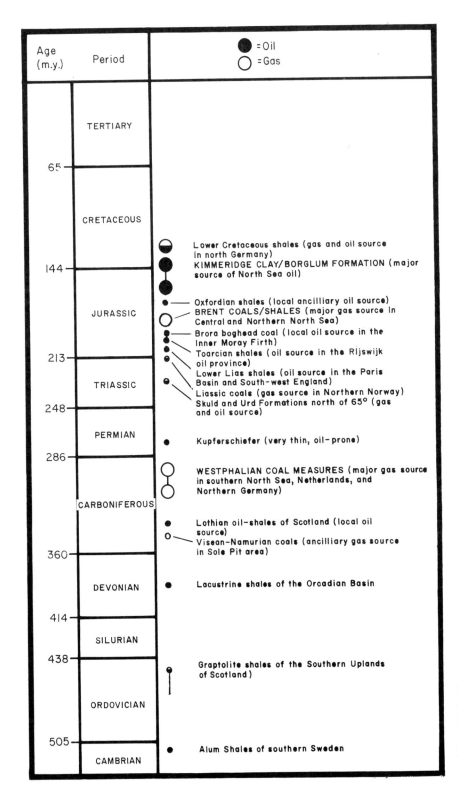

Fig. 9.11. Stratigraphic distribution of source-rocks in the North Sea area. Major, proven source-rocks are in upper case, less significant or putative source-rock intervals are in lower case text. Details are given in Section 9.5 for each stratigraphic interval.

3.27%R and 2.20%R respectively; Watson, 1976). Both Watson (1976) and Bergstrøm (1980) note a decrease in maturity to the west.

There is no current evidence of a Cambro-Ordovician source for any North Sea hydrocarbon occurrence, but generation from these rocks is known in the paleo-contiguous Appalachians of North America in the west and in Estonia to the east.

9.5.3 Devonian

Organic-rich Devonian (Old Red Sandstone) sediments

are found in the Orcadian basin of north-east Scotland and the Shetlands. These lacustrine shales (Donovan, 1980), constitute a minor lithology of an 18,000 ft thick sequence (Donovan *et al.*, 1974), and are often associated with oil staining, bitumens and seeps (Parnell, 1983). Hall and Douglas (1983) obtained TOC values of 0.6 to 5.2% for five Devonian samples. Kerogen type is poorly defined and appears variable from oil-prone to gas-prone, and onshore maturity is probably within the oil window based on estimates from Hall and Douglas' figured sterane and triterpane distributions.

The Orcadian Basin itself extends from eastern Scotland and the Orkneys over to the Hornelen Basin off the west coast of Norway (Ziegler, 1982). The extent to which the lacustrine source-rock facies continues under the North Sea is unknown.

9.5.4 Carboniferous

Carboniferous sediments (coals and associated carbonaceous shales) have sourced the major gas-fields of the Southern North Sea, onshore Netherlands and Germany (Barnard and Cooper, 1983; Bartenstein, 1979; Tissot and Bessereau, 1982), whilst the minor U.K. onshore oilfields of the East Midlands and Central Valley of Scotland are reputedly sourced from locally developed liptinite-rich sediments within the Carboniferous.

Many of the gas fields sourced by the Carboniferous

occur in areas subjected to either early Mesozoic subsidence and inversion (Fig. 9.12), as in the Sole Pit Trough (Glennie and Boegner, 1981) and the Broad Fourteens Basin (Oele *et al.*, 1981), or to areas of higher geothermal gradients (Kettel, 1983).

The Carboniferous source-rock story is not so much about kerogen quantity (which is huge), or quality (which is overwhelmingly gas-prone), but about maturity and the timing of gas generation relative to the development of reservoir structures.

Dinantian (Visean/Tournaisian)

The Visean oil shales of the Central Valley of Scotland constitute a high quality oil-prone source. TOC values of 11.2% (Bitterli, 1963) and oil yields of about 30 gals/ton (∿115 litres/tonne) (Duncan and Swanson, 1965), probably represent the richest beds. Algal

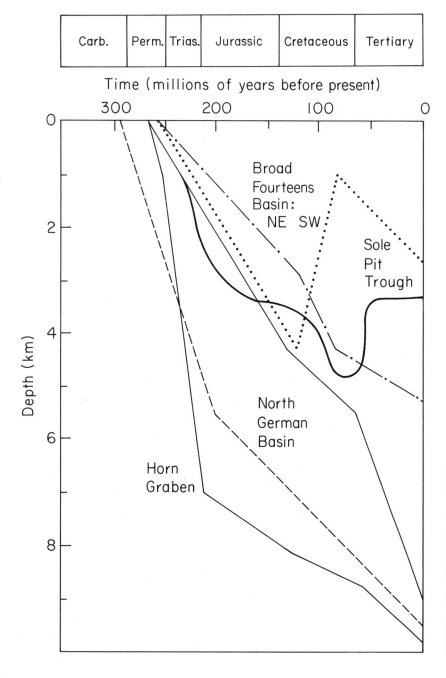

Fig. 9.12. Upper Paleozoic subsidence curves uncorrected for compaction from the deepest parts of the N.E. and S.W. Broad Fourteens Basin (Oele *et al.*, 1978), Sole Pit Trough (Glennie and Boegner, 1981), the Horn Graben (Day *et al.*, 1981), the Central Graben (Day *et al.*, 1981) and the North German Basin approximated from Teichmüller *et al.*, 1979; Day *et al.*, 1981; Ziegler, 1982). To estimate the burial of the base of the Westphalian from these curves add about 1200 m for the Stephanian/ Westphalian interval. Maturity gradients will differ from location to location, but onset of gas generation is expected after 4-5 km burial.

('boghead') coals or 'Torbanites' exist which consist of almost pure compressed bodies of the colonial algae *Botryococcus* (Allan *et al.*, 1980). At outcrop, these boghead coals and oil shales appear to be early to mid mature with respect to oil generation as estimated from the rank of the surrounding coals (high-medium volatile bituminous). They should produce a fairly high-wax oil judging from the n-alkane distribution recorded by Douglas *et al.* (1969) and Allan *et al.* (1980) and given the known oil-generative capacity of fresh-water algae. The Lothian oil shales may well be the source of the nearby Cousland oil-field.

The oil shale facies was probably laterally restricted, but outcrops are known as far apart as Linwood, south of Glasgow in the west, to the Lothians and the Firth of Forth in the east. The extension of this facies further east (offshore) is a matter of speculation (Ziegler, 1982), but its presence has not been reported in any published well information.

Coals of Visean age have been recorded over the southern part of the mid North Sea High, and may constitute a source for relatively small gas accumulations. Elsewhere in the North Sea area, the known Lower Carboniferous does not generally constitute a significant hydrocarbon source, given the high source quality of the overlying strata.

The Namurian

The Namurian rocks may locally possess oil as well as gas potential. The shales of the Visean/Lower Namurian Limestone and Scremerston Coal Groups of Northumberland are organic rich (1.08%-3.80% TOC), containing approximately equal quantities of exinite and vitrinite (Powell *et al.*, 1976). These rocks could generate some oil or condensate as well as gas. The bitumens of Windy Knoll in Derbyshire, though deposited in the Visean limestones, are believed to derive from the overlying Namurian Edale Shales (Pering, 1973) possibly under the influence of hydrothermal activity. Namurian shales may also have sourced the oil of the Eakring field of Nottinghamshire, which is reservoired in the associated Millstone Grit.

Namurian marine shales may have oil potential over a limited area to the north of the London-Brabant High. Further north, the Namurian is relatively rich in coal seams and will hence be an additional source for gas in the Southern North Sea.

The Coal Measures (*Westphalian*)

The Coal Measures, comprising coal seams and associated shales, were deposited in a large delta system covering much of North Western Europe (Ziegler, 1982). In the North Sea region, the area extended from about 57°N to the Variscan front in the South (Fig. 9.13), although coals are probably restricted to an area south of 55° 30′N. The Coal Measures are typically between 1000 m to 2500 m thick, of which in North-West Germany, coal seams comprise about 3% (Lutz

et al., 1975). It is still a matter of largely academic debate (e.g. Rigby and Smith, 1982) whether coals or the associated carbonaceous shales constitute the major source of gas: whatever the detailed source, Westphalian strata contain a large thickness of rich, dominantly gas-prone coals and shales.

Carboniferous coal rank in North Western Europe varies from sub-bituminous to meta-anthracite. Gas generation occurs between vitrinite reflectance values of 1%R and 3%R (Fig. 9.8), that is, from medium volatile bituminous coal to anthracite. The presence of Kupferschiefer at a sub-bituminous equivalent rank overlying Westphalian anthracites in North Germany and North-east England (Boigk *et al.*, 1971; Gibbons, in press) shows that at least some of the area attained its present maturity by early Zechstein time. In the Variscan tectonic episode, vast amounts of gas and possibly oil must have been generated but then rapidly lost during uplift and erosion. The major Variscan coalification occurred just to the north of the Variscan Front (Fig. 9.13) as evidenced by the anthracites of South Wales, Kent and the Ruhr district.

Many areas survived the Variscan Orogeny with their gas-generating potential intact (i.e. having vitrinite reflectance values of less than 2%R), and it is these areas that should be considered as being capable of generating more gas on subsequent burial (Fig. 9.13).

In the centre of the Sole Pit Trough, burial of the top Carboniferous up to about 4000 m occurred progressively from the Triassic to the late Cretaceous, Fig. 9.12 (Glennie and Boegner, 1981). Gas generation and migration will thus have occurred during the Jurassic and Cretaceous with migration to the basin flanks. Late Cretaceous inversion resulted in remigration of gas from the former basin margin structures (e.g. Indefatigable area) into Rotliegend reservoirs (e.g. West Sole and Leman Bank), that had already suffered diagenetic damage. The survival of reservoired hydrocarbons to this day is witness to an exceptionally good seal.

In contrast, Oele *et al.* (1981) have shown that generation in the Broad Fourteens Basin continued from late Jurassic to the present, with a north-easterly migration of the depo-centre.

Barnard and Cooper (1983) have noted that the Westphalian can be deeply eroded in areas of inversion as well as on regional highs. It is the areas of inversion, however that appear to have sourced the major gas fields. They suggest that in some areas only remnants of the Westphalian may be present, in which case the Namurian in Yoredale (coal-shale-limestone cyclothems) or Coal Measures facies may be a possible additional gas source.

A source-rock/reservoir mass balance can be attempted for the 693×10^9 m^3 (24.5×10^{12} cu. ft) of reserves in the gasfields surrounding the Sole Pit Trough (Barnard and Cooper, 1983). Using an estimate of the area of the Sole Pit Trough as defined in Glennie and Boegner (1981) of 10.12×10^3 km^2 (2.5×10^6 acres) and assuming generation of 3×10^9 m^3 gas/km^3 (0.14

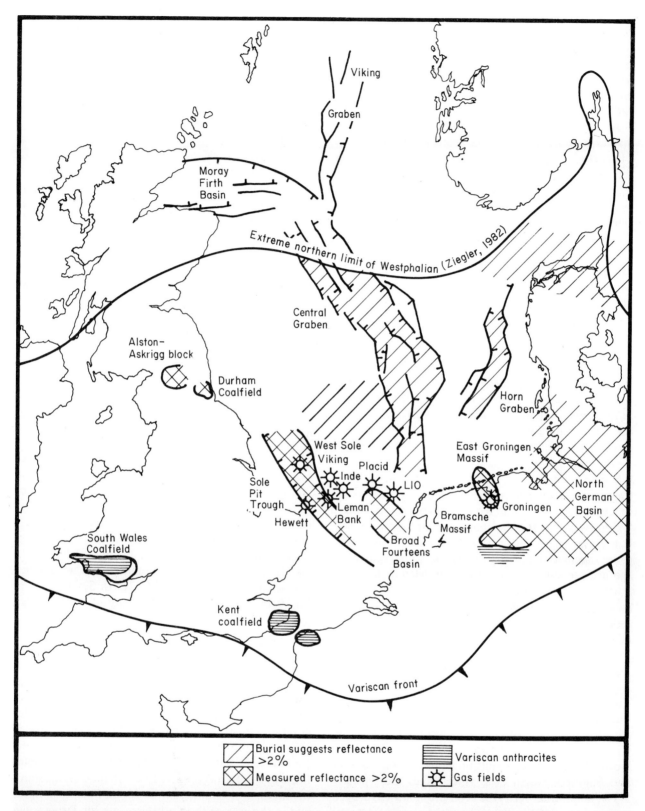

Fig. 9.13. Distribution and maturity of Westphalian (Upper Carboniferous) gas-prone source-rocks bounded to the south by the Variscan Front and to the north by erosion or non-deposition (after Ziegler, 1982, encl. 11). Cross-hatched areas have attained post-Variscan maturity >2%R (Barnard *et al.*, 1983; Teichmüller *et al.*, 1979; Bartenstein, 1979; Kettel, 1983) and diagonals signify areas where base Zechstein burial exceeds 4 km, and hence gas generation may have occurred. Some major gas fields are indicated: note that they are generally associated with areas of inversion (see Fig. 9.12).

mcf/acre ft) per 1% TOC in the shales and 160×10^9 m³/km³ (7 mcf/acre ft) of seam-coal (Fig. 9.8), indicates potential generation of 30.6×10^9 m³ gas per metre (0.364×10^{12} cu. ft of gas per ft) of 1% shale, or 1619×10^9 m³ gas per metre (18×10^{12} cu. ft of gas per ft) thickness of coal seam. Clearly there is adequate source potential to account for the reservoired gas even given a relatively incomplete Westphalian section and a low migration/remigration efficiency.

The second area of major post-Variscan burial is in

North Germany/Southern Denmark which may have sourced the massive Groningen field in north-east Netherlands. However, Lutz *et al.* (1975), suggest a source to the south-west of the field. A further source area may have been more precisely defined by Kettel (1983) as overlying the East Groningen massif.

A third area of gas generation from the Westphalian may exist under the Central Graben and Horn Graben, where Day *et al.* (1981) show local burial of up to 10 km for the base Zechstein. The Coal Measures, if preserved, are expected on regional grounds to be of bituminous coal rank (Eames, 1975). The timing of gas generation can be estimated as Triassic-Jurassic from Fig. 9.12.

Finally, some small areas where the Westphalian overlies post-Variscan intrusives (e.g. Bramische Massif, the Alston granite) could also have generated gas at a time when reservoirs, structures and seals existed.

No Carboniferous in source-rock facies is known from the area west of the Shetlands (Ridd, 1981). Carboniferous coals are mined in Ireland (Griffiths, 1983) and the Carboniferous is present in gas-prone Coal Measure facies in the Kish Bank Basin (Jenner, 1981), and possibly in the northern part of the Irish Sea (Barr *et al.*, 1981). Of the two gas finds west of England, the Morecambe Field is of unknown source, but compositional similarities with the Southern North Sea gas fields (Ebbern, 1981) suggest the Carboniferous, which is known to underlie the area, as the source. In contrast, the Kinsale Head field is believed by Colley *et al.* (1981) to have a complex origin from Mesozoic source-rocks.

Somewhat beyond the boundaries of the North Sea, but pertinent to current exploration, mature, gas-prone, organic-rich shales and coals of Carboniferous age are present on Bjornøya (Bear Island), offshore northern Norway (Bjorøy *et al.*, 1983). Carboniferous coals and coal measures are also present on Svalbard (Spitsbergen). This gas-prone source sequence is part of a separate North Greenland Basin (Rønnevik, 1981).

9.5.5 Permian

The Kupferschiefer/Marl Slate horizon—typically less than 1 m (3 ft) thick—is the only Permian sediment with a high-grade hydrocarbon source potential in the North Sea area. It outcrops in Durham, U.K., and North Germany and extends under much of the central and Southern North Sea. TOC values are typically 5-8%, and the kerogen type is oil-prone with hydrogen indices above 600 mg/g (Gibbons, in press). Any oil generated would be recognisable by its high porphyrin content. No oil of this type has been reported in the North Sea, where the oils generally have very low porphyrin contents. However, the high Vanadium and Nickel content of Buchan oil (Table 9.6) may suggest a Kupferschiefer contribution. This horizon is present north of the Buchan field (Taylor, 1981). The absence of effective generation is probably a function of the thinness of this unit. Locally, shows in the Zechstein may derive from the Kupferschiefer. Onshore Nether-

lands, Van den Bosch (1983) reports oil-prone Zechstein shales and carbonates.

9.5.6 Triassic

The Triassic sediments have no source potential in the North Sea area. However, in some regions to the north of latitude 62°N, marine sedimentation continued during the Triassic. Knowledge of the source potential is restricted to a limited number of outcrops. The marginal marine Skuld and Urd Formations of Bjornøya Island south of Svalbard (Spitsbergen) contains shales and silty shales with up to 2.0% TOC of dominantly gas-prone kerogen which is in the lower part of the oil window (Bjorøy *et al.*, 1983). On Svalbard itself, middle Triassic shales of the Botneheia Formation contain considerable quantities (2.9-5.5% TOC) of oil and gas-prone kerogen, which ranges from late immature to peak oil generating maturity (Forsberg and Bjorøy, 1983). This high quality oil and gas-prone source may extend as far south as about 65°N according to facies maps of Rønnevik (1981) and Rønnevik *et al.* (1983). The Rhaetic (Top Triassic) of Southern England may locally develop in an oil or gas-prone facies.

9.5.7 Jurassic

The Jurassic contains the source of the bulk of North Sea oil in the Upper Jurassic-basal Cretaceous Kimmeridge Clay and Borglum Formations, and a major source of gas in the Middle Jurassic coals. Ironically, relatively little detailed work has been published on the best oil source-rock, the 'hot shale' of the Kimmeridge Clay Formation (Fuller, 1975; Oudin, 1976; Brooks and Thusu, 1977). A typical summary of the Jurassic/Cretaceous source-rocks in the North Sea is given in Fig. 9.14, from Kirk's (1980) study of the Statfjord area.

Lower Jurassic (Liassic)

Within the North Sea area itself, the Lower Jurassic, where it develops source-rock quality, is dominantly gas-prone: only the Upper Lias (Toarcian) is locally developed in an oil-prone facies. The Lower Jurassic is absent, or present only as erosional remnants, in the Central and Witch Ground Grabens and the Outer Moray Firth Basin (Ziegler, 1982). In the extreme south of the Central Graben, in the Broad Fourteens Basin and in the area of the Yorkshire-Sole Pit Trough, the Lower Jurassic is preserved.

In the Yorkshire-Sole Pit Trough area, lean gas-prone shales (0.8% TOC average) and minor coals (jet) occur in the lower and middle Liassic sequence. An Upper Lias (Toarcian) rich oil-prone facies is variously known in north-west Europe as the Jet Rock and Bituminous shales of Yorkshire, the Posidoniaschiefer of Germany and Les Schistes Bitumineux in France, Belgium and Luxemberg. These rocks are rich in kerogen of sapropelic type, typically with 2% to 12% TOC

Period/Series Group/FM	Lithology	Thickness range (ft)	TOC[1]	EOM[2]	EOM/TOC	CPI[3]	Ro[4]
Paleocene Rogaland Group		548–820	0·43 (64)	141 (3)	3·3 (3)	1·45 (3)	0·41 (3)
Upper Cretaceous Shetland Group		1758–3998	0·78 (201)	172 (40)	2·64 (40)	1·71 (20)	0·43 (12)
Lower Cretaceous Cromer Knoll Group		0–482	1·04 (4)	78 (2)	1·41 (2)	–	0·71 (2)
Upper Jurassic Kimmeridge Clay FM (hot shale)		0–298	4·58 (4)	2160 (4)	4·00 (4)	–	–
Middle Jurassic Heather FM		0–994	2·24 (16)	350 (9)	1·71 (9)	1·68 (9)	0·40 (2)
Middle Jurassic Brent FM		0–1023	–	–	–	–	0·44 (5)
Lower Jurassic Dunlin FM		0–1430	1·47 (86)	234 (38)	1·95 (38)	1·68 (36)	0·59 (6)
Lower Jurassic / Triassic Statfjord FM		0–1016	0·44 (19)	–	–	–	0·68 (4)
Triassic Cormorant FM		Unknown. Max. Pen. 6032+ft	0·35 (56)	50 (6)	1·71 (6)	1·37 (3)	0·73 (8)

( ) Number of values

● Oil reservoir

1. Total organic carbon (% rock weight)
2. Extractable organic material (ppm)
3. Carbon Preference Index
4. Vitrinite reflectance

Fig. 9.14. Typical source-rock summary for a North Sea well (from Kirk's 1980 study of Statfjord field), identifying the Upper Jurassic Kimmeridge Clay Formation as the best source-rock interval (4.58% TOC and 47 mg extract/g TOC), with the Heather and Dunlin being leaner and more gas-prone. The sample density is somewhat puzzling, being highest in the Lower Jurassic, Upper Cretaceous and Paleocene. The gas-prone nature of the Lower Cretaceous Cromer Knoll sediments is highlighted by the presence of altered rather than primary vitrinite, as shown by the off-trend average vitrinite reflectance value of 0.71%R. Otherwise a consistant maturity trend is displayed by the vitrinite reflectance values, with the carbon preference indices being characteristically fickle.

(Barnard and Cooper, 1981; Tissot *et al.*, 1971; Huc, 1976; Brand and Hoffman, 1963).

The Hettangian/Sinemurian Blue Lias of Dorset, north Somerset and Glamorgan, U.K., is present as cyclic sequences of limestone-marl-shale. The shales, particularly when laminated, contain good yields (up to 18% TOC) of mixed oil and gas-prone kerogen, and are late immature to early mature with respect to oil generation (Cornford and Douglas, in preparation). The shales of Sinemurian Black Venn Marls of Dorset are also rich in mixed oil and gas-prone organic matter but here are immature. The Lias has been suggested as a possible source for the Wytch Farm oil, onshore Dorset (Colter and Havard, 1981).

To the east, in the Danish sub-basin, the open-marine shales of the Fjerritslev Formation have low TOC contents and are probably gas-prone, since they are dif-ferentiated by Hamar *et al.* (1983) from the anaerobic 'hot' shales of the Upper Jurassic/Lower Cretaceous. Apart from some 40 m of Upper Jurassic claystone (Olsen, 1983) the Jurassic is believed to be absent in the Horn Graben (Best *et al.*, 1983).

Since it is absent, or only present as remnants in the graben areas, the Lower Lias will not, without local heating (e.g. Altebäumer *et al.*, 1983), have reached the gas-generating stage (vitrinite reflectance level of 1.0%R) in the southern part of the North Sea. The Upper Liassic oil-prone shales between Yorkshire and the Netherlands may, however, have reached early maturity for oil generation in the pre-inversion Sole Pit Trough at the end Cretaceous (Glennie and Boegner, 1981), in the Cleveland Basin, Yorkshire (Barnard and Cooper, 1983), and in the Broad Fourteens Basin from the Cretaceous onwards (Oele *et al.*, 1981). The

Toarcian is believed to be the source for the oils in the Rijswijk oil province of the Netherlands (Bodenhausen and Ott, 1981).

Minor lean, probably gas-prone, shales of the Statfjord and Dunlin Groups are preserved in certain areas in the northern North Sea such as the East Shetland Basin (e.g. Magnus, Thistle, Heather). In comparison with the overlying Middle Jurassic this is probably not a significant gas-prone source-rock. Rønnevik *et al.* (1983) show mature (for oil) Toarcian marine shales of the Drake Formation, with an average of 2% TOC and oil-prone kerogen in the Norwegian Viking Graben and the Horda Basin, and gas-prone kerogen in the Norwegian sector of the Central Graben.

North of 62°, mature Lower Jurassic coals and, deeper in the basin, marine shales may be present in the Haltenbanken and Traenabanken areas as gas and oil-prone source-rocks respectively (Rønnevik *et al.*, 1983). Rich mixed gas and oil-prone shales are reported at the base of an undifferentiated Jurassic sequence on Svalbard (Bjorøy and Vigran, 1980).

Middle Jurassic

In the North Sea area, Middle Jurassic deposition is limited (Ziegler, 1982, Encl. 19). With the exception of the oil-prone, algal-rich shales and coals of the Inner Moray Firth Basin, the Middle Jurassic generally has gas-generating potential in the paralic/deltaic sands, shales and coals of the Yorkshire-Sole Pit Trough area (Hancock and Fisher, 1981), the Moray Firth Basin (e.g. Maher, 1981; Bissada, 1983), the Viking Graben (Eynon, 1981; Pearson *et al.*, 1983) and the Horda, Egersund and North Danish basins (Koch, 1983; Hamar *et al.*, 1983). Coals are shown as minor components in these sequences (Hancock and Fisher, 1981; Eynon, 1981; Parry *et al.*, 1981). Goff (1983) estimates a total thickness of 10 m of Brent coals in the drainage area of the Frigg field, but suggests that the isotopic composition of Frigg gas indicates generation from the shales rather than coals (see Rigby and Smith, 1982). The gas in the giant Sleipner field is believed to come from Middle Jurassic coals (Larsen and Jaarvik, 1981). Middle Jurassic coals and shales frequently contain considerable altered vitrinite and inertinite (Cope, 1980), which somewhat down-grades their gas-generating potential.

The Middle Jurassic shales commonly have TOC values as high as 5% and, like most coal-measure sequences, the TOC values fluctuate considerably. Their gas-prone nature is shown by the kerogen composition of the claystone facies, which is dominated by vitrinitic material (Parry *et al.*, 1981). Interestingly, minor amounts of oil/condensate prone algae, resin and cuticle are present in these shales within the Viking Graben. The Callovian shale of the Ninian field contains up to 4% TOC and oil-prone kerogen (Albright *et al.*, 1980). In addition, Bissada (1983) has reported fairly rich (1.4-2.7% TOC) oil-prone rocks together with coals in the Middle Jurassic of the Moray Firth in the area of the Piper field.

A concentration of the freshwater algae *Botryococcus* is found in the oil-prone Parrot coal and associated shales of the Brora section of the Moray Firth coast. Barnard and Cooper (1981) have suggested that this might be the source for the high wax Beatrice crude.

In terms of maturity, the Brent coals and shales will be mature for gas generation (>1%R) in the Viking Graben below about 3.9 km (Goff, 1983). Interpolating between the top Trias and base Cretaceous depth maps of Day *et al.* (1981), gas generation will be confined to the central part of the Viking and Witch Ground grabens and excluded from the East Shetland Basin, and the Horda, Egersund and North Danish basins. Measured onshore reflectance data (up to 0.87%R) from Barnard and Cooper (1983), and offshore subsidence shown by Glennie and Boegner (1981), suggest that the Middle Jurassic of the Yorkshire-Sole Pit Trough area would not have attained gas generating rank even at pre-inversion maximum burial. The Middle Jurassic of the Inner Moray Firth occurs at about 2000-2100 m (6500-6800 ft) in the Beatrice field area (Linsley *et al.*, 1980), but additional offstructure burial is possible according to the regional seismic maps of Day *et al.* (1981).

To the west of the North Sea, the Middle Jurassic is present as the arenaceous gas-prone coal-bearing Great Esturine Series in Skye and the Inner Isles, outcrops that may be representative of both the Minch Basins (Ziegler, 1982). Whilst a similar facies is reported over the Rona Ridge and in the West Shetland Basin, thick dark grey marine shales are present on the northwest flank of the Rona Ridge (Ridd, 1981).

To the north of 62°, the Middle Jurassic Hestberget member of the Andøya Island outlier, contains rich, immature oil-prone (1.0-11.5% TOC) as well as gas-prone (4.6-12.3% TOC) sediments (Bjorøy *et al.*, 1980). A study of an undifferentiated Jurassic-Cretaceous shale sequence from Spitsbergen (Bjorøy and Vigran, 1980) indicates good quantities (0.45-16.0% TOC) of oil and gas-prone kerogen.

The Upper Jurassic and basal Lower Cretaceous (Oxfordian to Ryazanian)

The Kimmeridge Clay Formation, which falls within this age range, is recognised as the dominant oil (and associated condensate and gas) source-rock of the North Sea. It overlies the Oxfordian, which, where developed in an argillaceous facies, is typically a fair to good gas-prone source, but locally also has good oil potential.

Oxfordian

The Oxfordian shales of the North Sea (i.e. the Heather Formation of the Viking Graben, the Haugesund Formation of the Central Graben, and the Egersund Formation of the Fiskebank sub-basin), are generally fair to rich gas-prone source-rocks (Barnard and Cooper,

1981). In the Statfjord field, Kirk (1980) reports an average of 2.24% TOC for the Heather Formation, which appears gas-prone (Figure 14). Larsen and Jaarvik (1981) have suggested the Heather Formation may be the source for the condensate in the Sleipner field. Locally, Oxfordian shales can have some oil potential (Bissada, 1983), and where thick, (e.g. in the East Shetland Basin) could augment the Kimmeridge Clay Formation (Goff, 1983). For example, Bissada (1983) found the Oxfordian ('non Kimmeridge Clay Upper Jurassic') of U.K. Quadrants 14 and 15 to be richer (1.4-8.8% average TOC values for eight wells) than, and to contain equally oil-prone kerogen as, the overlying Kimmeridge Clay Formation.

Onshore U.K., Fuller (1975) reports the Oxford Clay of Yorkshire to be a lean, gas-prone source. In southern England, the Oxford Clay is reported to be a rich but immature potential source for oil in Dorset (Colter and Havard, 1981). Duff (1975) reports mean TOC values of 2.9 and 4.1% for two intervals of the Oxford Clay of Central England.

In the Troms area of Northern Norway, the Upper Jurassic is argillaceous, with the lower part of this unit being as good as, if not better than, the overlying Kimmeridge Clay equivalent in terms of oil-source potential: it is locally mature on structure in at least some wells

(Bjorøy *et al.*, 1983). On Svalbard, the undifferentiated Jurassic section, presumably containing material representative of the Oxfordian, comprises mature oil-prone (Type II) kerogen (Bjorøy and Vigran, 1980).

The Kimmeridge Clay/Borglum Formations:

These formations contain the major oil source-rocks of the North Sea (Barnard and Cooper, 1981). They are developed as organic rich 'black' shales in most of the graben areas of the North Sea (Ziegler, 1982; Rønnevik *et al.*, 1983). As previously noted, the high natural gamma ray log response by which the formation is commonly characterised (Fig. 9.6) has earned it the name of 'hot shale'. Typical average TOC values are shown in Table 9.3, which also catalogues studies of the source-rock potential of the Kimmeridge Clay Formation. TOC contents vary from area to area, and also with lithology in a given section. In using values from Table 9.3, it should be noted that there is a natural tendency for source-rock reports to have analysed the richest lithologies. The richest lithologies are often characterised in well site descriptions as 'olive black' or 'dark olive grey' as opposed to 'dark grey', with sample selection being based on intervals with the highest natural gamma log response.

Table 9.3. TOC values for the Kimmeridge Clay/Borglum Formations of the North Sea

Area	TOC (%)	Reference	Comments
North Sea, unspecified	2.7 av.	Fuller, 1980	= 3.25% organic matter
North Sea, miscellaneous	7.1 av.	Fig. 9.5	2-12% range, richer samples
Unidentified N.Sea well	5.6, 4.9 av.	Brooks and Thusu, 1977	upper and lower intervals
Ekofisk	1.4-2.6 av.	Van den Bark and Thomas, 1980	NOCS well 2/4-19B
Outer Moray Firth Basin	1.0-3.8 av.	Bissada, 1983	U.K. Quad. 14 and 15 Piper wells
S. Viking Graben	2.5-4.5	Pearson *et al.* 1983	U.K. well 16/22-2
Inner Moray Firth	3-6	Pearson and Watkins, 1983	range of U.K. Quad. 12 wells
Brae area well	4.29 av.	Reitsema, 1983	U.K. well 16/7a-19N
E. Shetlands Basin	5.4	Goff, 1983	estimated average
Ninian	6-9	Albright *et al.* 1980	range
Tern Field	3.4-8.1	Grantham *et al.* 1980	U.K. well 210/25-3: 6.8% av.
Statfjord	4.58 av.	Kirk, 1980	Licence 037, av.
Southern Norwegian N.Sea	7	Hamar *et al.* 1983	7-17.5% in hot shales
S. Norway Shelf	2.1(5.0) av.	Fuller, 1975	NOCS 2/11-1 core av.
Danish N. Sea	3.85 av.	Thomsen *et al.* 1983	NW Central graben, well I-1
"	1.59 av.	"	SE Central graben, well M-8
Norwegian N. Sea	5 av.	Rønnevik *et al.* 1983	range 1-15%
Dorset type section (U.K.)	3.75 av.	Fuller, 1975	bituminous shale
"	1.6 av.	"	grey clay
"	15.3-40.0	Cosgrove, 1970	richest oil shale band
Yorkshire outcrop	7.95 av.	Fuller, 1975	bituminous shale
"	30.9	"	oil shale
"	3.4	"	dark grey shale
Sutherland outcrop	5.5 av.	"	black shale
Andøya, Svalbard	1.4, 4.3	Dypvik *et al.* 1979	Northern Norway, Spitsbergen

The 'hot shale' facies can be developed locally, and is diachronous: for example, the Tau 'hot shale' Member of the Borglum Formation of the Fiske sub-basin is of Volgian (uppermost Jurassic) age, whilst the 'hot shale' of the Norwegian sector of the adjacent Central Graben comprises the Ryazanian (basal Cretaceous) Mandal Formation (Hamar *et al.*, 1983). Thomsen *et al.* (1983) have shown that while the most oil-prone shales occur at the top of the Upper Jurassic in the deepest (north-west) part of the Danish Central Graben (well I-1), the best source quality in a generally poorer section occurs towards the base of the Upper Jurassic on the southeast flank of the graben (well M-8).

The typical kerogen type of the 'hot shales' (Table 9.4) is a mixture of bacterially degraded algal debris of marine planktonic origin (amorphous liptinite), and degraded humic matter of terrigenous origin (amorphous vitrinite). This amorphous component is mixed with variable amounts of particulate vitrinite (woody debris) and inertinite—highly altered (oxidised or burnt) material of land-plant origin. Land-plant spores, and marine algae such as dinoflagellates and hystrichospheres, are present in minor to trace amounts. Framboidal pyrite is common, attesting to the action of sulphate-reducing bacteria. Under the reflected-light microscope, the kerogen in a polished whole-rock preparation appears as a diffuse fluorescent background (Gutjahr, 1983), while isolated kerogen in transmitted light appears as a clumpy, amorphous, dully or blotchily fluorescent mass (Batten, 1983). Subjected to Rock Eval pyrolysis, this mixture generally comes out as a Type II oil-prone kerogen (Barnard *et al.*, 1981).

Characterising the kerogens by optical microscopy, Fisher and Miles (1983) have reported that in the Fladen Ground Spur area of the South Viking and Witch Ground Grabens, the kerogen types of the Kimmeridge Clay Formation vary spatially and temporally. Stratigraphically, the Volgian strata contain the best quality source material in this area, while spatially, the most oil-prone kerogen is deposited in the graben centre and away from positive features and the routes of sediment input.

This concept has been generalised by Barnard *et al.* (1981), using Rock Eval pyrolysis data to illustrate the variation of TOC and four kerogen types down through a single 180 m (600 ft) section of the Kimmeridge Clay Formation (Fig. 9.15a), and spatially in terms of the change of dominant kerogen type from platform edge to the graben centre (Fig. 9.15b). This model can be briefly summarised as an increase in (gas-prone) terrigenous kerogen, and a decrease in (oil-prone) marine planktonic/bacterial kerogen towards the paleo-coastline.

It is clear from the above discussion that the characteristics of the 'hot shales' of the Kimmeridge Clay Formation vary considerably. However, for volumetric calculations and broad source-rock/oil correlations, some average properties are summarised in Table 9.4. Such a summary is validated by the high compositional uniformity of North Sea crude oils from the south Central Graben to the north Viking Graben (see section 9.6).

The Upper Jurassic is also developed as a rich oil-prone source beyond the North Sea. North of 62° Rønnevik *et al.* (1983) predict marine shales with good source quality. More specifically, Bjorøy *et al.* (1980) report TOC contents of 0.98-6.80% but generally gas/condensate-prone kerogen for the Ryazanian to Kimmeridgian of Andøya Island. This section is immature.

In the undifferentiated Jurassic of Svalbard, TOC values are lower but the kerogen is of better quality, and fully mature (Bjorøy and Vigran, 1980). In the Troms area, the Upper Jurassic contains an organically-rich oil and gas-prone claystone, which on structure is

Table 9.4. Some average properties of typical Kimmeridge Clay Formation hot shales of the North Sea

Property	Value	Reference	Comment
TOC (%)	5	Table 9.3	Realistic average
H/C ratio (kerogen)	0.9-1.2		Immature kerogen
Rock Eval hydrogen index	450-600	Barnard *et al.* 1981	Immature kerogen
Kerogen type	II	"	Rock Eval
$\delta^{13}C‰$ (kerogen) pdb	−27.6 to −28.7 (−25)	Reitsema, 1983 Fuller, 1975	Brae, Moray-Statfjord areas
Organic petrography*			
% amorphous liptinite	30-80	—	Bacterially degraded algae
% particulate liptinite	1-10	—	Dinoflagellates, spores etc.
% vitrinite (am + partic)	20-70	—	(amorphous + particulate)
% inertinite	1-25	—	Fusinite and semi-fusinite
Oil yield (bbls/acre ft per 1% TOC)	50	Fig. 9.7	In the source rock at peak maturity

*Percentage values are visual estimates of area percent of slide occupied by kerogen components.

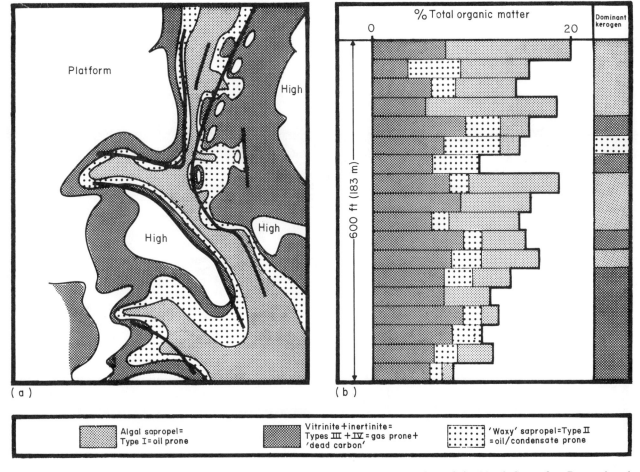

Fig. 9.15. Distribution of kerogen type in the Kimmeridge Clay/Borglum Formation of the North Sea, after Barnard and Cooper (1981, 1983).
(a) Idealised sketch of areal distribution.
(b) Vertical distribution through a 600 ft section of the Kimmeridge Clay Formation.
Note that kerogen quality improves away from structural highs and sediment sources (a), and is generally more oil-prone at the top than at the bottom of the formation (b) For definitions of kerogen types see Table 9.1 and Fig. 9.4.

late immature to early mature with respect to oil generation (Bjorøy *et al.*, 1983).

To the south, the Kimmeridge Clay Formation is not of optimum quality. In the Rijswijk oil province the equivalent Delfland Formation may have been a secondary source to the Toarcian (Bodenhausen and Ott, 1981). Onshore U.K., the Kimmeridge Clay contains variable quantities of oil-prone kerogen from Yorkshire to Dorset, being most mature in the North (Williams and Douglas, 1980; Douglas and Williams, 1981). Ridd (1981) notes that the Upper Jurassic west of the Shetlands is an organic-rich claystone with a high natural gamma ray response: it thickens to the north-west into the Faeroe Basin.

Depositional environment

The depositional environment of the organic-rich shales of the Kimmeridge Clay Formation has been much disputed. Worldwide, the Upper Jurassic is a time of transgression (Vail and Todd, 1981), which favours the deposition of organic-rich rocks (Hallam and Bradshaw, 1979), probably due to high bioproductivity and sedimentation rates and low concentrations of dissolved oxygen (Demaison and Moore, 1980 and refs. therein).

Published work on depositional environments is restricted to the U.K. onshore outcrops where limestones, marls, dolomitic limestones, oil shales, bituminous shale and clay are all recognised (Tyson *et al.*, 1979). Gallois (1976) suggested that blooms of phytoplankton—evidenced by the presence of coccolith limestones interbedded with the oil shales—produced an excess of organic matter over available dissolved oxygen, and hence preservation of organic-rich, anoxic sediments. It was then suggested (Tyson *et al.*, 1979) that the anoxicity was the cause and not the effect of the coccolith blooms. The model of Tyson *et al.*, envisaged a stratified water body (halocline or thermocline), anoxic at depth, which overturned periodically, liberating nutrients and giving rise to algal blooms. A modification of these two models has been suggested by Irwin (1979) who reported bioturbation within the oil-shale facies, and noted that biogenic calcite would be rapidly dissolved below any O_2-H_2S interface. Cornford *et al.* (1980) and Cornford and Douglas (in preparation) have suggested that the influx of terrigenous organic debris may also have reduced the background oxygen levels in extensive shelf seas.

It is clear that deposition occurred below wave base in an oxygen deficient environment with high planktonic

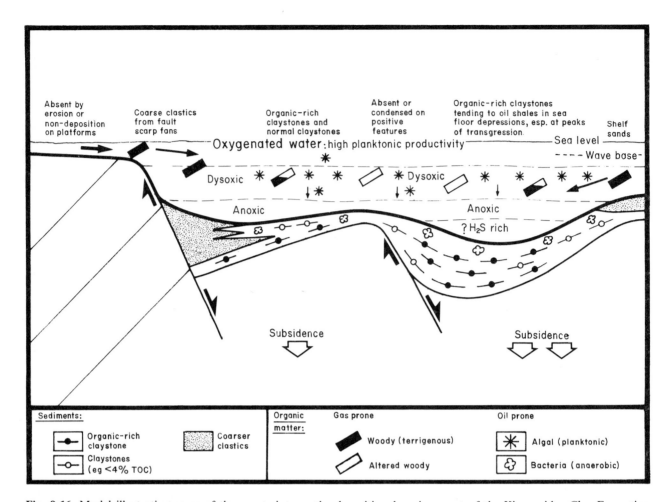

Fig. 9.16. Model illustrating some of the constraints on the depositional environment of the Kimmeridge Clay Formation organic-rich shales of the North Sea. Sedimentation rates are high (typically 20-30 m/m.y. decompacted) and locally can be much higher especially where allochthonous sands are injected into the basin.

bioproductivity (Fig. 9.16). The water depth was probably not great (e.g. ∿200 m?) since we see condensed sequences on 'highs' in the North Sea. Where coarser clastics are associated with the organic-rich facies such as in the Brae area, down-slope transport as indicated, but water depths still need not be great (Stow *et al.*, 1982). The association of these coarse clastics with fault scarps suggests syn-sedimentary fault movement. The overall picture is one of deposition in a relatively shallow sea with highly productive surface waters and anoxic water at depth, with occasional mixing. The highly organic 'oil shale' facies may have accumulated in deeper sea-floor depressions controlled by fault subsidence, where ponding of highly anoxic water and higher sedimentation rates would favour organic preservation. The surrounding land areas must have been rich in higher-plant vegetation, which contributed the vitrinitic components to the kerogen.

Oil-source correlations

The 'hot-shales' of the Kimmeridge Clay Formation have repeatedly been shown (e.g. Oudin, 1976) to give the best match with North Sea oil properties using oil/source-rock correlation techniques (Tissot and Welte, 1978). Considering only the more recent detailed studies, Reitsema (1983) has demonstrated that the Brae oils

correlate with the 'hot shales' using n- and isoprenoid alkane, sterane and triterpane distributions, as well as the stable carbon isotope curve-matching technique. He emphasised the need to correlate the properties of the different oils within the Brae complex with specific facies of the hot shale. Fisher and Miles (1983) have noted that in the Southern Viking and Witch Ground grabens, a correlation can be made between specific kerogen types, maturity, and oil properties.

Oil/source-rock correlations have latterly relied heavily on sterane and triterpane fingerprinting using the computerised gas chromatography—mass spectrometry technique (gc-ms). Cornford *et al.* (1983) using a suite of about 100 North Sea oils and 'hot shale' extracts, have demonstrated a detailed correlation, emphasising that the 'hot shale' generates a recognisable oil type at each maturity stage (Fig. 9.17). It is concluded that the bulk of reservoirs have received a full spectrum of maturity products from the source-rock on the flanks of the graben as shown in Fig. 9.17c. Grantham *et al.* (1981) have suggested an oil/source-rock correlation in the Tern area (U.K. Block 210/25) using the C_{27} and C_{28} triterpanes to identify a specific interval of the 'hot shales'. Fuller (1975) has shown a good correlation between North Sea oils and the Upper Jurassic source-rocks using optical rotation and stable-carbon isotope data.

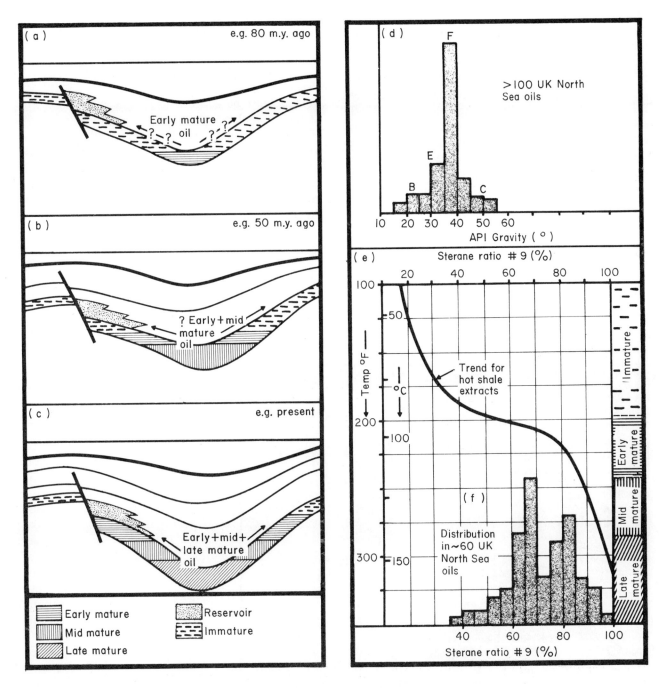

Fig. 9.17. North Sea oil properties controlled by source-rock maturity (Cornford *et al.*, 1983).

(a), (b), (c) Progressive oil generation from a single source-rock horizon being buried in a graben. Early mature oil, generated first in the graben centre (a), cannot migrate due to poor drainage. Effective expulsion from the graben centre occurs only when this location reaches the mid (b) or even late mature stage (c), by which time early mature generation is occurring on the basin flanks, where improved drainage may facilitate expulsion.

(d), (e), (f) show the result of this process on oil properties. In the distribution of API gravities of U.K. North Sea oil (d), four oil types are identified B = Biodegraded, C = Condensate/light oil, E = Early mature, F = Full maturity spectrum. (A full maturity spectrum oil is one that contains molecules characteristic of the expelled products of early, mid and late mature source-rocks). The maturity designation is derived, amongst other properties, from comparison of a sterane ratio trend measured on a maturity series of source-rocks (e), with the same ratio measured on a collection of about 60 oils (f). Early mature oils are rare, and are found on the flanks of the basin, generally in Upper Jurassic reservoirs. The bulk of the oils contain molecules characteristic of a full maturity spectrum, or just a late maturity fraction of source-rock products, consistant with the poor drainage of the hot shale source-rock. Full maturity spectrum oils are found in reservoirs that have received oil from a range of source-rock maturities as in the graben model c.

Volumetric modelling

The North Sea oil province, with a well defined single oil source-rock and a simple history of maturation, is ideal for developing and testing volumetric (mass balance) models for the generation and accumulation of hydrocarbons (see Section 9.4 and Welte *et al.*, 1983 and refs. therein). Fuller (1975) detailed a mass balance for one basin in the North Viking Graben area. Goff (1983) has published an attempt to account for the oils of the North Viking Graben (East Shetland Basin) in terms of the volumes of mature source-rock of known richness. Goff uses the equation:

$$\text{Oil generated} = \text{SRV} \times \text{OMV} \times \text{GP} \times \text{FOIH} \times \text{TR} \times \text{VI}$$

where

SRV = Source Rock Volume (m³ or acre ft).
OMV = Organic Matter content by Volume (from TOC, see below).
GP = Genetic Potential (fraction of oil-prone kerogen =0.7 for the Kimmeridge Clay Formation).

FOIH = Fraction of Oil In Hydrocarbon yield (0.8 according to Goff).
TR = Transformation Ratio (the extent to which the kerogen is converted—depends on the maturity of the source-rock).
VI = Volume Increase on oil generation (1.2 according to Goff).

To this must be added a migration and entrapment factor. From these calculations he obtained the overall efficiency of generation—migration and entrapment to be 20-30%, having allowed for some additional generation from the Heather Formation. A major factor not adquately discussed by Goff (1983), and important for prospect evaluation is the definition of the drainage area (see Section 9.4).

Maturity

The maturity of the Kimmeridge Clay/Borglum Formation can be estimated from a base Cretaceous seismic depth map such as that produced by Day *et al.* (1981) and reproduced in part in Fig. 9.18. Similar, more

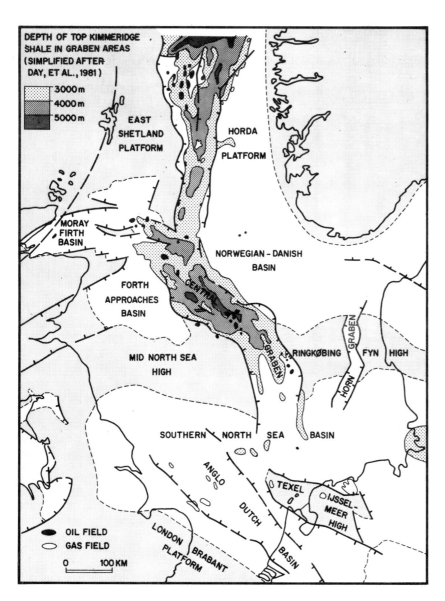

Fig. 9.18. Depth of burial at the top of the Kimmeridge Clay/Borglum Formations (simplified by Glennie from Day *et al.*, 1981). Major burial to oil-generating depths (e.g. >3000 m) occurs in the Central, Witch Ground and Viking grabens. The relationship between depth of burial of the source-rock and hydrocarbon type can be made by comparison with Fig. 9.21.

detailed maps have been produced by Goff (1983) for the East Shetland Basin and Viking Graben, and by Michelsen and Andersen (1983) for the Danish Central Graben area. Depth maps can be used to define the area of mature rocks using a scheme such as shown in Table 9.5, where the temperatures for generation boundaries taken from Cornford *et al.* (1983) are converted to depth using a range of geothermal gradients (i.e. 25, 30, 35°C/km; 1.4, 1.6, 1.9°F/100 ft), plus 4°C (40°F) at sea floor. This spread covers the majority of mapped variation in geothermal gradient in the North Sea (Harper, 1971; Cornelius, 1975; Carstens and Finstad, 1981) as shown in Fig. 9.9.

Maps of maturity *per se* have been published by Rønnevik *et al.* (1983) for the Norwegian North Sea, and for the Moray Firth-South Viking Graben area by Fisher and Miles (1983). Goff (1983) has produced a series of maps of the East Shetland Basin to show the progressive expansion of the areas of maturity with time from the basin centre to the basin edge (Fig. 9.20). Equating temperature with maturity in this way, however, ignores the effect of the cooking time or effective heating time. Within the North Sea region there are areas of both rapid and slow Neogene and recent subsidence, as well as of Tertiary uplift and erosion (Fig. 9.19).

It is well established that under conditions of rapid burial the increase in reflectance of vitrinite is retarded (e.g. Waples, 1980), but, it is not at all clear whether anything other than extremely rapid burial significantly retards the generation of hydrocarbons in terms of m³ (bbls) generated per km³ (acre ft) of source-rock (Cornford *et al.*, 1983). Since the majority of the North Sea suffered fairly uniform burial during the Tertiary, and the temperature zones used have been established within the North Sea Basin (Cornford *et al.*, 1983), these maturity intervals can probably be used with some confidence throughout the North Sea region except in areas of uplift and erosion such as the Inner

Moray Firth, or areas with very high Neogene burial rates as in certain areas of the Central Graben.

Finally, it is clear from Fig. 9.19 that the Kimmeridge Clay Formation in the deepest parts of the Viking and Central Grabens reached the early mature stage by the late Cretaceous/early Tertiary, and full maturity by the mid Paleogene. Generation will have continued from that time onwards with the locus of generation moving up the flanks of the basin (Fig. 9.17a-c).

As the North Sea Basin moves into a mature phase of exploration, it is of increasing importance to define the early mature generation phase as precisely as possible, and hence to recognise those geological situations under which the early mature oil will be effectively drained from the source-rocks. This knowledge will optimise the search for oil in the areas of marginal maturity on the flanks of the grabens, areas such as the West Forties Basin, the Egersund Basin, the inner Moray Firth Basin and the western margin of the East Shetland Basin.

9.5.8 The Cretaceous

The Lower Cretaceous (excluding the Ryazanian, see previous section) is developed as a dominantly shaley facies over much of the North Sea (Ziegler, 1982, encl. 21, 22; Hesjedal and Hamar, 1983). Major depocentres (>500 m thickness) occurred in the Moray Firth Basin, the Viking and Central Grabens, in the Horda, Egersund and North Danish basins, and in the Broad Fourteens and Lower Saxony basins (Ziegler, 1983, encl. 31). In all these basins the section is believed to contain fair to good amounts of gas-prone kerogen (Barnard and Cooper, 1981).

TOC values that average 0.6% to 1.4% were reported for seven wells in the Moray Firth Basin (Bissada, 1983), with the kerogens of most wells being of gas/condensate-prone land-plant type except in Block 14/20, where the kerogen is of lower-plant origin. In the

Table 9.5. Depths to specific maturity levels for areas of high, average and low geothermal gradient

Maturity level for oil (and associated gas)*	Temperature		Depth (km) to boundary at geothermal gradients (°C/km)† of:		
	°C	°F	25	30	35
Immature					
	80	175	3.0	2.5	2.2
Early mature					
	110	225	4.2	3.5	3.0
Mid mature					
	130	265	5.0	4.2	3.6
Late mature					
	155?	310?	6.0	5.0	4.3

* after Cornford *et al.* 1983
† = 1.4, 1.6, 1.9° F/100 ft respectively

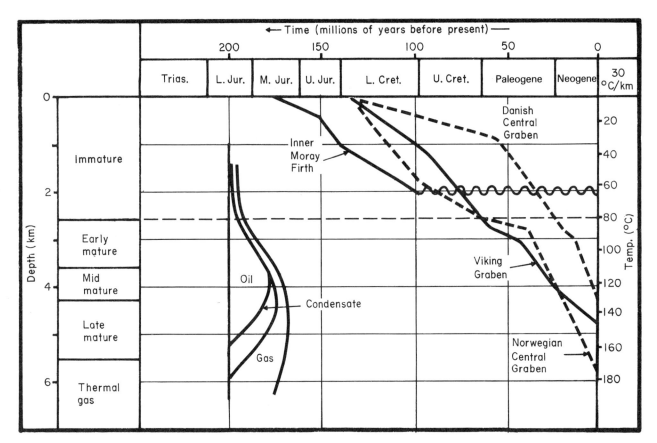

Fig. 9.19. The timing of generation from the Upper Jurassic of the North Sea as deduced from the subsidence curves (uncorrected for compaction) from the central parts of the Danish Central Graben (Hansen and Mikkelson, 1983), the Norwegian Central and Viking Grabens (Rønnevik *et al.*, 1983), and the Inner Moray Firth/Beatrice Field (Linsley *et al.*, 1980). Approximate maturity with respect to oil generation is shown using an average geothermal gradient of 30°C/km (Table 9.5). This shows that generation started in the late Cretaceous/Early Tertiary in the Graben centres, being earlier in the deepest parts of the Central than in the Viking Graben.

Statfjord area, the TOC values of the Lower Cretaceous Cromer Knoll Group average 1.04% (Kirk, 1980) and appear to be gas-prone.

Locally, particularly during the Albian and Aptian, a thinly developed oil-prone kerogen facies was deposited in the southern half of the North Sea and may be equivalent to the English onshore pyritous, black, Speeton Clay.

The Lower Cretaceous is, along with the Jurassic, the source for some of the North-West German gas fields, according to Tissot and Bessereau (1982). The Wealden facies of the Lower Cretaceous, as displayed on the South Coast of England, is in a gas-prone facies with thin lignite seams and stringers and carbonaceous shales. The high sand/shale ratios will favour efficient drainage of any generated hydrocarbons.

North of 62°, variable quantities of mature gas and oil-prone kerogen occur in the uppermost part of the Skarstein Formation (Aptian/Barremian) on Andøya Island (Bjorøy *et al.*, 1980). Rich gas and oil-prone shales also occur on Svalbard (Bjorøy and Vigran, 1980; Leythaeuser *et al.*, 1983).

The regional maturity of the Lower Cretaceous can be deduced from the base Cretaceous depth map (Fig. 9.18), which indicates that it will be mature for gas (>1%R, >4000 m) in the Central, Viking and Witch Ground grabens. It also reaches gas generating matu-

rities in the North German Basin, where it is believed to have contributed to commercial gas fields.

The Upper Cretaceous is devoid of source potential in the southern part of the North Sea, where it is developed mainly as chalk.

9.5.9 Tertiary

In the North Sea area Tertiary sediments are unlikely to be effective source-rocks because of their low degree of maturity. Day *et al.* (1981) show that the top Cretaceous is buried below 3000 m only in the Central Graben. Paleocene shales, however, are lean and gas and condensate-prone. They were suggested by Pennington (1975), as the source for the oil in the Argyll field, but this is not now thought probable on the grounds of regional maturity and kerogen type. TOC values average 0.43% in the Paleocene Røgland Group in the Statfjord area (Kirk, 1980), whilst two Paleocene intervals of Ekofisk well NOCS 2/4-12 gave average TOC values of 1.8 and 2.0% (Van den Bark and Thomas, 1980). In the Ekofisk area, the shales have reached maturities of 0.59%-0.62%R at 3030-3070 m (9930-10060 ft). Since this is in the area of deepest Paleocene burial, no significant gas or condensate generation is expected from this formation.

North of 62°, the Tertiary may be a potential source

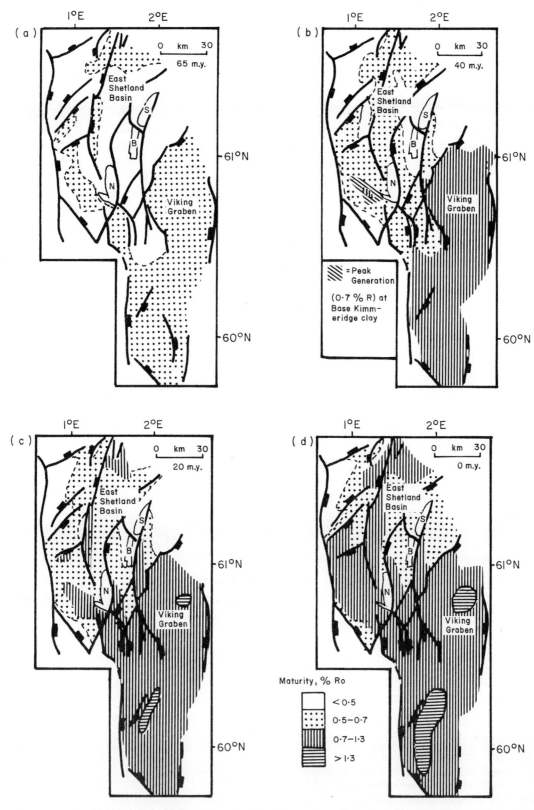

Fig. 9.20. The progressive maturation of the Kimmeridge Clay Formation of the East Shetland Basin with time (reproduced from Goff, 1983 with permission). (a) 65 million years ago, (b) 40 million years ago, (c) 20 million years ago, and (d) present day. Note that generation started in the Viking graben about 40 m.y. ago, and only in the last 20 m.y. has moved into the half grabens of the East Shetland Basin. Light oil and condensate generation, initiated about 20 m.y. ago, is restricted to the central portions of the Viking Graben. N = Ninian, B = Brent, S = Statfjord.

for gas where burial is sufficient. Paleocene/Eocene coals on Spitsbergen are immature with respect to gas generation (0.4-0.7% vitrinite reflectance), but are interbedded with black marine shales containing abundant amorphous kerogen (Manum and Throndsen, 1977).

9.6 North Sea Hydrocarbons

A typical North Sea oil sourced by the Kimmeridge Clay Formation is a low sulphur, medium gravity, naphtheno-paraffinic oil (Table 9.6). Gas sourced from the Westphalian Coal Measures of the southern North Sea is sweet and dry in composition (Table 9.7). Little is reported on the properties of the associated gas in the oil province: Frigg field gas (? sourced by Middle Jurassic coal measures) is dry and sweet (Héritier *et al.*, 1979; Goff, 1983). The gas of the Sleipner field, which also contains significant condensate, is wetter, with a high CO_2 content (Larsen and Jaarvik, 1981). Compi-

lations of the composition of the oils have been made by Aalund (1983 a and b), and Cornford *et al.*, (1983). Little information is available for gas (Barnard and Cooper, 1983). Boigk and Stahl (1970) and Boigk *et al.* (1971) have noted regional trends in the contents of methane, nitrogen and carbon dioxide within the southern North Sea and North German offshore and land areas, although a number of possible explanations for these trends were offered. Kettel (1982) presents evidence to show that gases with high nitrogen contents derive from the most mature source rocks. The nitrogen content is generally higher in the Bunter than in the Rotliegend reservoirs of the U.K. southern North Sea (Barnard and Cooper, 1983), which may suggest enrichment during (re)migration. Some key properties of the oils and gases are summarised in Tables 9.6 and 9.7. In addition to sweet medium gravity oils, condensates are known in the Beryl Embayment, the Witch Ground Graben and the Central Graben (Fig. 9.21) associated with the most deeply buried Kimmeridge

Table 9.6. Average properties of North Sea oils sourced from the Kimmeridge Clay Formation

Property	Range	Average ($\pm$sd)	Comments
API gravity (°)[1]	17-51	36($\pm$6.5)	Dead oils
Sulphur content (%) (a)[2]	0.13-0.55	0.32($\pm$0.12)	15 oils excl. (b) below
(b)	0.56-1.57	1.0	Piper, Claymore, Tartan Buchan
Gas/Oil ratio (sft^3/bbl) (a)[3]	216-1547	671($\pm$415)	Analytically heterogenous
(b)	216-952	562($\pm$271)	Excl. Ekofisk
Asphaltenes (%)[1]	0.1-5.1	1.2($\pm$1.2)	Excl. asphaltene-enriched oils up to 35%
Saturate/Aromatics[1]	0.62-8.0	2.02($\pm$1.2)	Topped oils
Pristane/nC_{17} ratio[1]	0.3-1.0	0.63($\pm$0.17)	Excl. biodegraded oils
Phytane/nC_{18} ratio[1]	0.2-1.1	0.56($\pm$0.18)	" "
Pristane/Phytane ratio[1]	0.6-1.9	1.24($\pm$0.25)	" "
$\delta^{13}C$ (‰)[4]	28.4-29.8	–	Brae, Statfjord-Moray oils
	–	$\sim$29	Piper area oils (av.)
V/Ni (ppm, wt)[2] V	0.53-6.0	3.1($\pm$1.8)	Av. of 9 oils, excl.
Ni	0.5-5.0	1.8($\pm$1.3)	Buchan (=26ppmV, 4.5ppmNi)
Wax (%) (a)[2]	4.0-7.7	6.3($\pm$1.1)	5 oils
(b)[3]	–	17	Beatrice[5]

(1) Cornford *et al.* 1983 (approx. 60 analyses of mainly exploration DST and RFT samples).
(2) Aalund, 1979, 1983 (Production Crudes).
(3) Compiled from field reports in Illing and Hobson, 1981, Woodland, 1975, and Halbouty, 1980.
(4) Reitsema, 1983; Bissada, 1983, Fuller, 1975.
(5) Beatrice field may have another source.
    sd = standard deviation

Fig. 9.21. *opposite.* Generalised map showing (a) the distribution of oil types in the North Sea by API gravity (from Barnard and Cooper, 1981) with insets (b) showing detailed distribution of oil according to gravity, and (c) predicted hydrocarbon product based on kerogen type and maturity for the Witch Ground Graben (Fisher and Miles, 1983). The general distribution of oil type is clearly controlled by the maturation of the Kimmeridge Clay Formation (Fig. 9.18), while local variation may also be influenced by kerogen type and source-rock drainage.

Table 9.7. Properties of gas from the North Sea and adjacent areas

	C_1(%)	C_2+(%)	N_2(%)	CO_2(%)	$\delta^{13}CH_4$‰
U.K. Southern N. Sea[1]					
average	91.2	5.2	3.6	0.27	-
range	83.2-95.0	3.7-8.2	1.0-8.4	0.1-0.5	-
Groningen[1]	81.6	2.7	14.8	0.9	−36.6
Kinsale Head[2]	99.1	0.2	0.4	0.3	−45.5 to −48.3
Morcombe[3]	Dry	-	7-8	variable	-
Sleipner[4]	78-80	12-16	-	7-9	-
Frigg[5]	95.5	3.6	0.4	0.3	−43.3
NOCS Block 31/2[6]	92.6	5.4	1.5	0.5	-
NW Germany[7]	65-91	8-30	1	-	−54 to −44
Holland, Waddenzee fields[8]	77.1-88.7	2.87-6.4	3.1-19.7	0.02-1.26	−31.1
Broad Fourteens (K/13)[9]	85.3	6.7	5.4, 7.1	1.7, 0.1	

(1) Barnard and Cooper, 1983.
(2) Colley *et al.*, 1981.
(3) Ebbern, 1981.
(4) Larsen and Jaarvik, 1981 (C_2+ figures include N_2).
(5) Héritier *et al.* 1981.

(6) Brekke *et al.* 1981.
(7) Tissot and Bessereau, 1982 (Jurassic to Lower Cretaceous source).
(8) Cottençon *et al.* 1975; Van den Bosch, 1983.
(9) Roos and Smits, 1983 (remigrated Westphalian gas).

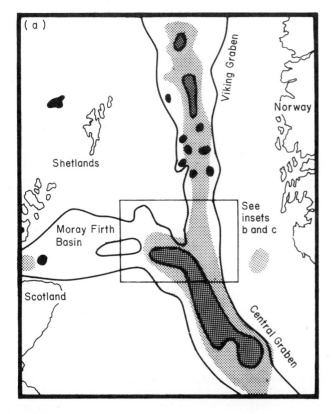

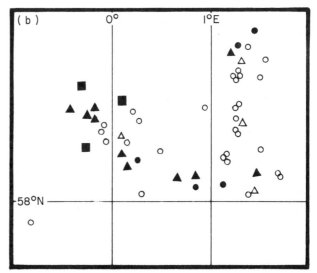

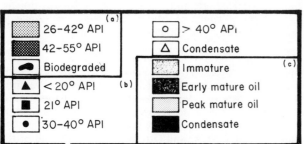

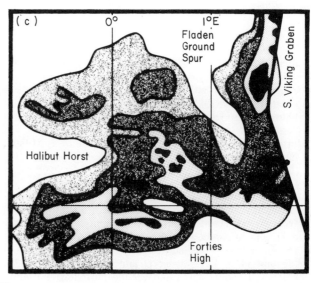

Clay Formation (Fig. 9.18), a relationship detailed by Barnard and Cooper (1981).

Heavy oils (API gravity <20°) are found over much of the North Sea, generally in reservoirs with present day temperatures less than 60°C (∿2000 m at 30°C/km) (Barnard and Cooper, 1981), and hence in Tertiary rocks. There is good evidence to show that they are produced within the reservoir by bacterial degradation of normal oil sourced by the Kimmeridge Clay Formation (e.g. Oudin, 1976). Bacterial degradation of reservoired oil is now readily recognised by its molecular and isotopic composition (Tissot and Welte, 1978; Cornford et al., 1983). There is no commercial production of heavy oil in the North Sea. Intermediate gravity oils (20-35° API gravity) are known from the Clair field, West of Shetland (22-25° API gravity), which Barnard and Cooper (1981) have classified as a biodegraded oil. The same authors have noted that early mature North Sea oils may fall in the 28-35° API gravity range, in agreement with analytical data presented by Cornford et al. (1983).

Higher sulphur oils are found in some of the Witch Ground Graben fields (e.g. Piper, Claymore, Tartan and Buchan). Beatrice oil, though light (∿38° API) and sweet, has a high wax content (17%) and pour point (65°F; 18°C), and low gas/oil ratio (Linsley et al., 1980). A similar high-wax, (16%), low sulphur (0.3%), 35° API, low-GOR oil is produced from the De Lier field of the Rijswijk oil province (Bodenhausen and Ott, 1981). Both oils are believed to come from non-marine source-rocks.

9.7 Acknowledgements

I am pleased to thank Ken Glennie and Gordon Speers for detailed comments on a draft version of the text, and my many former colleagues in Britoil who have influenced my understanding of the genesis of North Sea oil. Douglas Kelso and Sara Cornford deserve special thanks for drafting the figures and typing the text respectively.

9.8 References

Aalund, L.R. (1979) Guide to world crudes—2. *Oil and Gas Journal,* Dec. 3, pp. 99-113.

Aalund, L.R. (1983a) Guide to export crudes of the 80s—3. North Sea now offers 14 export crudes. *Oil and Gas Journal,* May **23**, pp. 69-76.

Aalund, L.R. (1983b) Guide to export crudes of the 80s—4. North Sea crudes: Flotta to Thistle. *Oil and Gas Journal,* June 6, pp. 75-79.

Albright, W.A., Turner, W.L. and Williamson, K.R. (1980) Ninian field, U.K. sector, North Sea. In: Halbouty, M.T. (Ed.) *Giant oil and gas fields of the decade: 1968-1978* (AAPG Memoir 30).

Allan, J., Bjorøy, M. and Douglas, A.G. (1980) A geochemical study of the exinite group maceral alginite selected from three Permo-Carboniferous torbanites. In: Douglas, A.G. and Maxwell, J.R. (Eds.) *Advances in organic geochemistry* 1979. Pergamon Press, pp. 599-618.

Altebäumer, F.J., Leythaeuser, D. and Schaefer, R.G. (1983) Effects of geologically rapid heating on maturation and hydrocarbon and generation in Lower Jurassic shales from N.W. Germany. In: Bjorøy, M.*et al.* (Eds.) *Advances in organic geochemistry* 1981. John Wiley, pp. 80-86.

Andersson, A., Dahlman, B. and Gee, D.G. (1982) Kerogen and uranium resources in the Cambrian Alum shales of Billingen-Falbygden and Närke areas, Sweden. *Geol. För. Stockh. Förh.* **104** (3), 197-209.

Barnard, P.C. and Cooper, B.S. (1981) Oils and source rocks of the North Sea area. In: Illing, L.V. and Hobson, G.D. (Eds.) q.v. pp. 169-175.

Barnard, P.C. and Cooper, B.S. (1983) A review of geochemical data related to the North-west European gas province. In: Brooks, J. (Ed.) q.v. pp. 19-33.

Barnard, P.C., Collins, A.G. and Cooper, B.S. (1981) Identification and distribution of kerogen facies in a source rock horizon—examples from the North Sea basin. In: Brooks, J. (Ed.) *Organic maturation studies and fossil fuel exploration.* Academic Press, pp. 271-282.

Barnard, P.C., Collins, A.G. and Cooper, B.S. (1981a) Generation of hydrocarbons—time, temperature and source rock quality. In: Brooks, J. (Ed.) *Organic maturation studies and fossil fuel exploration.* Academic Press, pp. 337-342.

Barr, K.W., Colter, V.S. and Young, R. (1981) The geology of the Cardigan Bay—St George's Channel basin. In: Illing, L.V. and Hobson, G.D. (Eds.) q.v. pp. 432-443.

Bartenstein, H. (1979) Essay on coalification and hydrocarbon potential of the N.W. European Paleozoic. *Geol. Mijnbouw* **59** (2), 155-168.

Batten, D.J. (1983) Identification of amorphous sedimentary organic matter by transmitted light microscopy. In: Brooks, J. (Ed.) q.v. pp. 275-287.

Bergstrøm, S.M. (1980) Conodants as paleotemperature tools in Ordovician rocks of the Caledonides and adjacent areas in Scandinavia and the British Isles. *Geol. För. Stockh. Förh.* **102** (4), 377-392.

Berstad, S. and Dypvik, J. (1982) Sedimentological evolution and natural radioactivity of Tertiary sediments from the Central North Sea. *J. Petrol. Geol.* **5** (1), 77-88.

Best, G., Kockel, F. and Schöneich, H. (1983) The geological history of the southern Horn Graben. *Geol. Mijnbouw.* **62** (1), 25-33.

Bissada, K.K. (1983) Petroleum generation in Mesozoic sediments of the Moray Firth Basin, North Sea area. In: Bjorøy, M. (Ed.) *Advances in geochemistry* 1981. John Wiley, pp. 7-15.

Bitterli, P. (1963) Aspects of the genesis of bituminous rock sequences. *Geol. Mijnbouw.* **42** (6), 183-201.

Bitterli, P. (1963) On the classification of bituminous rocks from Western Europe. In: *Proceedings of 6th World Petroleum Congress, Frankfurt,* Section 1, pp. 155-165.

Bjørlykke, K., Dypvik, H. and Finstad, K.G. (1975) The Kimmeridgian shale, its compositions and radioactivity. In: Finstad, K.G. and Selley, R.C. (Eds.) (1976) *Jurassic Northern North Sea Symposium.* pp. 12.1-12.20, NPS.

Bjorøy, M. and Vigran, J.O. (1980) Geochemical study of the organic matter in outcrop samples from Agardhfjellet, Spitzbergen. In: Douglas, A.G. and Maxwell, J.R. (Eds.) *Advances in organic geochemistry* 1979. Pergamon Press, pp. 141-147.

Bjorøy, M, Hall, K. and Vigran, J.O. (1980) An organic geochemical study of Mesozoic shales from Andøya, North Norway. In: Douglas, A.G. and Maxwell, J.R. (Eds.) *Advances in organic geochemistry* 1979. Pergamon Press, pp. 77-91.

Bjorøy, M., Mork, A. and Vigran, J.O. (1983) Organic geochemical studies of the Devonian to Triassic succession on Bjornøya and the implications for the Barent Shelf. In: Bjorøy, M. *et al.* (Eds.) *Advances in organic geochemistry* 1981. John Wiley, pp. 49-59.

Bodenhausen, J.W.A. and Ott, W.F. (1981) Habitat of the Rijswijk oil province, onshore The Netherlands. In:

Illing, L.V. and Hobson, G.D. (Eds.) q.v. pp. 301-309.

Boigk, V.H. and Stahl, W. (1970) Zum Problem der Entstehung nordwestdeutcher Erdgaslagerstätten. In: *Erdöl and Kohle—Erdgas—Petrochemie, 111* Brennstoff Chemie **23** (6), 325-333.

Boigk, H., Stahl, W., Teichmüller, M. and Teichmüller, R. (1971) Metamorphism of coal and natural gas. *Forschr. Geol. Rheinld. u. Westphal.* **19**, 104-111.

Bostick, N.H. (1974) Phytoclasts as indicators of thermal metamorphism, Franciscan assemblage and Great Valley sequence (upper Mesozoic), California. *Geol. Soc. Am. Spec. Paper* **153**, 1-17.

Bostick, N.H. (1979) *Microscopic measurement of the level of catagenesis of solid organic matter in sedimentary rocks to aid exploration for petroleum and to determine former burial temperatures: a review.* SEPM Special Publication No. 26. Society of Economic Paleontologists and Mineralotists, pp. 17-43.

Brand, E. and Hoffmann, K. (1963) Stratigraphy and facies of the North-west German Jurassic and genesis of its oil deposits. In: *Proceedings of 6th World Petroleum Congress, Frankfurt.* Section 1, Paper 7, pp. 223-246.

Brekke, T., Pegrum, R.M. and Watts, P.B. (1981) First exploration results in Block 31/2 offshore Norway. In: *Norwegian Symposium on Exploration, Bergen, Sept. 14-16.* Norwegian Petroleum Society, p. 16.1-16.34.

Brooks, J. (Ed.) (1983) *Petroleum geochemistry and exploration of Europe.* Blackwell Scientific Publications.

Brooks, J. and Thusu, B. (1977) Oil source rock identification and characterisation of the Jurassic sediments in the northern North Sea. *Chem. Geol.* **20**, 283-294.

Brooks, J., Cornford, C. and Archer, R. (in press). The role of marine source rocks in petroleum exploration. In: *Marine petroleum source rocks* meeting, 17-18 May 1983. Petroleum Geochemistry group of the Geological Society.

Bujak, J.P., Barss, H.S. and Williams, G.L. (1977) Offshore Canada's organic type and colour and hydrocarbon potential. *Oil and Gas J.* **4**, 198-202.

Carstens, H. and Finstad, K.G. (1981) Geothermal gradients of the northern North Sea Basin, 59-62° N. In: Illing, L.V. and Hobson, G.D. (Eds.) q.v. pp. 152-161.

Colley, M.G., McWilliams, A.S.F. and Myres, R.C. (1981) The geology of the Kinsale Head gas field, Celtic Sea, Ireland. In: Illing, L.V. and Hobson, G.D. (Eds.) q.v. pp. 504-510.

Colter, V.S. and Havard, D.J. (1981) The Wytch Farm oil field, Dorset. In: Illing, L.V. and Hobson, G.D. (Eds.) q.v. pp. 494-503.

Connan, J. and Cassou, A.M. (1977) Properties of gases and petroleum liquids derived from terrestrial kerogen at various maturation levels. *Proceedings of the International Palynological Congress, Leon, Spain.*

Cooper, B.S., Coleman, S.H., Barnard, P.C. and Butterworth, J.S. (1975) Paleotemperatures in the northern North Sea Basin. In: Woodland, A.W. (Ed.) q.v. pp. 487-492.

Cope, M.J. (1980) Physical and chemical properties of coalified and charcoalified phytoclasts from some British Mesozoic sediments: an organic geochemical approach to paleobotany. In: Douglas, A.G. and Maxwell, J.R. (Eds.) *Advances in organic geochemistry* 1979. Pergamon Press, pp. 663-677.

Cornelius, C.D. (1975) Geothermal aspects of hydrocarbon exploration in the North Sea area. *Norges Geol. Unders.* **316**, 29-67.

Cornford, C. and Douglas, A.G. (In preparation) Maturity, depositional environment and hydrocarbon source potential of Lower Lias limestone/shale sequences of south-west Britain.

Cornford, C., Morrow, J.A., Turrington, A., Miles, J.A. and Brooks, J. (1983) Some geological controls on oil composition in the U.K. North Sea. In: Brooks, J. (Ed.) q.v. pp. 175-194.

Cornford, C., Rullkötter, J. and Welte, D. (1980) A synthesis of organic petrographic and geochemical results from DSDP sites in the eastern central North Atlantic. In: Douglas, A.G. and Maxwell, J.R. (Eds.) *Advances in organic geochemistry* 1979. Pergamon Press, pp. 445-453.

Cosgrove, M.E. (1970) Iodine in the bituminous Kimmeridge Shales of the Dorset coast, England. *Geochim, Cosmochim. Acta* **34**, 830-836.

Cottençon, A., Parant, B. and Flacelière, G. (1975) Lower Cretaceous gas fields in Holland. In: Woodland, A.W. (Ed.) q.v. pp. 403-412.

Day, G.A., Cooper, B.A., Anderson, C., Burgers, W.F., Rønnevik, H.C. and Schöneich, H. (1981) Regional seismic structure maps of the North Sea. In: Illing, L.V. and Hobson, G.D. (Eds.) q.v. pp. 76-84.

Demaison, G.J. and Moore, G.T. (1980) Anoxic environments and oil source bed genesis. *Org. Geochem.* **2**, 9-31.

Donovan, R.N. (1980) Lacustrine cycles, fish ecology and stratigraphic zonation in the Middle Devonian of Caithness. *Scott. J. Geol.* **16**, 35-72.

Donovan, R.N., Foster, R.J. and Westoll, T.S. (1974) A stratigraphical revision of the Old Red Sandstone of north-eastern Caithness. *Trans. R. Soc. Edinb.* **69**, 167-201.

Douglas, A.G. and Williams, P.F.V. (1981) Kimmeridge oil shale: a study of organic maturation. In: Brooks, J. (Ed.) *Organic maturation studies and fossil fuel exploration.* Academic Press, pp. 256-269.

Douglas, A.G., Eglinton, G. and Maxwell, J.R. (1969) The organic geochemistry of certain samples from the Scottish Carboniferous formation. *Geochim. Cosmochim. Acta,* **33**, 579-590.

Dow, W.G. and O'Connor, D.I. (1982) Kerogen maturity and type by reflected light microscopy applied to petroleum exploration. In: Staplin, F.L. *et al.* (Ed.) *How to assess maturation and paleotemperature,* SEPM Short Course No. 7, pp. 133-157.

Duff, K.L. (1975) Paleoecology of a bituminous shale—the Lower Oxford Clay of central England. *Palaeontology* **18**, 443-482.

Duncan, D.C. and Swanson, V.E. (1965) Organic-rich shale of the United States and world land areas. *US Geol. Survey Circ.* No. 523, Washington.

Durand, B. (1983) Present trends in organic geochemistry in research in migration of hydrocarbons. In: Bjorøy, M. *et al.* (Eds.) *Advances in organic geochemistry* 1981. John Wiley, pp. 177-128.

Dypvik, H., Rueslätten, H.G. and Thondsen, T. (1979) Composition of organic matter from North Atlantic Kimmeridgian shales. *Am. Assoc. Petrol. Geol. Bull.* **63** (12), 2222-2226.

Eames, T.D. (1975) Coal rank and gas source relationships—Rotliegendes reservoirs. In: Woodland, A.W. (Ed.) q.v. pp. 191-203.

Ebbern, J. (1981) The geology of the Morecombe gas field. In: Illing, L.V. and Hobson, G.D. (Eds.) q.v., pp. 485-493.

Eynon, G. (1981) Basin development and sedimentation in the Middle Jurassic of the northern North Sea. In: Illing, L.V. and Hobson, G.D. (Eds.) q.v. pp. 196-204.

Fertl, W.H. (1976) Elucidation of oil shales using geophysical well logging techniques. In: Yen and Chilingarian (Eds), *Oil Shale* Elsevier, pp. 199-213.

Fisher, M.J. and Miles, J.A. (1983) Kerogen types, organic maturation and hydrocarbon occurrences in the Moray Firth and South Viking Graben, North Sea Basin. In: Brooks, J. (Ed.) q.v. pp. 195-201.

Forsberg, A. and Bjorøy, M. (1983) A sedimentalogical and organic geochemical study of the Botneheia Formation, Svalbard, with special emphasis on the effects of weathering on the organic matter in the shales. In: Bjorøy, M. *et al.* (Eds.) *Advances in organic geochemistry* 1981. John Wiley, pp. 60-68.

Fuller, J.G.C.M. (1975) Jurassic source rock potential—and

202   C. Cornford

hydrocarbon correlation, North Sea. In: *Proceedings of the Symposium on Jurassic—northern North Sea.* Norwegian Petroleum Society meeting.

Fuller, J.G.C.M. (1980) In: Jones, J.M. and Scott, P.W. *Progress report on fossil fuels—exploration and exploitataion.* Proc. Yorks. Geol. Soc., 42, 581-593.

Galois, R.W. (1976) Coccolith blooms in the Kimmeridge Clay, and origin of North Sea oil. *Nature* 259, 473-475.

Gibbons, M. (in press) The depositional environment and petroleum geochemistry of the Marl Slate/Kupferschiefer. In: Brooks, J. and Fleet, A. (Eds.) (1984) *Marine petroleum source rocks.* Blackwell Scientific Publications.

Glennie, K.W. and Boegner, P.L.E. (1981) Sole Pit inversion tectonics. In: Illing, L.V. and Hobson, G.D. (Eds.) q.v. pp. 110-120.

Goff, J.C. (1983) Hydrocarbon generation and migration from Jurassic source rocks in the E. Shetland Basin and Viking Graben of the northern North Sea. *J. Geol. Soc. Lond.* 140, 445-474.

Gold, T. and Soter, S. (1982) Abiogenic methane and the origin of petroleum. *Energy Exploration and Exploitation* 1 (2), 89-104.

Grantham, P.J., Posthuma, J. and De Groot, K. (1980) Variation and significance of the C_{27} and C_{28} triterpane content of a North Sea core and various North Sea crude oils. In: Douglas, A.G. and Maxwell, J.R. (Eds.) *Advances in organic geochemistry* 1979. Pergamon Press, pp. 29-38.

Gretener, P.E. and Curtis, C.D. (1982) Role of temperature and time on organic metamorphism. *Am. Assoc. Petrol. Bull.* 66 (8), 1124-1129.

Griffiths, A.E. (1983) The search for petroleum in Northern Ireland. In: Brooks, J. (Ed.) q.v. pp. 213-222.

Gutjahr, C.C.M. (1983) Incident light microscopy of oil and gas source rocks. *Geol. Mijnbouw* 62 (3), 417-425.

Halbouty, M.T. (Ed.) 1980 *Giant oil and gas fields of the decade 1968-1978.* Am. Assoc. Petrol. Geol., Tuisa. Memoir 30.

Hall, P.B. and Douglas, A.G. (1983) The distribution of cyclic alkanes in two lacustrine deposits. In: Bjorøy, M. *et al.* (Eds.) *Advances in organic geochemistry* 1981. John Wiley, pp. 576-587.

Hallam, A. and Bradshaw, M.J. (1979) Bituminous shales and oolitic ironstones as indicators of transgressions and regressions. *J. Geol. Soc. Lond.* 136, 157-164.

Hamar, G.P. (1975) A Jurassic structure complex in northern North Sea. In: Finstad, K.G. and Selley, R.C. (Eds.) (975) *Proceedings of the Jurassic Northern North Sea Symposium.* Norwegian Petroleum Society, p. 17.1-17.18.

Hamar, G.P., Fjaeran, T. and Hesjedal, A. (1983) Jurassic stratigraphy and tectonics of the south-southeastern Norwegian offshore. *Geol. Mijnbouw* 62 (1), 103-114.

Hancock, N.J. and Fisher, M.J. (1981) Middle Jurassic North Sea deltas with particular reference to Yorkshire. In: Illing, L.V. and Hobson, G.D. (Eds.) q.v. pp. 186-195.

Hansen, J.M. and Mikkelsen, N. (1983) Hydrocarbon geological aspects of subsidence curves: interpretation based on released wells in the Danish Central Graben. *Bull. Geol. Soc. Denmark* 31, 159-169.

Harper, M.L. (1971) Approximate geothermal gradients in the North Sea Basin. *Nature* 230, 235-236.

Héritier, F.E., Lossel, P. and Wathne, E. (1979) Frigg field—large submarine fan trap in Lower Eocene rocks of North Sea. *Am. Assoc. Petrol. Geol. Bull.* 63 (11), 1999-2020.

Héritier, F.E., Lossel, P. and Wathne, E. (1981) The Frigg gas field. In: Illing, L.V. and Hobson, G.D. (Eds.) q.v. pp. 380-391.

Hesjedal, A. and Hamar, G.P. (1983) Lower Cretaceous stratigraphy and tectonics of the south-southeastern Norwegian offshore. *Geol. Mijnbouw* 62 (1), 135-144.

Hood, A., Gutjahr, C.C.M. and Heacock, R.L. (1975) Organic metamorphism and the generation of petroleum. *Am. Assoc. Petrol. Geol. Bull.* 59, 986-996.

Hopkinson, E.C., Fertl, W.H. and Oliver, D.W. (1982) The continuous carbon/oxygen log—basic concepts and recent field experience. *J. Petrol. Technol.,* October 1983, 2441-2448.

Huc. A. (1976) Mise en evidence de provinces geochimiques dans les Schistes Bitumineux du Toarcien de l'est du Bassin de Paris, *Rev. Inst. Franç. Petrol.,* 31 (6), 933-953.

Illing, L.V. and Hobson, G.D. (Eds.) (1981) *Petroleum geology of the Continental Shelf of north-west Europe.* Heyden & Son.

Irwin, H. (1979) An environmental model for the type Kimmeridge Clay (comment and reply). *Nature* 279, 819-820.

Islam, S., Hesse, R. and Chagnon, A. (1982) Zonation of diagenesis and low grade metamorphism in Cambro-Ordovician Flysche of Gaspé peninsula, Quebec Appalachians. *Canadian Mineralogist* 20, 155-167.

Jenner, J.K. (1981) The structure and stratigraphy of the Kish Bank basin. In: Illing, L.V. and Hobson, G.D. (Eds.) q.v. pp. 426-431.

Juntgen, H. and Karweil, J. (1966) Gasbildung und Gasspeicherung in Steinkohlflözen. *Erdöl, Kohle, Erdgas, Petrochemie* 19, 251-258, 339-344.

Kettel, D. (1982) Norddeutsche Erdgase: Stickstoffgehalt und Isotopenvariationen als Reife-und Faziesindikatoren. *Erdöl und Kohle, Erdgas, Petrochemie mit Brenstoffchemie* 35 (12), 557-559.

Kettel, D. (1983) The East Groningen Massif; detection of an intrusive body by means of coalification. *Geol. Mijnbouw* 60 (1), 203-210.

Kirk, R.H. (1980) Statfjord Field, a North Sea giant. In: Halbouty, M.T. (Ed.) *Giant oil and gas fields of the decade 1968-1978.* (AAPG Memoir 30). pp. 95-116.

Koch, J.-O. (1983) Sedimentology of the Middle and Upper Jurassic reservoirs of Denmark. *Geol. Mijnbouw* 62 (1), 115-129.

Larsen, R.M. and Jaarvik, L.J. (1981) The geology of the Sleipner field complex. In: *Norwegian Symposium on Exploration, 1981, Bergen.* Norwegian Petroleum Society, p. 15.1-15.31.

Leythaeuser, D. Mackenzie, A.S., Schaefer, R.G., Altebäumer, F.J. and Bjorøy, M. (1983) Recognition of migration and its effects within two coreholes in shale/sandstone sequences from Svalbard, Norway. In: Bjorøy, M. *et al.* (Eds.) *Advances in organic geochemistry* 1981. John Wiley, pp. 136-146.

Leythaeuser, D., Schaefer, R.G. and Yukler, A. (1982) The role of diffusion in primary migration of hydrocarbons. *Am. Assoc. Petrol. Geol. Bull.* 66 (4), 408-429.

Linsley, P.N., Potter, H.C., McNab, G. and Racher, D. (1980) The Beatrice Field, Inver Moray Firth, U.K. North Sea. In: Halbouty, M.T.(Ed.) *Giant oil and gas fields of the decade 1968-1978* (AAPG Memoir 30), pp. 117-129.

Lutz, M., Kaasschieter, J.P.H. and van Wijke, D.H. (1975) Geological factors controlling Rotliegend gas accumulations in the mid-European basin. *Proceedings of 9th World Petroleum Congress.* Applied Science, 93-103.

Mackenzie, A.S. and McKenzie, D. (1983) Isomerisation and aromatisation of hydrocarbons in sedimentary basins formed by extension. *Geol. Mag.* 120 (5), 417-470.

Mackenzie, A.S., Patience, R.L., Maxwell, J.R., Vandenbroucke, M. and Durand, B. (1980) Molecular parameters of maturation in the Toarcian shales, Paris basin, France. 1. Changes in the configuration of the acyclic isoprenoid alkanes, steranes and triterpanes. *Geochim. Cosmochim. Acta* 44, 1709-1721.

Maher, C.E. (1981) The Piper oilfield. In: Illing, L.V. and Hobson, G.D. (Eds.) q.v. pp. 358-370.

Manum, S.B. and Throndsen, T. (1977) Rank of coal and dispersed organic matter and its geological bearing in the

Spitzbergen Tertiary. *Norsk. Polarinst. Arbok* 1977, pp. 159-177, pp. 179-187.

McIver, R. (1975) Hydrocarbon occurrences from JOIDES Deep Sea Drilling Project. *World Petroleum Congress, Tokyo*, Panel Discussion, vol. 1, p. 1.

Meissner, F.F. (1978) Petroleum geology of the Bakken Formation, Williston Basin, North Dakota and Montana. *Williston Basin Symposium*, Montana Geol. Soc., pp. 207-227.

Meyer, B.L. and Nederlof, M.H. (1984) Identification of source rocks on wireline logs by density/resistivity and interval velocity/resistivity cross-plots. **68** (2), 121-129.

Michelsen, O. and Andersen, C. (1983) Mesozoic structural and sedimentary development of the Danish Central Graben. *Geol. Mijnbouw* **62** (1), 93-102.

Müller, P.J. and Suess, E. (1979) Productivity, sedimentation rate and sedimentary organic matter in the oceans. 1. Organic carbon preservation. *Deep Sea Research* **26**A, 1347-1362.

Nordberg, H.E. (1981) Seismic hydrocarbon indicators in the North Sea. *Norwegian Symposium on Exploration, Bergen.* Norwegian Petroleum Society, p. 8.1-8.40.

Oele, J.A., Hol, A.C.P.J. and Tiemens, J. (1981) Some Rotliegend gas fields of the K and L blocks, Netherlands offshore (1968-1978)—a case history. In: Illing, L.V. and Hobson, G.D. (Eds.) q.v. pp. 289-300.

Olsen, J.C. (1983) The structural outline of the Horn Graben. *Geol. Mijnbouw* **62** (1), 47-50.

Oudin, J-L. (1976) Étude géochimique du Bassin de la Mere du Nord. *Bull. Centre Rech. Pau-SNPA* **10** (1), 339-358.

Oxburgh, E.R. and Andrews-Speed, C.P. (1981) Temperature, thermal gradients and heat flow in the south-west North Sea. In: Illing, L.V. and Hobson, G.D. (Eds.) q.v. pp. 114-151.

Parnell, J. (1983) The distribution of hydrocarbon minerals in the Orcadian Basin. *Scott. J. Geol.* **19**, 205-213.

Parry, C.C., Whitley, P.K.J. and Simpson, R.D.H. (1981) Integration of palynological and sedimentalogical methods in facies analysis of the Brent Formation. In: Illing, L.V. and Hobson, G.D. (Eds.) q.v. pp. 205-215.

Pearson, M.J. and Watkins, D. (1983) Organofacies and early maturation effects in Upper Jurassic sediments from the Inner Moray Firth Basin, North Sea. In: Brooks, J. (Ed.) q.v. pp. 147-160.

Pearson, M.J., Watkins, D., Pittion, J-L., Caston, D. and Small, J.S. (1983) Aspects of burial diagenesis, organic maturation and paleogeothermal history of an area in the southern Viking Graben, North Sea. In: Brooks, J. (Ed.) q.v. pp. 161-173.

Pennington, J.J. (1975) The geology of the Argyll field. In: Woodland, A.W. (Ed.) q.v. pp. 285-291.

Pering, K.L. (1973) Bitumens associated with lead, zinc and fluorite ore minerals in North Derbyshire, England. *Geochim. Cosmochim. Acta.* **37**, 401-417.

Powell, T.G., Douglas, A.G. and Allan, J. (1976) Variations in the type and distribution of organic matter in some Carboniferous sediments from Northern England. *Chemical Geol.* **18**, 137-148.

Pratsch, J-C. (1983) Gasfields, N.W. German Basin: secondary gas migration is a major geologic parameter. *J. Petrol. Geol.* **5** (3), 229-244.

Price, L.C. (1983) Geologic time as a parameter in organic metamorphism and vitrinite reflectance as an absolute paleogeothermometer. *J. Petrol. Geol.* **6** (1), 5-38.

Reitsema, R.H. (1983) Geochemistry of North and South Brae areas, North Sea. In: Brooks, J. (Ed.) q.v. pp. 203-212.

Ridd, M.F. (1981) Petroleum geology west of the Shetlands. In: Illing, L.V. and Hobson, G.D. (Eds.) q.v. pp. 414-425.

Rigby, D. and Smith, J.W. (1982) A reassessment of stable carbon isotopes in hydrocarbon exploration. *Erdöl, Kohle, Erdgas, Petrochemie* **35** (9), 415-417.

Roberts III, W.H. and Cordell, R.J. (1980) *Problems of petroleum migration.* AAPG Studies in Geology No. 10. Am. Assoc. Petrol. Geol., Tulsa.

Rønnevik, H.C. (1981) Geology of the Barents Sea. In: Illing, L.V. and Hobson, G.D. (Eds.) q.v. pp. 395-406.

Rønnevik, J., Eggen, S. and Vollset, J. (1983) Exploration of the Norwegian Shelf. In: Brooks, J. (Ed.) q.v. pp. 71-93

Roos, B.M. and Smits, B.J. (1983) Rotliegend and main Buntsandstein gas fields in Block K/13—a case history. *Geol. Mijnbouw* **62** (1), 75-82.

Schmoker, J.W. (1979) Determination of organic content of Appalachian Devonian shales from formation-density logs. *Am. Assoc. Petrol. Bull.* **63** (9), 1504-1509.

Schmoker, J.W. (1981) Determination of organic matter content of Appalachian Devonian shale from gamma ray logs. *Am. Assoc. Petrol. Bull.* **65**, 1285-1298.

Smith, P.M.R. (1983) Spectral correlation of spore colour standards. In: Brooks, J. (Ed.) q.v. pp. 289-294.

Snowdon, L.R. and Powell, T.G. (1979) Families of crude oils and condensates in the Beaufort-Mackenzie Basin. *Bull. Canadian Petrol. Geol.* **27** (2), 139-162.

Snowdon, L.R. and Powell, T.G. (1982) Immature oil and condensate—modification of hydrocarbon generation model for terrestrial organic matter. *Am. Assoc. Petrol. Geol. Bull.* **66** (6), 775-788.

Stow, D.A.V., Bishop, C.D. and Mills, S.J. (1982) Sedimentology of the Brae field North Sea: fan models and controls. *J. Petrol. Geol.* **5** (2), 129-148.

Taylor, J.C.M. (1981) Zechstein facies and petroleum prospects in the central and northern North Sea. In: Illing, L.V. and Hobson, G.D. (Eds.) q.v. pp. 176-185.

Teichmüller, M. and Teichmüller, R. (1979) Diagenesis of coal (coalification) In: Larsen, G. and Chilingar, G.V. (Eds.) *Diagenesis in sediments and sedimentary rocks.* Elsevier, pp. 207-246.

Teichmüller, M., Teichmüller, R. and Bartenstein, H. (1979) Inkohlung und Erdgas in Nordwestdeutchland. Eine Inkohlungskarte de Oberflacher des Oberkarbons. *Fortschr. Geol. Rheinld. u. Westf.* **27**, 137-170.

Thomsen, E. Lindgreen, H. and Wrang, P. (1983) Investigation on the source rock potential of Denmark. *Geol. Mijnbouw* **62** (1), 221-239.

Tissot, B.P. and Bessereau, G. (1982) Géochimie des gaz naturels et origine des gisements de gaz en Europe occidentale. *Rev. Inst. Français de Petrol.* **37** (1), 63-77.

Tissot, B.P. and Welte, D.H. (1978) *Petroleum formation and occurrence.* Springer Verlag.

Tissot, B., Chalifet-Debyser, Y., Deroo, G. and Oudin, J.L. (1971) Origin and evolution of hydrocarbons in early Toarcian shales, Paris Basin, France. *Am. Assoc. Petrol. Geol. Bull.* **55** (12), 2177-2193.

Tixier, M.P. and Curtis, M.R. (1967) Oil shale yield predicted from well logs. *Proceedings of 7th World Petroleum Congress, Mexico City*, vol. 3, pp. 713-715.

Toth, D.J., Lerche, I., Petroy, D.E., Meyer, R.J. and Kendall, C.G. St. C. (1983) Vitrinite reflectance and the derivation of heat flow changes with time. In: Bjorøy, M. et al. (Eds.) *Advances in organic geochemistry* 1981. John Wiley, pp. 588-596.

Tyson, R.V., Wilson, R.C.L. and Downie, C. (1979) A stratified water column environmental model for the type Kimmeridge Clay. *Nature* **277**, 377-380.

Vail, P.R. and Todd, R.G. (1981) Northern North Sea Jurassic unconformities, chronostratigraphy and sea-level changes from seismic stratigraphy. In: Illing, L.V. and Hobson, G.D. (Eds.) q.v. pp. 216-235.

Van den Bark, E. and Thomas, O.D. (1980) Ekofisk: first of the giant oil fields in Western Europe. In: Halbouty, M.T. (Ed.) *Giant oil and gas fields of the decade 1968-1978* (AAPG Memoir 30), pp. 195-224.

Van den Bosch, W.J. (1983) The Harlingen gas field, the

only gas field in the Upper Cretaceous chalk of the Netherlands. *Geol. Mijnbouw* **62** (1), 145-156.

van Wijhe, D.H., Lutz, M. and Kaasschieter, J.P.H. (1980) The Rotliegend in The Netherlands and its gas accumulations. *Geol. Mijnbouw* **59**, 3-24.

Waples, D. (1980) Time and temperature in petroleum formation; application of Lopatin's method to petroleum exploration. *Am. Assoc. Petrol. Geol. Bull.* **64**, 916-926.

Waples, D.W. (1983) Reappraisal of anoxia and organic richness with emphasis on Cretaceous of North Atlantic. *Am. Assoc. Petrol. Geol. Bull.* **67** (6), 963-978.

Watson, S. (1976) *Sedimentary geochemistry of the Moffat Shales; a carbonaceous sequence in the Southern Uplands of Scotland*. Unpublished Ph.D., University of St Andrews, Scotland.

Welte, D.H. and Yukler, M.A. (1981) Petroleum origin and accumulation in basin evolution—a quantitative model. *Am. Assoc. Petrol. Geol. Bull.* **65** (8), 1387-1396.

Welte, D.H., Yukler, M.A., Radke, M., Leythaeuser, D., Mann, U. and Ritter, U. (1983) Organic geochemistry and basin modelling—important tools in petroleum exploration. In: Brooks, J. (Ed.) q.v. pp. 237-254.

Williams, P.F.V. and Douglas, A.G. (1980) A preliminary organic geochemical investigation of the Kimmeridge oil shales. In: Douglas, A.G. and Maxwell, J.R. (Eds.) *Advances in organic geochemistry* 1979. Pergamon Press, pp. 531-545.

Woodland, A.W. (Ed.) (1975) *Petroleum and the Continental Shelf of North-west Europe,* vol. 1, Geology. Applied Science Publishers.

Ziegler, P.A. (1982) *Geological atlas of Western and Central Europe*. Shell Int. Pet. Maatsch. B.V.

Chapter 10 — North Sea Hydrocarbon Plays

A.J. PARSLEY

10.1 Introduction

A hydrocarbon play is a set of petroleum geological circumstances which combine to create the conditions necessary for the accumulation of oil and/or gas. A single play may contain a number of discoveries and prospects but the favourable combination of the controlling geological parameters usually occurs over a limited geographic area, often referred to as a 'fairway'. This is particularly true of the Northern North Sea, where almost all the major oil discoveries, including those in the Tertiary, are restricted to the

centres and margins of the main Mesozoic rift basins (Fig. 10.1).

In the competitive atmosphere of North Sea exploration, the petroleum geologist is continuously required to make predictive judgements regarding the acquisition of new acreage in licensing rounds, or the drilling of wildcat wells on existing acreage. To do this, extrapolations are made from proven plays or new plays are invented and conceptually evaluated. In either case, the geologist relies heavily upon the acquisition of the most up-to-date seismic and well data, and the interpretation of this new data set by means of

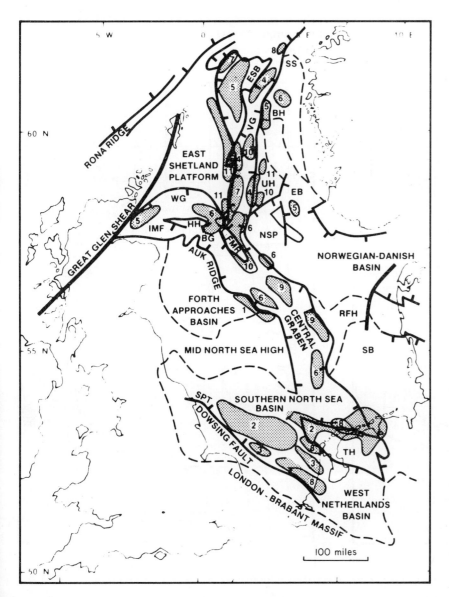

Fig. 10.1. North-Sea hydrocarbon plays.
Key: 1 Palaeozoic oil plays in graben margins. 2 Rotliegend gas play. 3 Bunter gas play. 4 L. and M. Jurassic gas/condensate play. 5 L. and M. Jurassic oil play. 6 U. Jurassic shallow marine sands oil and gas play. 7 U. Jurassic fan/turbidite sands oil and gas/condensate play. 8 L. Cretaceous oil and gas/condensate play. 9 U. Cretaceous/Danian chalk oil and gas/condensate play. 10 L. Tertiary sands oil and gas plays. 11 L. Tertiary sands heavy (and medium) oil plays.
BG Buchan Graben, BH Bergen High, EB Ergersund Basin, ESB East Shetlands Basin, FMR Forties-Montrose Ridge, FS Fladen Ground Spur, HH Halibut Horst, IMF Inner Moray Firth Basin, NSP Norwegian Salt Platform, RFH Ringkøbing-Fyn High, SB Schillbank High, SPT Sole Pit Trough, SS Sogn Shelf, TH Texel High, UH Utsira High, WG Witch Ground Graben, VG Viking Graben.

regional structural and stratigraphic models. The preceeding chapters in this book have presented in some detail such a set of models. While at the large scale at which they have been presented these models are fairly robust, at the more detailed level of current exploration work in the North Sea they are subject to constant revision.

New plays in the North Sea tend to arise rapidly, and the acreage over which the fairway may extend is quickly taken up in subsequent licensing. An example is provided by the 4th Round of U.K. licensing in the East Shetland Basin. Fig. 10.2a shows the wells drilled in 1971 on the few 3rd Round blocks awarded in the area and the Brent discovery made by Shell; Fig. 10.2b shows the tremendous upsurge in drilling on this Middle Jurassic play that occurred in 1972-74 on 4th Round blocks awarded in 1972, and the discoveries made. The development of a new play is, of course, controlled by the evolution of petroleum geological thinking in the area, but this in itself is often tied to the acquisition of new fragments of stratigraphic information or to improvements in seismic data quality. For example, the establishment of the Brent sandstone play followed limited improvements in seismic data quality which suggested that the previously recognised base Cretaceous structure could be related to rotated Jurassic fault blocks similar to those of the East Greenland

Basin (Bowen, 1975). However, some plays have emerged as a result of exploration for other primary reservoir objectives; for example, the discovery of oil in the Upper Jurassic sandstones of the Piper Field resulted from an exploration programme aimed at extending the Forties/Montrose Palaeocene sandstone play to the north (Williams *et al.*, 1975).

10.2 Hydrocarbon play definition

The uncertainty associated with an undrilled prospect in the North Sea can be described by evaluation of the four primary geological parameters which together define the risk of a dry hole:

—the likelihood of the existence of a sealed trap at some point in geological time
—the likelihood of the presence of a porous and permeable reservoir
—the likelihood of the migration of oil and/or gas through the reservoir at some point in geological time
—the likelihood that trap formation predated hydrocarbon migration and the likelihood of preservation of an accumulation to the present day

Some of these parameters are unique to the prospect, for example the geometrical configuration of the trap, but others are controlled on a broader regional scale

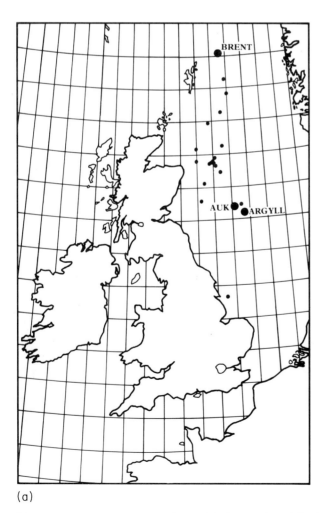

(a)

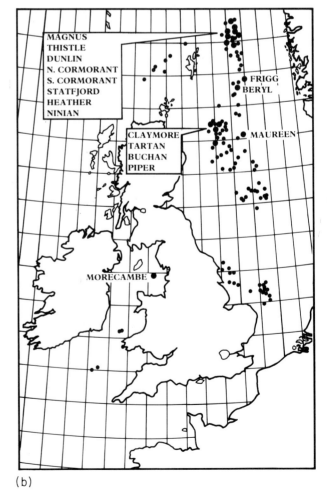

(b)

Fig. 10.2. Drilling activity associated with the opening of the Lower-Middle Jurassic sandstone play for 1971 (a) and 1972-74 (b).

and are related to the total play of which the prospect forms only part. An example is the likelihood of occurrence of the potential reservoir, which can only be assessed from a consideration of palaeogeographic, depositional and diagenetic models.

Many of the structures which form traps in the Northern North Sea are related to extensional faulting associated with rifting, as shown by Ziegler's (1982) sections illustrated in Fig. 10.3. Extensional faulting was periodically active from the Triassic to the Cretaceous but had almost entirely ceased by the Tertiary, when a period of broad basinal subsidence was established which has continued to the present day. Generation and migration of hydrocarbons from the dominant Upper Jurassic source rocks, north of the Mid-North Sea High (Fig. 10.4), began at the end of the Cretaceous and has continued until the Recent. The relative timing of trap formation and migration is therefore clearly favourable for most Mesozoic structural traps. This is not so for Upper Cretaceous and Tertiary fields, of which there are many, where trap formation began after migration had started. Apart from tilting due to basinal subsidence during the Tertiary, there has been little tectonic activity to rupture early-formed accumulations so that preservation is also generally favourable. There is evidence, however, of Laramide fault reactivation in the Northern North Sea as far north as Sleipner (Fagerland, 1983).

The primary control on North Sea play fairways is that the Upper Jurassic source rocks have reached maturity only within the deeply buried Mesozoic grabens, and that hydrocarbon migration has been dominantly vertical with only rare evidence of extensive lateral migration (refer to later discussion of Lower Tertiary sandstone plays). The location of the Mesozoic grabens is shown on the tectonic elements map of the North Sea in Fig. 10.4. The other major controlling parameter is the distribution of the principal reservoirs, which range in age from the Devonian to the Eocene, and it is on this basis that North Sea hydrocarbon plays can be most readily classified and discussed. The main North Sea reservoirs are illustrated in Fig. 10.5.

It should be noted that in the Southern North Sea Basin the controlling parameters on the plays are quite different. The principal source rocks are located in the Westphalian and the Lower Jurassic, and the area has been strongly affected by Late Cretaceous and Early Tertiary (Laramide) tectonics; the resulting inversions have locally led to rupture and spillage of hydrocarbons from older traps.

10.3 Lower Tertiary sandstone plays

The locations of the principal Lower Tertiary sandstone plays are shown in Fig. 10.1. Plays for light oils and gas or lean gas-condensate are situated above the axes of the buried Mesozoic grabens, whereas the heavy oil plays are generally perched above the upthrown graben margins. The principal constraints on these plays are the distribution of sandstone reservoirs and the development of traps against the unfaulted basinward dip of the Tertiary. These controls are illustrated in Fig. 10.6.

The uplift and erosion of the East Shetland Platform in the Early Tertiary resulted in large quantities of sand being shed eastwards and southeastwards into the depositional axes of the developing basins. These depocentres are approximately coincident with the buried Viking and Central grabens. To the north, shallow marine and deltaic environments in the western part of the fairway give way to deeper water submarine fans to the east (Lilleng, 1980). Depositional trends directed eastwards off the shelf are deflected north and south along the axis of the basin (Rochow, 1981), with the result that Lower Tertiary sandstones are generally lacking from the eastern basin flank. Sands being shed to the southeast, however, were transported by submarine processes far to the southeast along the axis of the basin. Turbidite sands reach as far south as approximately 56°N (Parker, 1975), being replaced beyond by distal basin argillites. It should be noted that in the northern North Sea, Mudge and Bliss (1983) have alternatively interpreted all these depositional patterns in terms of shallow water environments.

The broad basinal subsidence pattern in the Tertiary (Fig. 10.6) creates, in the absence of major faults this high in the sequence, an almost uninterrupted migration pathway for hydrocarbons moving through these reservoirs until they reach surface outcrop to the west or northwest. Closure along this pathway is a critical control on these plays. Three main trapping styles are recognised in the Lower Tertiary of the North Sea:

—dip-closure resulting from drape over mid-graben horsts
—dip-closures above salt domes
—stratigraphic traps

The two major oil discoveries made in Lower Tertiary sandstones (the Forties and Montrose fields) are formed by dip-closures, at the base Upper Palaeocene top seal, located above the prominent southeasterly-plunging Forties-Montrose Ridge (Fig. 10.4). The causal relationship may have been either renewed uplift of this Jurassic high in the Early Tertiary or draping (or both). In either case, trap development was not completed until the Miocene (Walmsley, 1975 and Fowler, 1975), which is late compared with the end Cretaceous onset of oil generation in the Central Graben (Cornford, this volume, Fig. 9.19). The mechanism of migration of Jurassic oil into these traps is not well understood since, although the horst blocks themselves are fault controlled at depth, there is little evidence that these faults propagate upwards in the section through the Chalk to the Palaeocene reservoir.

Development drilling on the Forties Field has demonstrated the complexity of the depositional geometry of these submarine sands (Carman and Young, 1981). Whilst this has not had a marked affect on the development of these giant fields, it has been a major factor in

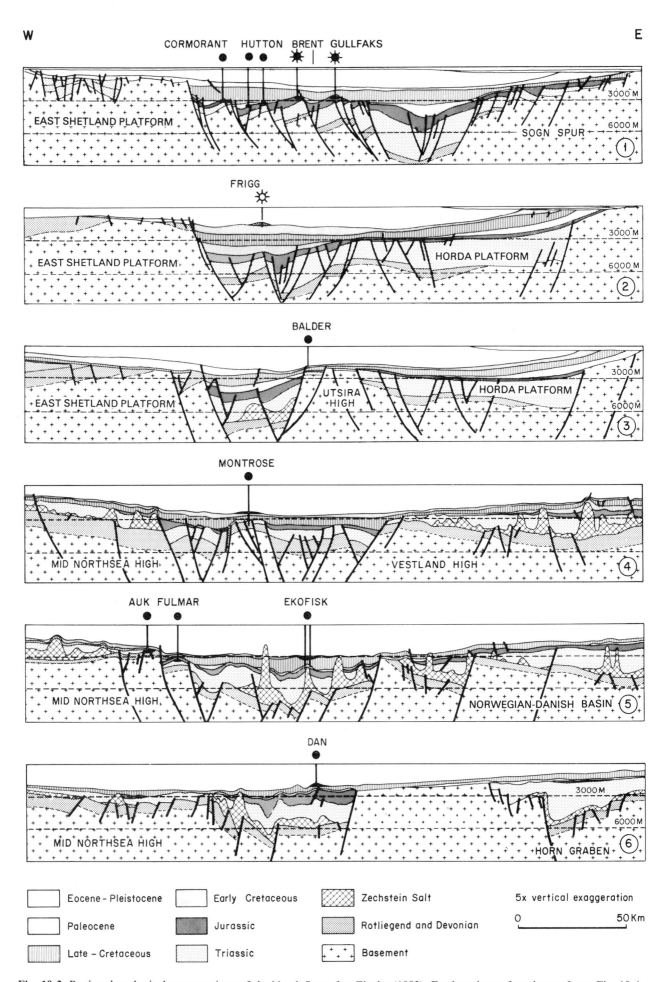

Fig. 10.3. Regional geological cross-sections of the North Sea; after Ziegler (1982). For locations of sections refer to Fig. 10.4.

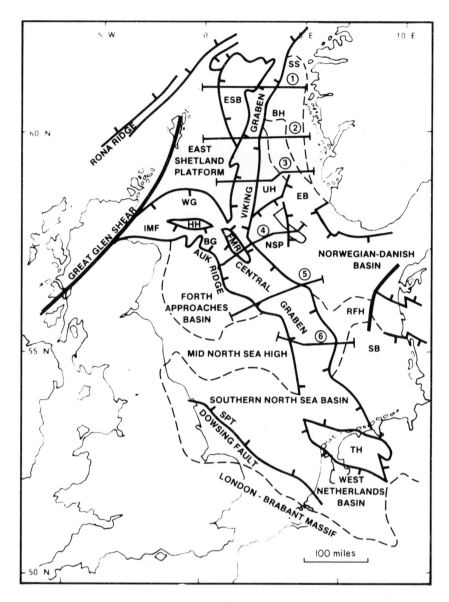

Fig. 10.4. North Sea tectonic elements. See Fig. 10.1 for Key. For geological cross-sections see Fig. 10.3.

limiting the commerciality of Lower Tertiary reservoirs in dip-closures above salt domes, where the gross trapped volume is relatively small. Within the area of development of the Permian salt (Fig. 10.7), there are numerous dip-closures above diapirs but so far only Maureen (U.K.) and Cod (Norway) have proved commercial.

The appraisal history of the Cod Field (Kessler *et al.*, 1980) illustrates this problem. The field was discovered in 1968 but appraisal was enigmatic because it has proved difficult to predict reservoir development. Nine of the eleven wells (Fig. 10.8) found gas-condensate with varying pressures and hydrocarbon-water contacts; one well found oil and gas-condensate in separate reservoirs, and two structurally high wells were dry. It is now believed that there are at least nine separate accumulations within this apparently simple domal feature. The hydrocarbons are trapped in reservoirs associated with three separate submarine-channel and fan-lobe progradations, and the complexity of the field has probably been enhanced by syndepositional structural growth.

Beyond the northern limit of salt diapirism, it is this depositional complexity which provides the principal trapping mechanism, although drape folds also occur. Two contrasting examples are the Frigg and Balder fields (Fig. 10.9). Trapping in the Frigg Field is related to the deposition of thick submarine channel and levee deposits of Eocene age, and the structural elevation of channel margins through differential compaction. In the Balder Field, approximately 60 metres of vertical closure relies on the irregular upper surface of the Middle to Upper Palaeocene sands, which were deposited in a submarine fan. The irregularity at the base of the top seal is the direct result of a depositional topography generated by the mounding of suprafan lobes, amplified by submarine erosion both by southwards-flowing axial currents and east-west currents, which may have been tidally influenced.

Balder contains a heavy (25° API) oil, and numerous heavy oil discoveries (10°-20° API typically) have been made along the margin of the East Shetlands Platform (Fig. 10.1). Closure to the west, against the predominant easterly regional dip, is created by depositional topography associated with shallow-marine and deltaic bar sands, enhanced by differential compaction. Draping of sheet sands over buried bars can also create closure. In some cases, the size of traps created by

Northern North Sea

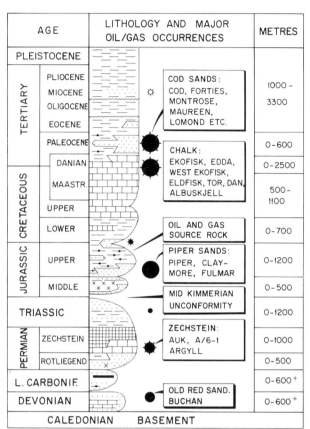

Central North Sea

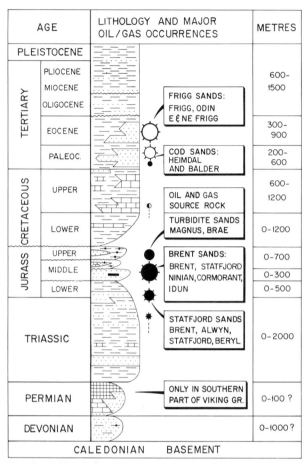

Southern North Sea

Fig. 10.5. Principal North Sea reservoir rocks; after Ziegler (1977).

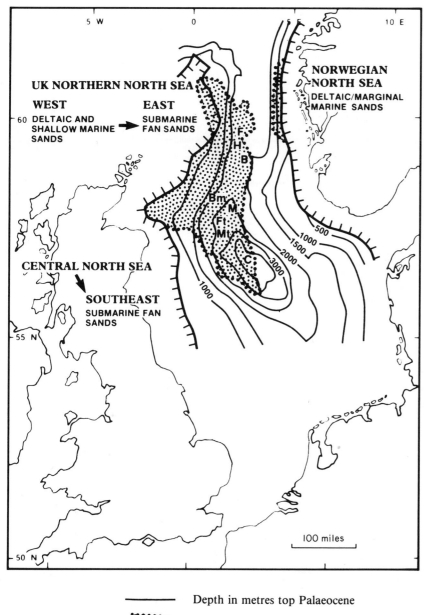

Fig. 10.6. Lower Tertiary sandstone play map; compiled from Hamar *et al.* (1980), Rochow (1981) and Parker (1975).
Key: **F** Frigg, **B** Balder, **Ft** Forties, **Mt** Montrose, **C** Cod, **M** Maureen, **H** Heimdal, **Bm** Balmoral.

—————— Depth in metres top Palaeocene

Areas of deposition of Lower Tertiary sands

structural closure may be enhanced by stratigraphic trapping due to facies changes. The commercial exploitation of the large volumes of trapped oil is prevented by the generally shallow depth of burial, the friable nature of the reservoir sandstones and the unfavourable relative mobility in the reservoir of water and viscous oil.

The mechanisms for the migration of hydrocarbons from the deeply buried Upper Jurassic source rocks into Palaeocene and Eocene sandstone reservoirs is not well understood. Permian salt diapirs which puncture the Upper Jurassic and reach up into the Lower Tertiary provide an obvious mechanism in the Central Graben (see discussion under chalk plays) but there is no evidence of salt diapirism in the northern Viking Graben. A comparison of the distribution of oil gravities (Barnard and Cooper, 1981) with the depth of Tertiary burial of Upper Jurassic source rocks, within the Mesozoic Grabens, is shown in Fig. 10.10. The lightest

oils and condensates in all reservoirs, including the Tertiary, tend to occur in areas of maximum Tertiary burial of Upper Jurassic source rocks, thus inferring that migration has been predominantly vertical. There is little evidence in the North Sea for long distance lateral migration of hydrocarbons into traps. In a recent study of hydrocarbon generation and migration in the northern North Sea, Goff (1983) concluded that the gas and heavy oil (24° API) of the Frigg Field had been both generated from the Upper Jurassic and had migrated vertically through 2000 metres of unfaulted Cretaceous and Palaeocene mudstones (Fig. 10.11). Migration is suggested to have occurred in the gas phase in microfractures in overpressured Upper Jurassic to Upper Cretaceous mudstones below 3500 metres, and thereafter at shallower depths in aqueous solution to the reservoir. Oil migration through microfractures at depth is also possible but at shallower depths, a tortuous path through siltstones and larger mudstone

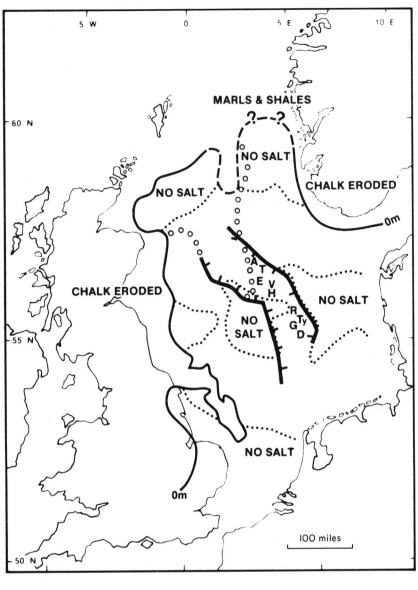

Active Late Cretaceous faults
——**0m** Erosional limit of chalk
– – – – Effective northern limit of chalk deposition
··········· Approximate limits of thick Permian salt
ooooooo Approximate eastern and southern limits of Lower Tertiary sand development

Fig. 10.7. Chalk reservoir play map; compiled from Hancock and Scholle (1975), Taylor (1981) and Day *et al.* (1981).
Key: **T** Tor, **A** Albuskjell, **E** Ekofisk & Eldfisk, **V** Valhall, **H** Hod, **R** Roar, **Ty** Tyra, **G** Gorm, **D** Dan.

pores has to be proposed. Goff (1983) also concludes that the Kimmeridge Clay was at peak maturity for oil generation below Frigg for 10–20 million years before the trap was sealed. Gas generation began at about the time of trap formation. Light oil may, therefore, have migrated up into the reservoir whilst it was still open to sea-water circulation, allowing *in situ* alteration to a heavy oil by water washing and biodegradation, which removed almost all paraffinic hydrocarbons (Barnard and Cooper, 1981). Alternatively, Heritier *et al.*, (1981) have suggested that bacteria were introduced by an eastwards invasion of meteoric waters, presumably from outcrop on the East Shetlands Platform. Above 60°C bacterial activity is reduced, and where Tertiary reservoirs are present

below 2000 metres, the light fractions are preserved and substantial well flow rates can be achieved.

10.4 Upper Cretaceous and Lower Tertiary chalk plays

Chalk rock is widespread in the Upper Cretaceous and Danian of the North Sea but commercial production is limited to two narrow fairways in the Central Graben (Fig. 10.1), despite the fact that wells elsewhere often encounter significant hydrocarbon shows. The reason is simply that chalk is not a good reservoir under normal conditions. Although depositional porosities are estimated to be as high as 70 per cent, and the low magnesian calcite composition of the coccolith plate-

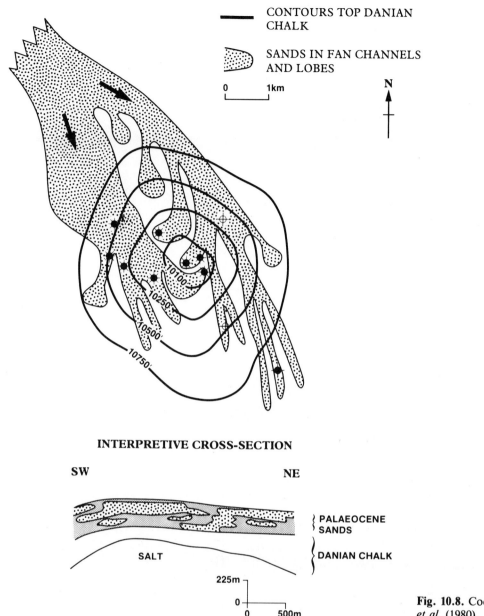

CONTOURS TOP DANIAN CHALK

SANDS IN FAN CHANNELS AND LOBES

0   1km

N

INTERPRETIVE CROSS-SECTION

SW   NE

PALAEOCENE SANDS

DANIAN CHALK

SALT

225m
0
0   500m

Fig. 10.8. Cod Field; after Kessler *et al.* (1980).

lets is relatively resistant to early diagenetic effects, mechanical compaction and solution and reprecipitation of calcite normally rapidly reduce porosity with increasing depth of burial. The chalk reservoir of the Ekofisk Field, however, has an average porosity of about 32 per cent compared with an expected range of 2-25 per cent at these depths (10,400 ft). What is more striking is that permeabilities at these depths would be expected to be in the range 0 to 0.5 mD (Hancock and Scholle, 1975) compared with effective permeabilities of about 12 mD derived from production data at Ekofisk (Thomas, 1980).

The chalk reservoir oilfields in the Norwegian sector of the North Sea are illustrated in Fig. 10.12: the accompanying cross section through Ekofisk (after Thomas, 1980) shows that these domal features are caused by diapiric intrusions of Permian salt at depth. Salt movements probably began in the Ekofisk area in the Triassic, slowed during the Cretaceous and Danian and were reactivated by Early Tertiary clastic sediment

loading (Blair, 1975). Salt intrusion was, therefore, most active at about the same time as peak oil-generation from the underlying Upper Jurassic source rocks (refer Cornford, this volume). Early spot-welding of coccoliths and platelets may have resulted in the formation of a rigid framework for the chalk, which led to the rock being microfractured in 'tensional' areas immediately above the intruding diapir (D'Heur, 1980). Matrix blocks created by fracturing are typically only 1 to 2 cm across where the chalk is thin or structurally weak (e.g. Tor Formation) but up to 50 cm or 1 m in zones where the clay content is higher (Hardman and Kennedy, 1980). It is now widely believed that oil migrated up through these fractures, and that when the column of oil in the fracture system was of the order of 50-100 m, the bouyancy force was sufficient to overcome the high entry pressure of the very narrow matrix pore throats. Hardman and Kennedy (1980) have observed that, possibly as a result of this process, high relief features have better oil saturations and that valid low

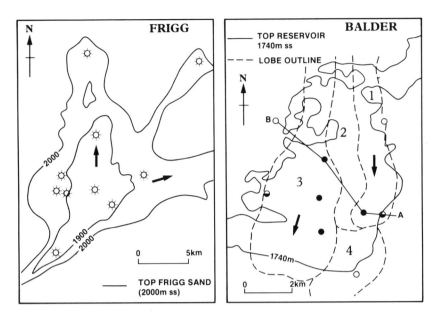

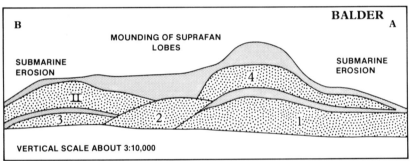

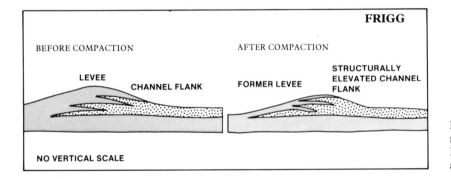

Fig. 10.9. Comparison of trapping mechanisms in Frigg and Balder Fields; after Heritier *et al.* (1981) and Sarg and Skjold (1982).

relief traps can have water-wet matrix porosity. The early entry of oil into the matrix pores and the associated inflated overpressures then retarded the normal progress of lithification with burial, by inhibiting chemical diagenesis and reducing the overburden pressure on the matrix grains. D'Heur (1980) has suggested that these processes were most effective in the centres of the uplifted domes, migrating laterally outwards towards the flanks: this could explain the reduction of porosity and increase of water saturation radially, and the concave-up oil-water contact in West Ekofisk.

The microfractures have remained open and in fluid continuity with matrix porosity to the present day, thereby contributing largely to the unexpectedly high observed bulk permeability.

The chalk oilfields of Denmark are broadly similar to those of Norway, although less productive. The major established fields are broad domal features

(Hurst, 1983) which are presumably less intensely fractured: examples are Dan (Childs and Reed, 1975), Gorm (Hurst, 1983) and Tyra (Fig. 10.12). In contrast, Skjold is located over a piercement and is more intensely fractured. It is noteworthy that Dan has a free gas cap whilst Gorm oil is undersaturated; the difference is thought to be caused by the sealing capability of the caprock which, in the case of Gorm, was insufficient to prevent gas escaping to shallower levels.

The mechanism of vertical migration of oil from the Jurassic source rocks through Lower Cretaceous claystones is not well understood. Faulting, which might create pathways, is common (e.g. Dan) but not ubiquitous, and West Ekofisk is reported by D'Heur (1980) to be unfaulted.

The primary controls on the fairway, therefore, appear to be the distribution of the Chalk itself and thick Permian salt at depth (Fig. 10.7). The require-

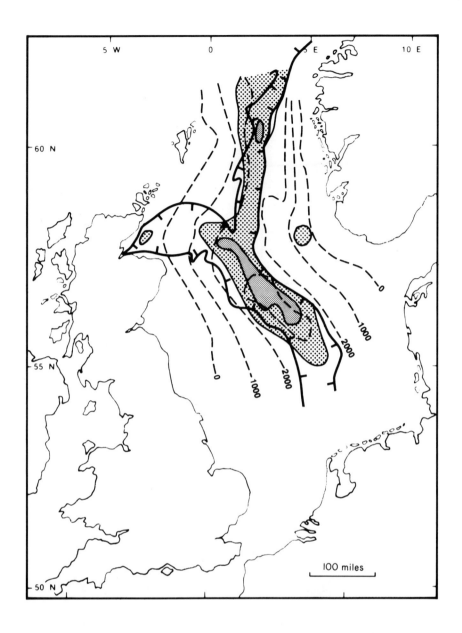

- - - - - 1000m Base Tertiary Metres

├──┴──┤ Mesozoic Graben Margin

▨ Oils 42°-55° API

▦ Oils 28°-42° API

Fig. 10.10. The location of the main Mesozoic Grabens and their Tertiary burial related to the distribution of oil gravities; partly from Barnard and Cooper (1981).

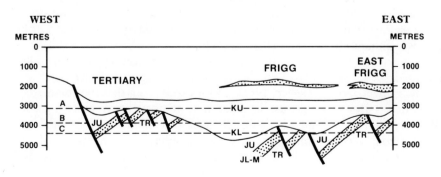

Fig. 10.11. Cross-section of the North Viking Graben through Frigg showing present-day maturity levels; from Heritier *et al.* (1981) and Goff (1983). A = Top peak oil generation, B = Top gas generation, C = Base oil window. Key: **TR** Triassic, **JL-M** Lower-Middle Jurassic, **JU** Upper Jurassic, **KL** Lower Cretaceous, **KU** Upper Cretaceous.

EKOFISK

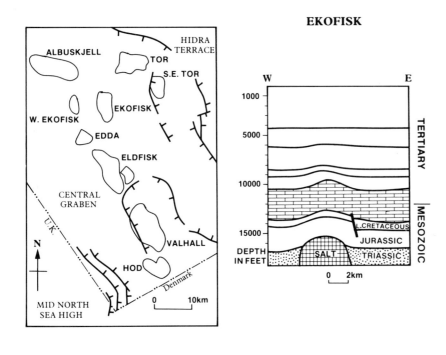

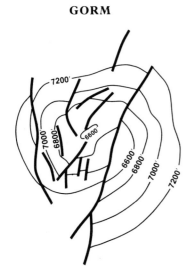

STRUCTURE MAP TOP MAASTRICHTIAN

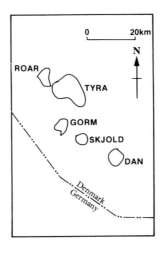

Fig. 10.12. Locations of oil and gas fields with Chalk reservoirs in Norway and Denmark with details on Ekofisk and Gorm; from Kennedy (1980), Thomas (1980) and Hurst (1983).

ment for vertical migration of oil into the traps further restricts the play to the locus of the Mesozoic grabens. However, it is still surprising that this play has not proved more extensive and it is possible that other parameters are exerting controls.

Firstly, the relative timing of trap formation and hydrocarbon migration is critical. Migration after trap formation could be unfavourable because microfractures could become annealed, preventing entry of oil into matrix pores. A second control may be the extent to which primary porosity in the chalk was enhanced by syndepositional or early post-depositional movements. In both Norway and Denmark, it is believed that the productive chalk reservoirs show evidence in cores of slump deposits or mass slides, debris flows and turbidite flows (Kennedy, 1980; Watts *et al.*, 1980; Hurst, 1983). Nygaard *et al.* (1983) have concluded that allochthonous units have higher porosities and oil

saturations, possibly as a result of the disintegration of early chalk rock structure by redeposition. The slopes required to trigger these gravity-induced movements could either be related to seismic fault movements, on the broad scale, or locally caused by halokinesis (Watts *et al.*, 1980). Skovbro (1983) has suggested that bathymetrically deeper areas of chalk deposition may have received a higher proportion of exotic material (e.g. Albuskjell, Ekofisk and Tor) explaining contrasts in reservoir development with shallower areas (e.g. Eldfisk, Valhall and Hod).

Fig. 10.7 also illustrates the locations of major faults active during the Late Cretaceous (Ziegler, 1975 and Ziegler, 1981), and it can be seen that the commercial fairways lie adjacent to the reactivated eastern graben margin fault. The lack of commercial Chalk production in the U.K. sector, adjacent to the western fault, may be related to the presence of sands in the basal

Tertiary clastic sequence overlying the Chalk (Fig. 10.6; refer also to Fig. 10.1). Estimates of the depth of burial of the Chalk at the time of fracturing, oil migration and overpressuring are in the range from 450 metres (Hurst, 1983; Gorm Field) to 1050 metres (Hardman and Kennedy, 1980; Hod Field), and it is surprising that this top seal was not ruptured when the chalk reservoir pressure was inflated. It is possible that these Tertiary claystones may themselves have been overpressured due to rapid deposition. Obviously, the presence of significant amounts of sandstone in the Lower Palaeocene would reduce the critical sealing capacity of the interbedded shales.

10.5 Lower Cretaceous sandstone plays

Lower Cretaceous reservoirs in the North Sea contain only a small proportion of the total volume of oil and gas discovered to date. The reason appears to be that reservoir development is generally poor in the major grabens where the Upper Jurassic source rocks are mature.

The Lower Cretaceous is very sandy in the Inner Moray Firth Basin (Fig. 10.13) but here the Kimmeridge Clay is too shallow for the source rocks to reach maturity. Further east into the Witch Ground and Buchan grabens, where source-rock maturity is attained, the Lower Cretaceous rapidly shales-out into deeper water claystones. Locally, reworking of sediment from uplifted fault-blocks gave rise to better reservoir development, from which substantial oil and gas-condensate flow rates have been reported.

Far to the north on the eastern margin of the Møre Basin, gas-condensate has been discovered in Aptian-Albian stacked-channel sequences of the middle and outer zones of a submarine fan complex (Myrland *et al.*, 1981). Appraisal drilling of the so-called Idun Field has demonstrated the discontinuity of the indi-

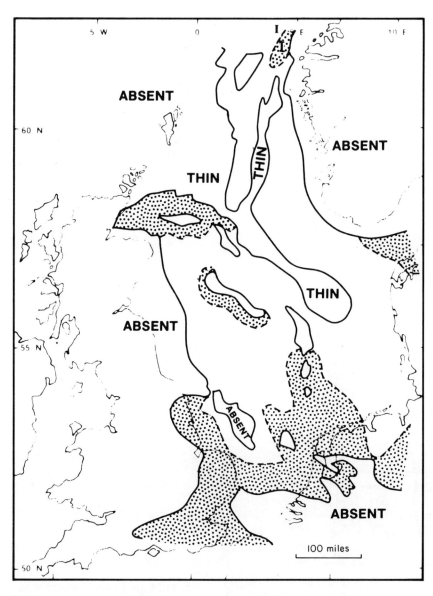

Significant sand development   **I**   Idun

Fig. 10.13. Distribution of significant Lower Cretaceous sand development in the North Sea; from Ziegler (1981) and Day *et al.* (1981).

vidual reservoir units and commercial production is doubtful.

Commercial production has been established, however, in shallow marine and deltaic sandstones to the south (Fig. 10.1) in the offshore extension of the West Netherlands Basin. Little has been published on these discoveries but it is likely that they are similar to the onshore fields of the Rijswijk Oil Province described by Bodenhausen and Ott (1981). The structures in which oil is trapped are Laramide in origin, with growth initiating in the Santonian-Campanian and reaching its peak in the Early Palaeocene phase of inversion of the West Netherlands Basin. The traps are formed by northwest-southeast trending flower struc-

tures (Fig. 10.14), associated with renewed (transpressional) wrench movements along Late Jurassic-Early Cretaceous transtensional faults, subsidence across which originally formed the West Netherlands Basin. The principal source rock is the Lower Jurassic Posidonia Shale. Reconstructions by Bodenhausen and Ott (1981) indicate that, in the northern part of the basin, this source rock reached maturity in the late Early Cretaceous to Late Cretaceous prior to inversion, and that early-formed traps were probably subsequently breached (Fig. 10.14). Further south, Tertiary and Quaternary burial has been sufficient to bring the Posidonia Shale to maturity after Laramide trap formation.

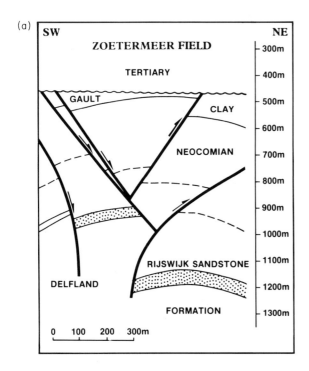

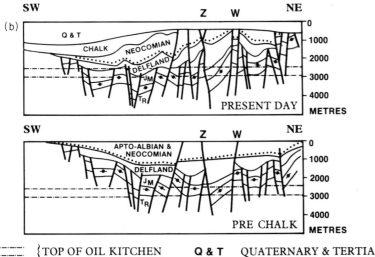

===== {	TOP OF OIL KITCHEN	**Q & T**	QUATERNARY & TERTIARY
Z	ZOETERMEER	··········	RIJSWISK SANDSTONE
W	WOUBRUGGE	**JM**	MIDDLE JURASSIC
TR	TRIASSIC	—●—	LOWER JURASSIC (POSIDONIA SHALE)

Fig. 10.14. (a) Lower Cretaceous oil play, West Netherlands Basin; (b) Reconstructed history of Rijswijk oil province. From Bodenhausen and Ott (1981).

Although reservoir development in the staged transgression of the London-Brabant Massif during the Neocomian can be good, the complexity of this type of trap offshore can be expected to place limits on field sizes and to make commercial development difficult.

In the north of the Netherlands, where the Dutch Central Graben terminates against the Texel High (Fig. 10.4), gas has been discovered in Valanginian sandstones in partly structural (drape domes) and partly stratigraphic traps. The source for the gas is thought to be Wealden coals, locally present between the Valangian and the truncated Permo-Triassic and older section, although the Westphalian may also have contributed (Cottencon *et al.*, 1975).

10.6 Upper Jurassic sandstone plays

Upper Jurassic sandstone plays are widespread in the North Sea and continue to be actively explored (Fig. 10.1). The sandstones are generally interbedded with or lie immediately below the Kimmeridge Clay source rock, so that migration paths are relatively short and simple and migration efficiency can be high.

10.6.1 Upper Jurassic shallow-marine sandstone play

The widespread Late Jurassic transgression in the North Sea led to the localised development of littoral and shallow marine depositional environments on the flanks of some major basins, where relatively higher wave or tidal energy, and suitable sediment supply, led to the accumulation of good quality reservoir sandstones.

In the outer part of the Moray Firth Basin, high-energy sands were deposited in the Witch Ground Graben during the Callovian-Oxfordian, and form the reservoir for the Piper Field (refer Brown, Figure 6.5). These Piper Sandstones were probably deposited as offshore bars located over actively growing fault blocks (Maher, 1980). Fault movements may have begun as early as the Triassic but significant activity is not recorded until the Late Jurassic, when depositional patterns were influenced by tectonics (Maher, 1980). In the outer Moray Firth Basin, fault activity did not generally cease at the end of the Jurassic, as in the East Shetland Basin, but continued through the Early Cretaceous, probably terminating in the Cenomanian. Upper Cretaceous sediments onlap the structure, which was not top-sealed until the Campanian. If the structure had continued growth thereafter, or perhaps had developed greater topographic relief, the top seal would have been Maastrichtian Chalk rather than Campanian Marl; Maher (1981) speculates that a similar structure in an adjacent block was dry because it depends on chalk for a top seal, the overlying Tertiary being sand-prone in this area with poor sealing potential. Source-rock maturity was first reached in the Early Tertiary (probably Palaeocene), so that the timing factor is favourable for this play.

The primary control on this fairway, however, is the distribution of the Piper Sandstone, which shales-out towards the basin centres and wedges-out onto the adjacent basement highs. To the west, further into the Moray Firth Basin, the fairway is limited by the lack of maturity of the Kimmeridge Clay (Fig. 10.10).

Similar sandstones are developed along the flanks of the Central Graben (Fig. 10.1), and provide the reservoir for the Fulmar and Clyde Fields to the west and the Ula Field to the east. In the Ula Field (Fig. 10.15), the sandstone reservoir is present over and to the east of the South Vestland Arch but attains maximum thickness (ca 150 metres) on the downthrown side of the main eastern graben fault, shaling-out westwards into the centre of the basin (Bailey *et al.*, 1981). Curiously, the reservoir is thickest over the culmination of the structure, which is thought to be formed by salt pillowing at depth, indicating the uplift of a small scale depocentre. According to Bailey *et al.* (1981), structural development occurred from an unspecified date in the Mesozoic to the Miocene. The reservoir sandstones are thought to have been accumulated by the repeated capture of migrating tidal sand bars in deeper waters along the downthrowing side of the active graben-margin fault (the Ula trend). The Clyde Field (Fig. 10.15) is, in many ways, similar to Ula except that structural development is reported to be related to growth-faulting rather than halokinesis (Gibbs, 1983). It is suggested that detached listric faulting of the Jurassic and Triassic was initiated by deep-seated extensional faulting associated with the development of the Central Graben. Gravity sliding down these detachment surfaces was accommodated by horizontal compaction and uplift and erosion of the toe of the slide. Clyde is top-sealed by the Kimmeridge Clay except in this area of erosion, where the reservoir is sealed by the Chalk (Fig. 10.15).

Ziegler (1978) has suggested that syndepositional fault activity along the margin of the Central Graben caused these shallow-marine sands to be mobilised and redeposited as sediment gravity flows.

Fig. 10.1 shows the locations of the Upper Jurassic fairways in the Central Graben based on published data. It is likely that these fairways will be extended along the graben margins as exploration drilling proves additional areas of deposition of these sands. Further south, in the North Netherlands Trough, discoveries have also been made in Upper Jurassic sandstones.

Deltaic and coastal sandstones of the Middle to Upper Jurassic, which prograded westwards across the Horda Platform, provide the reservoir for the giant Troll gas-field (Fig. 10.16) in the far north of the Norwegian North Sea (Brekke *et al.*, 1981). The trap is a rotated fault block of Late Jurassic age in which the reservoir is sealed by claystones ranging in age from the Upper Jurassic to the Paleocene. Gas is considered to have migrated eastwards for some distance from the North Viking Graben, and this may be one of the few documented examples in the North Sea of substantial lateral migration of hydrocarbons.

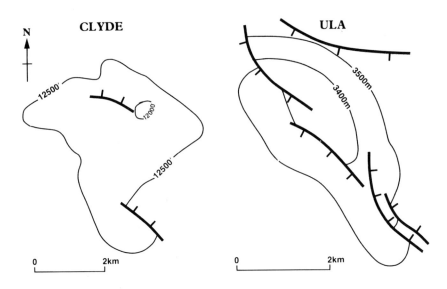

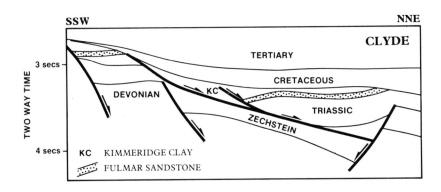

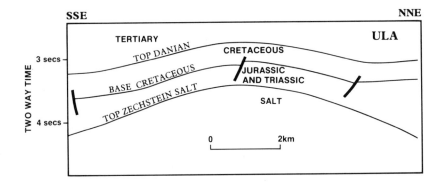

Fig. 10.15. The Clyde and Ula Upper Jurassic oil fields; from Gibbs (1983) and Bailey *et al.* (1981).

10.6.2 Upper Jurassic fan/turbidite sandstone play

The Late Jurassic was a period of intensification of fault movements leading to strong basin differentiation and the polarisation of the North Sea rift systems (Ziegler, 1981). The flanks of some of the grabens were uplifted and underwent intense subaerial erosion. Coarse clastics were deposited as aprons in the grabens close to the boundary faults, shaling-out rapidly into typical organic-rich Upper Jurassic clays.

The best published example of this play is the South Brae Field (Harms *et al.*, 1981), although similar discoveries have been made elsewhere in the South Viking Graben (Fig. 10.1), notably in Phillips' block 16/17, and the same type of facies development is known to be present elsewhere in the North Sea. Onshore examples

of Upper Jurassic fault-controlled sedimentation occur between Brora and Helmsdale on the Sutherland coast. The South Brae trap (Brown, Fig. 6.11) is primarily controlled by fault and dip closure; a seal must be present along the major western boundary fault against upthrown Devonian rocks. The trap was initially formed in the Late Jurassic to Early Cretaceous and is everywhere top-sealed by Kimmeridge Clay of Volgian age. Fault activity along the margin declined in intensity at the end of the Jurassic but continued until the Early Tertiary; however, the younger faults tend to step back towards the horst leaving the early formed trap relatively undisturbed. The Upper Jurassic reservoir at South Brae consists of conglomerates and sandstones which were deposited in a fan-delta building eastwards into a shallow sea. The combination of continuing uplift of the horst block, a rising sea level and a high

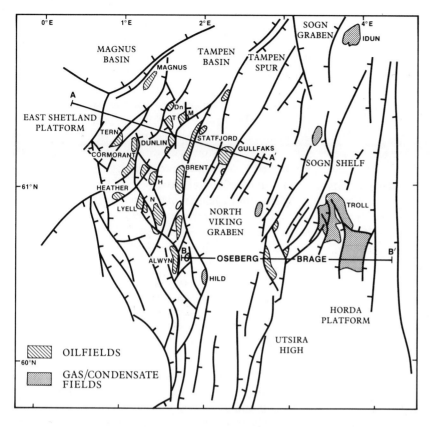

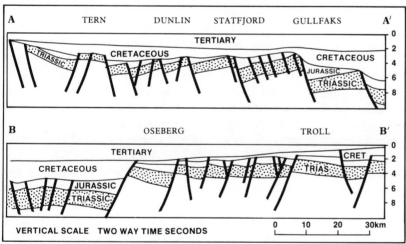

Fig. 10.16. North Viking Graben and East Shetlands Basin: major fault trends and location of main oil and gas fields; from Challinor and Outlaw (1981), Brekke *et al.* (1981) and other various sources. Key: **Dn** Don, **H** Hutton, **N** Ninian, **T** Thistle, **M** Murchison.

rate of sediment supply, led to the accumulation of the very thick reservoir sequence, which is so critical to the commercial potential of such prospects. Elsewhere in the South Viking Graben a different balance between these controlling parameters has led to the deposition of these coarse clastics in submarine fans. In fact, Stow *et al.* (1982) have suggested that South Brae itself may have been deposited in a shallow submarine environment.

Further north, on the north-western margin of the East Shetland Basin, the Magnus Field reservoir consists of the preserved middle-fan turbidite sandstones deposited in a Late Jurassic submarine fan (De'Ath and Schuyleman, 1981); the more proximal parts of the fan have been removed by post-Jurassic erosion associated with the development of the North Shetland Trough. It seems likely that the Magnus reservoir

sands were reworked from a shelf area to the west into deeper waters and this may also be the depositional model for some of the more sand-prone, rather than conglomeratic, reservoirs in the South Viking Graben.

The fairway for this play is obviously controlled by the location of intense activity on Late Jurassic faults but the highly variable reservoir characteristics along the fault-trend depend upon the specific sedimentological processes in operation, which in turn are controlled by the balance between the rates of horst uplift and basin subsidence. A further critical parameter is the timing of cessation of fault activity and its impact on the validity of the top seal. In view of the relatively low intensity of exploration along such faulted margins, Harms' (1981) prophecy that similar accumulations to Brae await discovery elsewhere in the North

Sea may well be fulfilled. A logical extension of this play would also be to explore the deeper basinal areas for the distal turbidite sandstone equivalents of these coarse clastic aprons.

10.7 Lower and Middle Jurassic sandstone plays

The discovery of the Brent Field in 1971 opened up the most prolific hydrocarbon play in the North Sea (Fig. 10.1). The principal reservoirs are the alluvial to coastal Statfjord Formation sandstones (Rhaetian to Sinemurian) and the deltaic to shallow-marine Brent Group sandstones (Bajocian to Bathonian) separated by marine shales of the Dunlin Group (Sinemurian to Aalenian), which also locally contain thin sandstones (e.g. Cook Formation).

Fig. 10.16 illustrates the location of the principal oil and gas-condensate discoveries. Most are located along the North Viking Graben and in the broad East Shetlands Basin, although discoveries have also been made in the Inner Moray Firth (Beatrice; Linsley *et al.*, 1980) and in the Ergersund Basin (Bream and Brisling). Apart from the constraint imposed by the distribution of mature Upper Jurassic source rocks, the principal control on the fairway is the distribution of reservoir sandstones. The Statfjord Formation is hydrocarbon-bearing only in the Brent, Statfjord and Gullfaks fields, and it is the Brent Group sandstones which contain the bulk of the reserves.

It is generally accepted that the Rannoch, Etive and Ness formations, of Bajocian age in the Brent Group, were deposited in a northerly-prograding delta, with the Tarbert Formation sandstones, of Bathonian age, being deposited as a result of renewed marine transgression (Budding and Inglin, 1981). The best reservoir development is usually in the Rannoch and Etive formations. A snapshot reconstruction of Middle Bajocian and Bathonian palaeogeography has been made by Skarpnes *et al.* (1980) and is illustrated in Fig. 10.17. The intersection of the generally east-west trend of deposition of the Rannoch and Etive formations with the northerly trend of the Viking Graben and East Shetlands Basin circumscribes the play. Transport of Bajocian sediments away from the main volcanic centre (Fig. 10.17) to the west, into the Inner Moray Firth Basin, and to the east, into the Norwegian-Danish Basin, is also recognised but in neither area was the balance between sediment accumulation and subsidence rate such as to result in the development of a major delta. An interesting assertion in this interpretation is that the Bathonian to Callovian transgression (Fig. 10.17) submerged the East Shetlands Platform, and brought an open-marine shoreline into the South Viking Graben and Inner Moray Firth Basin, creating the higher energy depositional environment needed for improved reservoir development. This is consistent with the reservoir geology of Sleipner, where Larsen and Jaarvik (1981) have interpreted the Callovian reservoir sands (Hugin Fm) to have been deposited in

shallow-marine to deltaic environments following a sea-level rise. Skarpnes *et al.* (1980) have also suggested that the Middle Jurassic reservoirs of the Beryl Field were deposited during this transgression. It should be noted, however, that quite different palaeogeographic models may be constructed for the Middle Jurassic. For example, Proctor (1980) has proposed that the dominant depositional element of the Brent Group reservoirs is the Murchison-Statfjord fan-delta, which prograded southwards into the East Shetlands Basin from a northerly landmass.

Apart from Sleipner and possibly Gudrun, which are salt structures, the trapping mechanism for these Middle Jurassic reservoirs is provided by rotated fault blocks. Cross sections of fields shown elsewhere in this volume (Statfjord, Fig. 6.10 and Brent, Fig. 5.12) illustrate typical traps. Fault movements started in the Triassic (or earlier) and continued until the end of the Jurassic. At the end of the Middle Jurassic the crests of the fault blocks rose above wave base and were truncated. Submergence during the Later Jurassic transgression was followed by renewed uplift and erosion at the end of the Jurassic, leaving truncated topographic highs at the sea bed, which were progressively onlapped and buried by Lower and Upper Cretaceous shales. The traps are thus partly stratigraphic, with the top seal being provided by Upper Jurassic and Lower and Upper Cretaceous shales.

The principal oil fields in the East Shetlands Basin are located along major north and northeast fault trends that hade east (Fig. 10.16), although west-hading trends are also developed, as are north-north-west fault trends. To the east of the graben on the Sogn Shelf, complementary west-hading fault sets provide closure for the Oseberg Field (Larsen *et al.*, 1981) and the Troll Field (Brekke *et al.*, 1981). Closure along the fault terraces is provided by dip or second order cross faults. The Oseberg Field is illustrated in Fig. 10.18 as an example of this play. Traps were formed in the Late Jurassic and sealed by the Late Cretaceous, predating the onset of oil generation at the end of the Cretaceous. Goff (1983) has estimated that peak oil generation was reached in the North Viking Graben during the Palaeocene and in the half-grabens of the East Shetlands Basin by the Oligocene. All the play parameters are very favourable within the fairway and, as a result of the close proximity of source rock and reservoir, migration efficiency is estimated to be high with 20-30 per cent of the oil generated now trapped in Jurassic sandstones. This is surprising, in that over the culminations of the major structures the Kimmeridge Clay only reached maturity recently, and downflank the Heather Shale intervenes between mature source rock and reservoir (Fig. 10.18). Goff (1983) has suggested that the high pressure gradients set up between generating source rocks and lower pressured Brent Group reservoirs would drive oil migration laterally even through the minute effective (connected) porosity of these intervening shales. Oil migration from the Brent Group reservoirs to the Statfjord

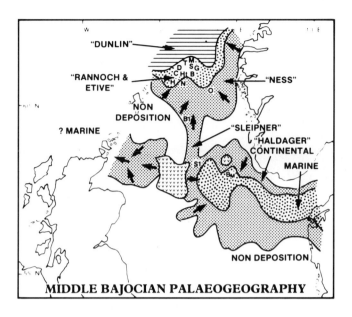

MIDDLE BAJOCIAN PALAEOGEOGRAPHY

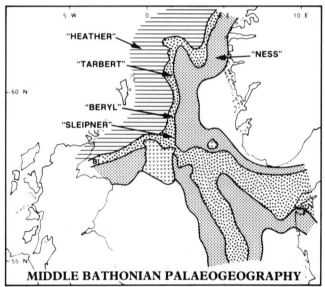

MIDDLE BATHONIAN PALAEOGEOGRAPHY

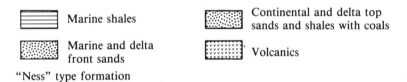

Marine shales

Marine and delta
front sands

"Ness" type formation

Continental and delta top
sands and shales with coals

Volcanics

Fig. 10.17. Snapshot palaeogeo-
graphic sketches of the Middle
Jurassic of the North Sea for the
Middle Bajocian and Middle Batho-
nian; from Skarpnes *et al.* (1982).
Key: **B** Brent, **Bl** Beryl, **Bm** Bream
and Brisling, **Bt** Beatrice, **C** Cor-
morant, **D** Dunlin, **G** Gullfaks,
H Heather, **Ht** Hutton, **M** Murchison,
N Ninian, **O** Oseberg, **S** Statfjord,
Se Sleipner, **T** Thistle.

Formation sandstones cannot, however, be explained
in this way but the frequency of faulting provides a
mechanism for secondary migration between these two
principal reservoirs. This more complex migration
path may be the reason that the Statfjord Formation is
frequently water-bearing beneath a productive Brent
reservoir.

Within the Viking Graben, progressive Tertiary
burial brought the Kimmeridge Clay to the phase of
wet gas generation in the Eocene. There is no evidence
that gas migrated in significant volumes westwards
into the East Shetlands Basin but the presence of a gas-
cap in Oseberg (Fig. 10.18) and the gas in Troll probably attest to migration towards the east. It is interesting to speculate that Troll may orginally have been an
oil field, with biodegradation resulting from the late

(Palaeocene) top seal, which was later flushed with gas
from the Viking Graben. Deep Middle Jurassic sandstones in the Viking Graben are gas or gas-condensate
bearing from Sleipner (Larsen and Jaarvik, 1981) northwards to Hild and onwards towards Idun (Fig. 10.16).
Condensate ratios vary widely, reflecting both differing
maturity histories and kerogen compositions of the
Kimmeridge Clay (Fisher and Miles, 1983).

Away from the main grabens, oil has been discovered
at Bream and Brisling, where the thin Borglum Formation source rocks locally reach maturity in the
Ergersund Basin (Fig. 10.4). To the west, in the Inner
Moray Firth, the Kimmeridge Clay is generally immature. However, Barnard and Cooper (1981) have
suggested that Bathonian algal oil shales could be the
source of the high-wax-content oil discovered in the

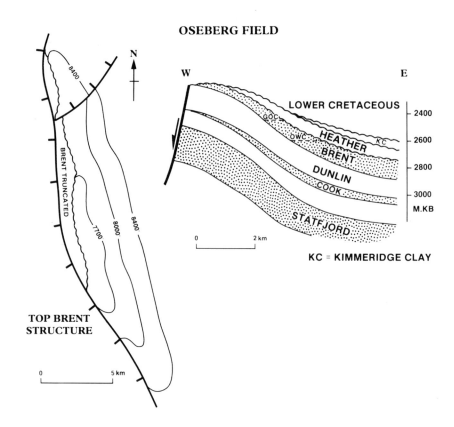

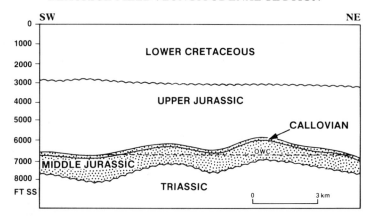

Fig. 10.18. The Oseberg and Beatrice Middle Jurassic oil fields; from Larsen *et al.* (1981) and Linsley *et al.* (1980).

Beatrice Field (Linsley *et al.*, 1980), the very thick Upper Jurassic section (Fig. 10.18) providing sufficient depth of burial for the Middle Jurassic to reach maturity.

10.8 Rotliegend and Bunter gas plays in the Southern North Sea

The Rotliegend gas play (Fig. 10.1) has been described by Glennie (Chapter 3, this volume) and is illustrated by the sections in Fig. 10.19. The widespread although variable quality Rotliegend Sandstone, blanketed by the near perfect top-seal of the Zechstein carbonates and evaporites, overlies with angular unconformity the coal-bearing sequences of the Westphalian, which are the gas source. Migration efficiency is high because the coal-measures sandstones, which can have large contact areas with the coal-seams, can act as pathways to

conduct the migrating gas to the Rotliegend reservoir at the unconformity.

These are the main favourable elements of this play, which is otherwise complicated by the relative timing of migration and trap formation. The Westphalian coals reached maturity for gas generation at different times in different parts of the basin as a result of different subsidence histories; for example, maturity was reached in the Broad Fourteens Basin and in the Sole Pit Basin in the Late Jurassic (Lutz *et al.*, 1975; Glennie and Boegner, 1981) but elsewhere maturity was not generally attained until the Late Cretaceous. Gas generation and migration in some parts of the basin was thus in progress at about the same time as the Cimmerian formation of the faulted anticlines which provide the Rotliegend traps. The timing aspects of this play are further complicated by basin inversion occurring during the Middle to Late Cretaceous and

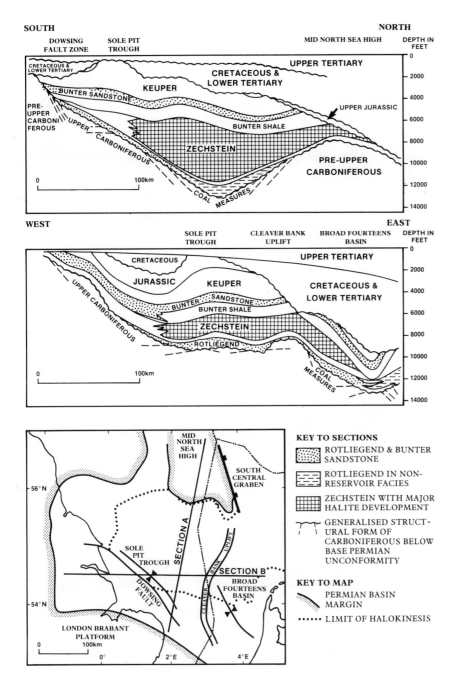

Fig. 10.19. Stratigraphic cross-sections of the southern North Sea Basin.

Tertiary in different parts of the southern North Sea (Fig. 10.19), which may have resulted in the escape of gas from early-formed accumulations, and remigration towards the elevated inversion axes, where reservoir parameters had been adversely affected by burial diagenesis. In some areas, where the generative capacity of the Westphalian coals was not exhausted by pre-inversion burial, later Cretaceous and Tertiary burial has led to a second young phase of gas generation. In northern Holland and in most of the Dutch offshore areas where this effect is seen (Van Wijhe *et al.*, 1980), this favourable timing aspect of the play, in combination with the other favourable parameters, has resulted in most accumulations being full to spill-point (Lutz *et al.*, 1975).

The fairway for commercial Rotliegend gas production is thus controlled in part by basin inversions but the other principal controlling factors are the distribution of the Westphalian coals beneath the unconformity, the reservoir quality of the Rotliegend Sandstones and the limits of the Zechstein top seal. A play-map defining this fairway is illustrated in Fig. 10.20.

Gas discoveries in the Bunter (Fig. 10.1) tend to occur only where the Zechstein top seal becomes less effective. Towards the margins of the Zechstein basin, thinning and facies changes have allowed the vertical migration of gas into the Bunter and Hewett Sandstone reservoirs of the Hewett Field (Cumming and Wyndham, 1975). However, in the Dutch sector of the North Sea, where the Zechstein salt is thickly developed, gas is also occasionally found in the Bunter Sandstone (Roos and Smits, 1983) in areas of strong Late Cretaceous to Early Tertiary inversion. It is suggested

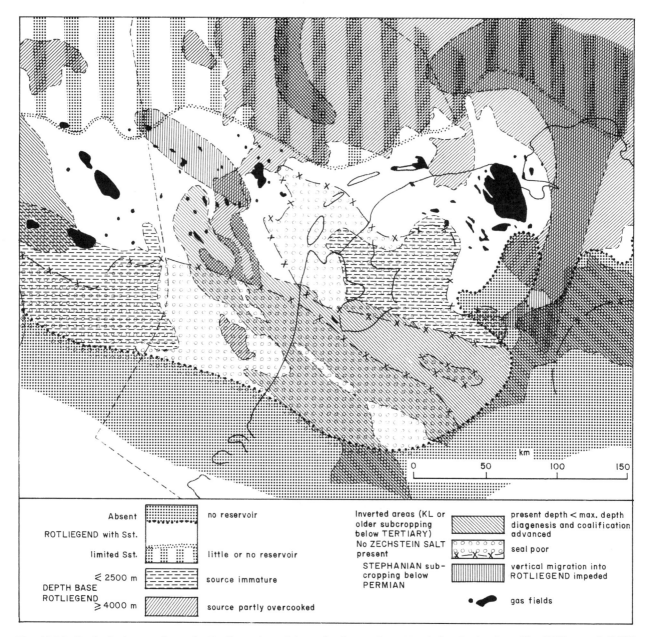

Fig. 10.20. Geological controls on the Rotliegend gas fairway in the southern North Sea Basin; from Van Wijhe *et al.* (1980).

that early formed Rotliegend gas accumulations (Oele *et al.*, 1981) leaked gas across active faults into the Bunter during uplift associated with inversion.

10.9 Palaeozoic plays on graben margins

Oil has been discovered in Permian and Devonian reservoirs along the upthrown margins of the grabens and in mid-graben horst blocks. In this play, oil has migrated from Upper Jurassic source rocks into stratigraphically older but structurally elevated reservoirs. In the Argyll Field, the principal oil-bearing reservoirs are vugular Zechstein dolomites and aeolian and fluviatile Rotliegend sandstones separated by a thin Kupferschiefer (Pennington, 1975). The same reservoirs are present in Auk, although the Rotliegend is largely water-bearing (Brennand and Van Veen, 1975). The Late Jurassic faulting, which resulted in the renewed differentiation of the Central Graben from the

Auk Ridge, was responsible for the main phase of development of both structures. However, further reactivation of fault movements in the Late Cretaceous resulted in local erosion, so that in both fields the top-seal is provided by Upper Cretaceous and Danian Chalk (Figure 3.19). Late movements, probably reactivation, along the faults at the extreme margins of the Central Graben and the Viking Graben are well illustrated in Fig. 10.3. Trap formation for this play is thus a relatively young feature with the same requirement for late generation and migration of hydrocarbons.

The fairway for this play is constrained by the development of good connected porosity in the Zechstein dolomites. At the Auk and Argyll fields, a relatively narrow zone is preserved between the truncated margin of the Zechstein and original tight shelf carbonate and anhydrite sequences, where leaching of the anhydrite has created good porosity and permeability. Taylor (1981) concludes that this exacting requirement may result in similar reservoirs being relatively un-common.

Since the Upper Jurassic shales on the platform to the west of the Auk Ridge are only marginally mature, the oil trapped at Auk and Argyll is believed to have migrated westwards across the marginal faults, from the mature Kimmeridgian source rocks in the Central Graben. In Argyll, the graben margin fault also forms the side-seal to the trap, so that it is necessary to postulate that the fault acted as both a migration path-way and a seal. This phenomenon could have arisen in two ways. The margin of the Central Graben is not formed by a single fault but by a discontinuous series of en-echelon faults (Fig. 3.19); the scissor move-ments along these faults must therefore create windows, where the source rock and reservoir are juxtaposed and cross-fault migration can occur. In itself, how-ever, this explanation is inadequate in that the small area of contact between the Kimmeridge Clay and the reservoir would be too small to provide for efficient migration, and it is necessary to postulate that Upper Jurassic sandstones, which are known around the Argyll Field (Pennington, 1975), have acted as carrier beds. Alternatively, the marginal faults, which ter-minate vertically in the Lower Tertiary shales, may have provided open surfaces for vertical hydrocarbon migration.

In both fields, the Rotliegend Sandstone rests with little angular disconformity on Devonian rocks. To the north, in the Buchan Graben, the Devonian Old Red Sandstone itself forms the main oil-bearing reser-voir in a horst block which is top-sealed by Late Aptian mustones (Buchan Field; Burnhill and Ramsay, 1981). It is reported that the relatively poor matrix permeabilities of the reservoir sandstones are enhanced by fracturing.

Perplexingly, Harms *et al.* (1981) have reported that Devonian rocks, in the intense fault zone between the South Viking Graben and the Fladen Ground Spur, provide the side seal for the South Brae oilfield.

10.10 Concluding remarks

In this brief review of the North Sea, only the major proven plays have been described, and there are a num-ber of discoveries which cannot be simply categorised in this way. However, the plays described do account for the bulk of proven reserves of oil and gas. Fig. 10.21 shows the breakdown of North Sea reserves by reservoir for fields of commercial significance.

Original reserves of gas in Permian and Triassic reservoirs in the Southern North Sea Basin amount to approximately 31×10^{12} ft^3 (878×10^9 m^3). Declining production from these older fields is being compensated for by the development of gas reserves in Tertiary and Middle Jurassic (associated gas) reservoirs in the Northern North Sea, which amount to approximately 22×10^{12} ft^3 (623×10^9 m^3). The giant Troll discovery,

with proven reserves of at least 12×10^{12} ft^3 (340×10^9 m^3), will have a significant impact on gas supplies to Europe when it is brought on stream. Total North Sea gas reserves are estimated at approximately 79×10^{12} ft^3 (2237×10^9 m^3) excluding discoveries of uncertain commercial potential.

The total oil reserves in the North Sea are estimated to be approximately 22×10^9 bbls (3500×10^6 m^3) of which the major part (62 per cent) is contained in Lower to Middle Jurassic sandstones of the Northern North Sea. An analysis of the sequence of oil dis-coveries in the U.K. sector of the North Sea during the 1970s shows a steady decline in average field size from about 500 to about 300×10^6 bbls (80 to 50×10^6 m^3), and more recent results indicate the trend to be con-tinuing.

The U.K. sector of the North Sea is now well into a mature stage of exploration, and there are few blocks within the areas of the main producing basins that have not already been explored to some extent. This does not mean that relinquished acreage in the U.K. lacks potential, and a number of significant discoveries (Clyde in the U.K. as well as Ula in Norway) have been made on previously licenced blocks. The emergence of new play concepts or improvements in economic con-ditions can open opportunities for drilling which were previously overlooked or discarded as unattractive. For example, the proposed U.K. gas-gathering system could have spurred drilling for deep Jurassic gas in the axes of the main grabens where exploration is still at a relatively early stage. Another example is that seismic interpreters are now beginning to give credence to the extensional origin of the North Sea grabens (Beach, 1983; Gibbs, 1983) with the result that new but small traps may be defined within the main fairways.

The situation in Norway is different in that the slower pace of licencing has resulted in a number of blocks within the areas of the main fairways remaining unlicenced and undrilled. There is, therefore, potential in the Norwegian North Sea, particularly between 60° and 62° N, for major discoveries of oil of the magni-tude not seen in the U.K. since the early 1970s. Other-wise, as in the U.K., the main potential is for the discovery of hydrocarbons in smaller and increasingly complex fields.

Appraisal of early discoveries has revealed that there is often an element of stratigraphic trapping in pros-pects which were originally considered to be structur-ally closed. Exploration for stratigraphic traps in the North Sea will become more important but in this high-cost offshore environment the commerical suc-cess of exploration programmes will depend upon the development of reliable structural and stratigraphic models.

10.11 Acknowledgements

This chapter is published by permission of Britoil plc.

OIL   Total = 21.8 billion bbls

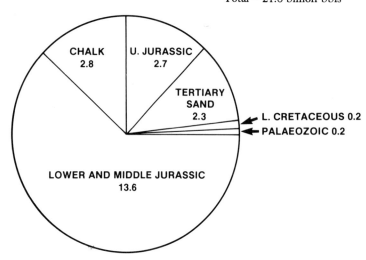

Oilfields in production or under development plus Oseberg and Gullfaks
note billion = 10⁹

GAS   Total = 79.4 trillion scf

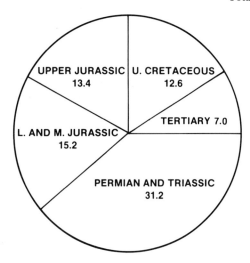

Dry Gas and Associated Gas in Gasfields and Oilfields in
production or under development including Sleipner and
Heimdal plus Troll.
note trillion = 10¹²

Fig. 10.21. Distribution of oil and gas reserves in the North Sea, classified by reservoir.

10.12 References

Barnard, P.C. and Cooper, B.S. (1981) Oils and source rocks of the North Sea area. In: Illing, L.V. and Hobson, G.D. (Eds.) q.v.

Bailey, C.C., Price, I. and Spencer, A.M. (1981) The Ula Oil Field, Block 7/12, Norway. In: *Norwegian Symposium on Exploration, Bergen.* Norsk Petroleumsforening. pp. 18/1-26.

Beach, A. (1984) *The structural evolution of the Witch Ground Graben.* In press. J. geol. Soc. London.

Blair, D.G. (1975) Structural styles in North Sea oil and gas fields. In: Woodland, A.W. (Ed.) q.v. 327-335.

Bodenhausen, J.W.A. and Ott, W.F. (1981) Habitat of the Rijswijk oil province, onshore The Netherlands. In: Illing, L.V. and Hobson, G.D. (Eds.) q.v. pp. 301-309.

Bowen, J.M. (1975) The Brent oil field. In: Woodland, A.W. (Ed.) q.v. pp. 353-362.

Brekke, T., Pegrum, R.M. and Watts, P.B. (1981) First exploration results in Block 31/2 (offshore Norway). In: *Norwegian Symposium on Exploration, Bergen.* Norsk Petroleumsforening. pp. 16.1-16.34.

Brennand, T.P. and Van Veen, F.R. (1975) Auk oil field. In: Woodland, A.W. (Ed.) q.v. pp. 295-311.

Budding, M.C. and Inglin, H.F. (1981) A reservoir geological model of the Brent Sands in Southern Cormorant. In: Illing, L.V. and Hobson, G.D. (Eds.) q.v. pp. 326-334.

Burnhill, T.J. and Ramsay, W.V. (1981) Mid-Cretaceous palaeontology and stratigraphy, Central North Sea. In: Illing, L.V. and Hobson, G.D. (Eds.) q.v. pp. 245-265.

Carman, G.J. and Young, R. (1981) Reservoir geology of the Forties Oilfield. In: Illing, L.V. and Hobson, G.D. (Eds.) q.v., pp. 371-379.

Challinor, A. and Outlaw, B.D. (1981) Structural evolution

of the North Viking Graben. In: Illing, L.V. and Hobson G.D. (Eds.) q.v. pp. 104-109.

Childs, F.B. and Reed, P.E.C. (1975) Geology of the Dan Field and the Danish North Sea. In: Woodland, A.W. (Ed.) q.v. pp. 429-438.

Cottencon, A., Parant, B. and Flaceliere, G. (1975) Lower Cretaceous gas fields in Holland. In: Woodland, A.W. (Ed.) q.v., pp. 403-412.

Cumming, A.D. and Wyndham, C.L. (1975) The geology and development of the Hewett gas field. In: Woodland, A.W. (Ed.) q.v. pp. 313-325.

Day, G.A., Cooper, B.A., Anderson, C., Burger, W.F.J., Rønnevik, H.C. and Schoneich, H. (1981) Regional seismic structure maps of the North Sea. In: Illing, L.V. and Hobson, G.D. (Eds.) q.v., pp. 76-84.

De'Ath, N.G. and Schuyleman, S.F. (1981) The geology of the Magnus oil field. In: Illing, L.V. and Hobson, G.D. (Eds.) q.v. pp. 342-351.

D'Heur, M. (1980) Chalk reservoir of the West Ekofisk Field. In: *The Sedimentation of the North Sea Reservoir Rocks, Geilo*. Norsk Petroleumsforening. Section X.

Fagerland, N. (1983) Tectonic analysis of a Viking Graben border fault. *AAPG Bull* 67, 2125-2136.

Fisher, M.J. and Miles, J.A. (1983) Kerogen types, organic maturation and hydrocarbon occurrences in the Moray Firth and South Viking Graben, North Sea Basin. In: Brooks, J. (Ed.) Petroleum geochemistry and exploration of Europe. Blackwell Scientific Publications, pp. 195-202.

Fowler, C. (1975) The geology of the Montrose Field. In: Woodland, A.W. (Ed.) q.v. pp. 467-477.

Gibbs, A. (1984) *The Clyde Field growth fault secondary detachment above basement faults in the North Sea*. In press. AAPG Bull.

Glennie, K.W. and Boegner, P.L.E. (1981) Sole Pit inversion tectonics. In: Illing, L.V. and Hobson, G.D. (Eds.) q.v. pp. 110-120

Goff, J.C. (1983) Hydrocarbon generation and migration from Jurassic source rocks in the E. Shetland Basin and Viking Graben of the Northern North Sea. *J. geol. Soc. London*, **140**, 445-474.

Hamar, G.P., Jakobsson, K.H., Ormaasen, D.E., and Skarpness, O. (1980) Tectonic development of the North Sea north of the Central Highs. In: *The Sedimentation of the North Sea reservoir rocks, Geilo*. Norsk Petroleums forening. pp. 3/1-11.

Hancock, J.M. and Scholle, P.A. (1975) Chalk of the North Sea. In: Woodland, A.W. (Ed.) q.v., pp. 413-427.

Hardman, R.F.P. and Kennedy, W.J. (1980) Chalk reservoirs of the Hod Fields, Norway. In: *The sedimentation of the North Sea reservoir rocks, Geilo*. Norsk Petroleumsforening. Section XI.

Harms, J.C., Tackenberg, P., Pickles, E. and Pollock, R.E. (1981) The Brae oil field area. In: Illing, L.V. and Hobson, G.D. (Eds.) q.v. pp. 352-357.

Heritier, F.E., Lossel, P. and Wathne, E. (1981) The Frigg gas field. In: Illing, L.V. and Hobson, G.D. (Eds.) q.v. pp. 380-391.

Hurst, C. (1983) Petroleum geology of the Gorm Field, Danish North Sea. *Geol. en Mijnbouw*, **62**, 157-168.

Illing, L.V. and Hobson, G.D. (Eds.) (1981) *Petroleum geology of the Continental Shelf of north-west Europe*. Heyden & Son. 521 p.

Kennedy, W.J. (1980 Aspects of chalk sedimentation in the Southern Norwegian offshore. In: *The sedimentation of the North Sea reservoir rocks, Geilo*. Norsk Petroleumsforening.

Kessler, L.G., Zang, R.D., Englehorn, J.A. and Eger, J.D. (1980) Stratigraphy and sedimentology of a Palaeocene submarine fan complex, Cod Field, Norwegian North Sea. In: *The Sedimentation of North Sea Reservoir Rocks, Geilo*. Norsk Petroleumsforening.

Larsen, R.M. and Jaarvik, L.J. (1981) The geology of the

Sleipner Field complex. In: *Norwegian symposium on exploration, Bergen*. Norsk Petroleumsforening. pp. 15.1-15.31.

Larsen, V., Aasheim, S.M. and Masset, J.M. (1981) 30/6-Alpha structure: a field case study in the silver block. In: *Norwegian symposium on exploration, Bergen*. Norsk Petroleumsforening. pp. 14/1-34.

Lilleng, T. (1980) Lower Tertiary (Danian-Lower Eocene) Reservoir sand developments between 59° 30′ N and 61° 30′ N of the northern North Sea. In: *The sedimentation of the North Sea reservoir rocks, Geilo*. Norsk Petroleumsforening.

Linsley, P.N., Potter, H.C., McNab, G. and Racher, D. (1980) The Beatrice Field Inner Moray Firth, U.K. North Sea. In: Halbouty, M.T. (Ed.) *Giant oil and gas fields of the decade 1968-1978*. AAPG Memoir 30. pp. 117-129.

Lutz, M., Kaasschieter, J.P.H. and Van Wijhe, D.H. (1975) Geological factors controlling Rotliegende gas accumulation in the mid-European Basin. *Proceedings of 9th world petroleum congress*, vol. 2, pp. 93-97.

Maher, C.E. (1980) Piper oil field. In: Halbouty, M.T. (Ed.) *Giant oil and gas fields of the decade 1968-1978*. AAPG Memoir 30. pp. 131-172.

Mudge, D.C. and Bliss, G.M. (1983) Stratigraphy and sedimentation of the Palaeocene sands in the North Sea. In: Brooks, J. (Ed.) *Petroleum geochemistry and exploration of Europe*. Blackwell Scientific Publications, pp. 95-112.

Myrland, R., Messell, K. and Raestad, N. (1981) Gas discovery in 35/3: an exploration history. In: *Norwegian symposium on exploration, Bergen*. Norsk Petroleumsforening. Section NSE/17.

Nygaard, E., Lieberkind, K. and Frykman, P. (1983) Sedimentology and reservoir parameters of the Chalk Group in the Danish Central Graben. *Geol. en Mijnbouw 62*, 117-190.

Oele, J.A., Hol, A.C.P.J. and Tiemans, J. (1981) Some Rotliegend gas fields of the K and L blocks, Netherlands offshore (1968-1978)—a case history. In: Illing, L.V. and Hobson, G.D. (Eds.) q.v., pp., 289-300.

Parker, J.R. (1975) Lower Tertiary sand development in Central North Sea. In: Woodland, A.W. (Ed.) q.v., pp. 447-453.

Pennington, J.J. (1975) The geology of the Argyll Field. In: Woodland, A.W. (Ed.) q.v. pp. 285-291.

Proctor, C.V. (1980) Distribution of Middle Jurassic Facies in the East Shetlands Basin and their control on reservoir capability. In: *The Sedimentation of the North Sea reservoir rocks, Geilo*. Norsk Petroleumsforening. pp. 15/1-22.

Rochow, K.A. (1981) Seismic stratigraphy of the North Sea 'Palaeocene' deposits. In: Illing, L.V. and Hobson, G.D. (Eds.) q.v. pp. 255-266.

Roos, B.M. and Smits, B.J. (1983) Rotliegend and Main Buntsandstein gas fields in block K/13—a case history. *Geol. en Mijnbouw 62*, pp. 75-82.

Sarg, J.F. and Skjold, L.D. (1982) Stratigraphic traps in Palaeocene sands in the Balder area, North Sea. In: Halbouty, M.T. (Ed.) *The deliberate search for the subtle trap*, AAPG Memoir 32.

Skarpnes, O., Briseid, E. and Milton, D.I. (1982) The 34/10 delta prospect of the Norwegian North Sea: exploration study of an unconformity trap. In: Halbouty, M.T. (Ed.) *The deliberate search for the subtle trap*, AAPG Memoir 32, pp. 207-216.

Skarpnes, O., Hamar, G.P., Jakobsson, K.H. and Ormaasen, D.E. (1980) Regional Jurassic setting of the North Sea north of the Central Highs. In: *The sedimentation of the North Sea reservoir rocks, Geilo*. Norsk Petroleumsforening. pp. 13/1-18.

Skovbro, B. (1983) Depositional conditions during Chalk sedimentation in the Ekofisk area, Norwegian North Sea. *Geol. en Mijnbouw, 62*, 169-176.

Stow, D.A.V., Bishop, C.D. and Mills, S.J. (1982) Sedimentology of the Brae oil field, North Sea: fan models and controls. *Petrol. Geol.,* **5**, 129-148.

Taylor, J.C.M. (1981) Zechstein facies and petroleum prospects in the Central and Northern North Sea. In: Illing, L.V. and Hobson, G.D. (Eds.) q.v. pp. 176-185.

Thomas, O.D. (1980) North Sea petroleum: past and future. *Proceedings of 10th world petroleum congress,* vol. 2, pp. 177-182.

Van den Bark, E. and Thomas, O.D. (1980) Ekofisk: first of the giant oil fields in Western Europe. In: Halbouty, M.T. (Ed.) *Giant oil and gas fields of the decade 1968-1978.* AAPG Memoir 30. pp. 195-224.

Van Wijhe, D.H., Lutz, M. and Kaaschieter, J.P.H. (1980) Rotliegend in The Netherlands and its gas accumulations. *Geol. en Mijnbouw,* vol. 59, pp. 3-24.

Walmsley, P.J. (1975) The Forties Field. In: Woodland, A.W. (Ed.) q.v. pp. 477-487.

Watts, N.L., Lapre, J.F., Van Schijndel-Goester, F.S. and Ford, A. (1980) Upper Cretaceous and Lower Tertiary chalks of the Albuskjell area, North Sea: deposition in a slope and a base-of-slope environment. *Geology,* **8**, 217-221.

Williams, J.J., Conner, D.C. and Peterson, K.E. (1975) The Piper oil field, U.K. North Sea: a fault-block structure with Upper Jurassic beach bar reservoir sands. In: Woodland, A.W. (Ed.) q.v. pp. 363-377.

Woodland, A.W. (Ed.) (1975) *Petroleum and the Continental Shelf of north-west Europe.* Applied Science Publishers. 501 p.

Ziegler, P.A. (1977) Geology and hydrocarbon provinces of the North Sea. *Geojournal* **1**, 7-32.

Ziegler, P.A. (1978) North-western Europe: tectonics and basin development. *Geol. en Mijnbouw* **53**, 43-50.

Ziegler, P.A. (1981) Evolution of sedimentary basins in North-west Europe. In: Illing, L.V. and Hobson, G.D. (Eds.) q.v. pp. 3-42.

Ziegler, P.A. (1982) *Geological atlas of Western and Central Europe.* Shell Internationale Petroleum Maatschappij B.V. 130 p.

Ziegler, W.H. (1975) Outline of the geological history of the North Sea. In: Woodland, A.W. (Ed.) q.v., pp. 165-190.

Index